Advanced Structured Materials

Volume 247

Series Editors

Andreas Öchsner, Faculty of Mechanical and Systems Engineering, Esslingen University of Applied Sciences, Esslingen, Germany

Lucas F. M. da Silva, Department of Mechanical Engineering, Faculty of Engineering, University of Porto, Porto, Portugal

Holm Altenbach, Faculty of Mechanical Engineering, Otto von Guericke University Magdeburg, Magdeburg, Germany

Common engineering materials are reaching their limits in many applications, and new developments are required to meet the increasing demands on engineering materials. The performance of materials can be improved by combining different materials to achieve better properties than with a single constituent, or by shaping the material or constituents into a specific structure. The interaction between material and structure can occur at different length scales, such as the micro, meso, or macro scale, and offers potential applications in very different fields.

This book series addresses the fundamental relationships between materials and their structure on overall properties (e.g., mechanical, thermal, chemical, electrical, or magnetic properties, etc.). Experimental data and procedures are presented, as well as methods for modeling structures and materials using numerical and analytical approaches. In addition, the series shows how these materials engineering and design processes are implemented and how new technologies can be used to optimize materials and processes.

Advanced Structured Materials is indexed in Google Scholar and Scopus.

Shiv Kumari Panda · P. M. Visakh
Editors

Bionanocomposites from Marine Origin

Preparation and Applications

Springer

Editors
Shiv Kumari Panda
Udayanath Autonomous College of Science and Technology,Adaspur
Cuttack, Odisha, India

P. M. Visakh
Natural Bioactive Materials Laboratory
Ege University
Bornova, Izmir, Türkiye

ISSN 1869-8433 ISSN 1869-8441 (electronic)
Advanced Structured Materials
ISBN 978-3-032-10748-0 ISBN 978-3-032-10749-7 (eBook)
https://doi.org/10.1007/978-3-032-10749-7

This Springer imprint is published by the registered company Springer Nature Switzerland AG
The registered company address is: Gewerbestrasse 11, 6330 Cham, Switzerland

Preface

The book *Bionanocomposites from Marine Origin: Preparation and Applications* brings together recent research advances in the field of marine-derived bionanocomposites, highlighting their preparation methods, functional properties and wide-ranging applications. It explores both fundamental aspects and applied perspectives, offering readers a comprehensive view of how marine resources can be transformed into sustainable nanocomposites for diverse technological, environmental and biomedical uses. In this book, we try to discuss many topics such as bionanocomposites from marine origin: preparation and application: scope and future prospects, marine nanoscience and nanotechnology, nanocellulose for sustainable water purification, the fate of plastic in the ocean environment, the interaction of toxic chemicals with micro/nanoplastics, advanced functional nanomaterials for marine antifouling technology, enhancing bioremediation of oil spills using nanotechnology, chitosan and cellulose smart bionanocomposite for cardiovascular devices, applications of nanocomposites from marine waste and applications of marine nanoscience and nanotechnology. This book will serve as a valuable reference for university and college faculty members, professionals, postdoctoral researchers, senior graduate students as well as researchers from marine-based industries and academic institutions working in the field of marine-derived biocomposites and bionanocomposites. So various chapters in this book are contributed by prominent researchers from industry, academia and government/private research laboratories across the globe. It covers an up-to-date record on the major findings and observations in the field of bionanocomposites from marine origin. Chapter 1 discusses the bionanocomposites from marine origin: preparation and application: scope and future prospects. Chapter 2 in this book entitled marine nanoscience and nanotechnology. This chapter author discusses many topics such as background of nanoscience and nanotechnology-introduction, key applications of marine nanoscience, the importance of nanotechnology and nanoscience in marine environments, marine nanoscience and nanotechnology: traditional and advancements, challenges and future directions, advantages of marine nanoscience and nanotechnology, disadvantages of marine nanoscience and nanotechnology and future prospects for marine nanoscience and nanotechnology. Nanocellulose for sustainable water purification discussed in Chap. 3. In this chapter, authors discuss with mainly different parts, such as classification and properties, structure property relationship, enabled water purification technologies and perspectives summary and outlook.

Chapter on the fate of plastic in the ocean environment is the Chap. 4 of this book, with some better topics such as the scientific basis of plastic pollution, bioaccumulation of micro- and nanoplastics (MNPs), ecological consequences of plastic pollution in marine environments and effect of MNPs on fishes and other marine organisms. Chapter 5 based on the interaction of toxic chemicals with micro/nanoplastics discusses with many topics such absorption behavior of micro/nanoplastics and influencing factors, types of pollutants adsorbed in micro/nanoplastics and impact of pollutant-micro/nanoplastics interactions. The following chapter (Chap. 6) provides a better idea about advanced functional nanomaterials for marine antifouling technology. Enhancing bioremediation of oil spills using nanotechnology is discussed in Chap. 7. In this chapter, authors explained about the conventional oil spill response techniques, bioremediation of oil spills, challenges in bioremediation of oil spills, nanomaterials for oil spill remediation, production of nanoparticles and nano-enhanced bioremediation of oil spills. Chapter 8 discusses chitosan and cellulose smart bionanocomposite for cardiovascular devices; authors of this chapter explained with manly four parts such as introduction to chitosan and cellulose bionanocomposite, fabrication of chitosan cellulose bionanocomposite, chitosan and cellulose properties and modifications, modification of chitosan and cellulose bionanocomposite, smart functionalities in bionanocomposite for cardiovascular applications, application in cardiovascular system, bionanocomposite for designing cardiovascular devices

Chapter 9 in this book discusses applications of bionanocomposites from marine waste; in this chapter, authors discuss many different topics such as fabrication and properties of bionanocomposites from marine waste, applications of bionanocomposites from marine waste, biobased applications, engineering applications, medical applications and packing applications. Final chapter (Chap. 10) in this book discusses the applications of marine nanoscience and nanotechnology with different topics such as applications of marine nanoscience, applications of marine nanotechnology, future prospects and challenges.

Finally, we would like to express our sincere gratitude to all the contributors of this book, who made excellent support to the successful completion of this venture. We are grateful to them for the commitment and the sincerity they have shown toward their contribution in the book. Without their enthusiasm and support, the compilation of a book could have not been possible. We would like to thank all the reviewers who have taken their valuable time to make critical comments on each chapter. We also thank the publisher Springer for recognizing the demand for such a book and for realizing the increasing importance of the area of *Bionanocomposites from Marine Origin: Preparation and Applications* and for starting such a new project, in which not many other publishers put their hands on.

Cuttack, India — Dr. Shiv Kumari Panda
Bornova, Türkiye — Dr. P. M. Visakh

Contents

About the Editors

Dr. Shiv Kumari Panda earned her Ph.D. from KIIT University in 2018, following her M.Sc. from Sambalpur University, Odisha, in 2011. She has been serving as Assistant Professor at Udayanath Autonomous College of Science and Technology, Odisha, since 2018. Dr. Shiv Kumari Panda is actively engaged in research, with a primary focus on the synthesis and transformation of waste materials into high-performance biomaterials with novel applications. In recognition of her contributions, she was achieved a four-year research grant from the Odisha State Higher Education Council under the Odisha University Research and Innovation Incentivisation Plan (OURIIP). She has published extensively in reputed Scopus-indexed journals and contributed over 15 research articles and more than 10 book chapters. She is also actively engaged in scholarly publishing and editorial activities, currently collaborating with Dr. P. M. Visakh on more than ten book projects, with three volumes already in press. She remains highly active in the global research community, contributing as a peer reviewer for leading scientific journals, including *Bioresource Technology Reports* (Elsevier), *Fibers and Polymers* (Springer Nature) and *ACS Sustainable Chemistry and Engineering* (American Chemical Society). Additionally, she has served as a reviewer for the International Conference on Materials Engineering and Manufacturing Systems, published in IOP Science: Materials Science and Engineering.

Dr. P. M. Visakh (M.Sc., M.Phil., Ph.D.) is a prolific editor with more than **50** books already published. Currently, he is working in Natural Bioactive Materials Laboratory Department of Bioengineering, Ege University, Bornova/Izmir, Turkey. He was working as Associate Professor in Tusur University, Tomsk, Russia, from 2017 till 2023.He did his Post doc. Research in Tomsk Polytechnic University, Tomsk, Russia (2014 2017). He obtained his Ph.D., M.Phil. and M.Sc. degrees from School of Chemical Sciences, Mahatma Gandhi University, Kottayam, Kerala, India. He edited **50 books** from Scrivener (Wiley), Springer, Royal Society of Chemistry, Elsevier and more than **25 books** in press, (from Wiley, Springer, Royal Society of Chemistry and Elsevier). He has been invited as a visiting researcher in Russia (2014 to 2023), Portugal (2013, 2014), Czech Republic (2012, 2013) Italy (2009, 2012), Argentina (2010) Sweden (2010, 2011, 2012), Switzerland (2010), Spain (2011, 2012), Slovenia (2011), France (2011), Belgium (2012) and Austria (2012) for his research work. He visited 12 countries. He visited 15 universities in Europe. He published 20 publications, 4 reviews and more than **50** book chapters. He has attended and presented more than 28 conferences; he has **3797 citations** and his **h-index is 29**. He acts as a guest editor for 4 international journals. Listed in the "**World Ranking of Top** 2% **Scientists**-2023 and 2024" Source-**Stanford University, USA**. **His interest of research:** polymer sciences, polymer nanocomposites, material sciences, bio-nanocomposites, rubber-based nanocomposites, fire retardant polymers and liquid crystalline polymers, and silicon sensors.

Chapter 1
Bionanocomposites from Marine Origin: Preparation and Application: Scope and Future Prospects

Shiv Kumari Panda and P. M. Visakh

Abstract In this chapter providing a short version of all chapters, here we are writing about the different chapter topics such introduction to marine nanoscience and nanotechnology, nanocellulose for sustainable water purification, fate of plastics in the ocean environment: the plastic odyssey, the interaction of toxic chemicals with micro/nanoplastics, advanced functional nanomaterials for marine antifouling technology, enhancing bioremediation of oil spills using nanotechnology, marine polysaccharides as nano-drug delivery systems, applications of bionanocomposites from marine waste, applications of marine nanoscience and nanotechnology.

Keywords Marine nanoscience · Nanotechnology · Nanocellulose · Fate of plastics · Ocean environment · Plastic odyssey · Toxic chemicals · Micro and nanoplastics · Marine antifouling technology

1.1 Introduction to Marine Nanoscience and Nanotechnology

The twentieth century's scientific and technological advancements, which demonstrated the undeniable worth of humankind, established theoretical and practical underpinnings that made it feasible to examine the structure of matter and observe objects at the nanoscale (10–9 m). The size of made things, which must have at least one dimension no larger than 100 nm, is one of the criteria currently used to define nanotechnology (Antczak 2012). The physical and chemical characteristics of such objects should be controllable during creation, and they should be able to be used to construct larger objects. Because nanotechnology is interdisciplinary, it is frequently

S. K. Panda (✉)
Udayanath Autonomous College of Science & Technology, Adaspur, Cuttack, Odisha, India
e-mail: shivkumaripanda@gmail.com

P. M. Visakh
Natural Bioactive Materials Laboratory, Department of Bioengineering, Ege University, Bornova, Izmir, Turkey

S. K. Panda and V. P.M. (eds.), *Bionanocomposites from Marine Origin*, Advanced Structured Materials 247, https://doi.org/10.1007/978-3-032-10749-7_1

unnecessary and occasionally impossible to categorize it by prefixing the name of a certain scientific discipline with nano- (e.g., nanochemistry, nano-biotechnology). As a result, this field of study and technology ought to be regarded as a single nanotechnology. Nanotechnology serves as a platform that unites many fields to create profitable, easily marketable products. Marine nanoscience and nanotechnology represent an interdisciplinary field poised to address a wide range of challenges in marine environments. By leveraging the unique properties of nanomaterials, this field offers innovations in environmental monitoring, marinc biotechnology, energy harvesting, and water purification. However, the potential applications of marine nanoscience are not without risks, particularly regarding the environmental impact of nanoparticles on marine ecosystems (Corsi et al. 2014). Nanotechnology, the manipulation of materials at the nanoscale (typically 1–100 nm), has revolutionized numerous scientific and industrial fields. In the marine environment, nanomaterials are being explored for applications that range from pollutant detection to energy production and drug delivery. Marine nanoscience, an emerging subdiscipline, takes advantage of the unique chemical, physical, and biological properties of nanomaterials to solve complex problems in marine ecosystems (Hochella 2002).

Despite its potential, concerns over the environmental impact and safety of nanomaterials in delicate maritime environments are also brought up by the field's quick development. The purpose of this review is to give a summary of the latest findings in the subject, describe the uses of marine nanoscience, and talk about nanotechnology in marine settings. Innovative approaches to contaminant removal in maritime settings are provided by nanotechnology. The cost, effectiveness, and scalability of traditional remediation techniques are frequently constrained, particularly in large-scale operations like oil spill response. Oil may be efficiently captured and removed from the water's surface using nanomaterials like magnetic nanoparticles and nanostructured sponges. Cleaning can be made simpler and more effective by engineering these materials to be extremely absorbent and magnetically separable (Singh et al. 2020). Removal of Heavy Metals: By removing dangerous heavy metals from seawater, nanomaterials can lower the possibility that they will bioaccumulate in marine food chains. Studies on nanocomposites and nanofiltration systems have demonstrated the potential to extract toxic metals from water with higher efficiency than conventional methods. For instance, nanoparticles with magnetic properties have been employed to clean oil and chemical contaminants from water surfaces, offering a cleaner and faster alternative to traditional methods like dispersants and skimmers (Mehmood et al. 2021).

1.2 Nanocellulose for Sustainable Water Purification

Nanocellulose (NC) is a nanomaterial characterized by a diameter measuring less than 100 nm, obtained from natural cellulose through specialized processing techniques. Its specific surface area can attain approximately 500 m^2/g, in stark contrast to soft cellulose pastes, which typically exhibit a surface area ranging from 1 to

4 m^2/g (Banavath et al. 2011). This substantial increase in surface area is associated with a greater availability of hydroxyl groups on the surface of nanocellulose, facilitating the grafting of functional groups or molecules via various methods, including carboxylation, etherification, silicification, sulfonation, esterification, 2,2,6,6-tetramethylpiperidin-1-yloxy radical (TEMPO)-mediated oxidation, phosphorylation, and amidation (Isogai et al. 2011). The field of biomaterials science currently leverages the structural and functional properties of nanocellulose to develop innovative materials with multifunctional characteristics, addressing contemporary demands across diverse sectors such as biomedicine, environmental science, energy, pharmaceuticals, agriculture, and food technology. Notably, it has been established that the incorporation of carboxyl, sulfonate, and phosphonate functional groups onto nanocellulose can effectively improve the selective absorption of contaminants in water remediation processes (Sehaqui et al. 2016). Furthermore, nanocellulose-based membranes and filters have proven effective in the removal of metal ions, dyes, and microorganisms, while modified nanocellulose matrices with hydrophobic or lipophilic functional groups can also eliminate organic contaminants, including oil and cyclohexene. The high crystallinity of sea squirt nanocellulose presents challenges for its solubility in water, as it tends to swell rather than dissolve, thereby limiting its potential applications. Consequently, extracted sea squirt nanocellulose is often processed into a more functional form. This tunicate-derived nanocellulose is characterized by exceptional properties, including high strength, purity, and Young's modulus, as well as notable optical, electrical, magnetic, and rheological attributes. The high concentration of hydroxyl groups present on the surface of nanocellulose enables various potential functional modifications. The applications of nanocellulose extracted from tunicates span various fields, including biomedicine, polymer-reinforced materials, papermaking, food packaging, and environmental engineering, which have attracted considerable scholarly interest. Nonetheless, the extraction and processing of cellulose from tunicates pose significant challenges that hinder its wider application. The methodology employed for extraction has a significant impact on the quality of the cellulose produced. A variety of research efforts have effectively isolated nanocellulose from the outer integument of tunicates. As it reported, Belton et al. (1989) utilized Dyson's enzyme to eliminate non-cellulosic constituents from the external layer of periphyton, subsequently subjecting the material to reflux with sodium hydroxide to obtain crude cellulose, albeit with residual colored components. At present, the primary techniques for cellulose extraction from tunicates consist of alkali treatment and the kraft process.

1.3 Fate of Plastics in the Ocean Environment: The Plastic Odyssey

The sources of marine plastic pollution are multifaceted, primarily stemming from land-based activities. Research indicates that approximately 80% of marine plastic debris is of terrestrial sources, including improper waste management and industrial runoff (Yap et al. 2022). Rivers play a pivotal function in delivering plastic waste to the oceans, with studies revealing that over 1000 rivers account for 80% of global riverine plastic emissions (Meijer et al. 2021). This transboundary nature of plastic pollution complicates mitigation efforts, as it requires coordinated action across multiple jurisdictions and stakeholders (Zhou and Luo 2024). Moreover, the economic drivers behind plastic production and consumption, such as the agricultural and fishing sectors, contribute significantly to the problem, with practices like plastic mulching and abandoned fishing gear being key contributors (Abalansa et al. 2020). Plastic pollution in the marine environment is significantly influenced by domestic usage of plastic products, which contributes to the vast amounts of plastic waste entering the marine environments. A primary source of this pollution stems from single-use plastics (SUPs), which have gained widespread popularity due to their convenience and low cost. Items such as plastic bags, food packaging, and straws are often discarded improperly, leading to their eventual entry into waterways and, subsequently, the ocean (Canning-Clode et al. 2020). The COVID-19 pandemic has exacerbated this issue, as the increased use of personal protective equipment (PPE) and other SUP items has resulted in a surge of plastic waste, further complicating waste management efforts and increasing marine litter (Canning-Clode et al. 2020). Marine plastic pollution is a pressing environmental issue characterized by the widespread chemical composition of marine plastics, which is complex, comprising various synthetic polymers and additives that can pose significant risks to marine life and human health. Plastics' physical and chemical characteristics contribute to their persistence in the environment. The primary constituents of marine plastic pollution are synthetic polymers, which are derived from petrochemical sources. The prevalent plastic polymers identified within marine ecosystems include polyethylene (PE), polypropylene (PP), polystyrene (PS), and polyvinyl chloride (PVC) (Rochman 2015). These materials are favored for their durability, lightweight nature, and versatility, making them prevalent in consumer products, packaging, and industrial applications. Microplastics, defined as plastic particles less than 5 mm in size, dominate marine debris and are derived from the fragmentation of larger plastic items or primary microplastics used in products like cosmetics. The composition of these microplastics varies, with common types including fragments of PE, PP, and PS, which have different densities affecting their buoyancy and their dissemination throughout marine environments. PE and PP are considered the most abundant groups of microplastics and nanoplastics found in marine waters around the globe. It is not unusual for PE as it is the most widely produced and plastic material in the world (Expósito et al. 2021). From day-to-day items to personal care products to medical utensils, the abundance of PE is widely recognizable (Lei et al. 2017). Not only that, the low-density nature and therefore their tendency to float in water are also the reasons behind the abundance of both PE and PP in marine waters.

1.4 The Interaction of Toxic Chemicals with Micro/Nanoplastics

Although there is a wide variety of plastics, with a notorious spectrum of physical and chemical properties and the presence of various additives, these materials commonly have low density, modulation of their physicochemical properties, as well as mechanical, chemical, and biological resistance, which allows for a broad versatility that results in their usage in various everyday applications, including storage systems, textile materials, construction elements, and medical supplies (Komorowska-Kaufman and Marciniak 2024). As a consequence of these characteristics, over 320 million tons of plastic are produced globally annually, approximately 40% of which is used for single-use packaging, and about 8 million tons are deposited in the environment (Ajithkumar et al. 2024; Palencia et al. 2016; Thompson et al. 2009; Sudarsan et al. 2024). The latter is an aspect of considerable concern because, despite the numerous benefits of these materials, there are serious environmental considerations related to every stage of their life cycle. Regarding their production, synthetic plastics are typically generated from raw materials such as fossil fuels, petroleum, and natural gas, whose derivatives generate monomers such as ethylene and styrene; in this conversion, pollutants and greenhouse gases such as carbon dioxide and nitrogen oxides are formed (Kim et al. 2023). Regarding their final disposal, synthetic plastics can be reused, recycled, and predominantly deposited in landfills; nevertheless, due to inadequate management practices, a significant amount ends up in the environment, leading to contamination processes. A typical example of this issue is the presence of low-density plastics, such as polystyrene (PS) and polypropylene (PP), in marine and coastal ecosystems (Agboola and Benson 2021; Bodus et al. 2024; Liu et al. 2021; Um et al. 2023). Additionally, MNPs exposed to environmental conditions undergo aging processes, such as oxidation, photodegradation, and erosion, which alter their crystalline structure, thermal stability, mechanical strength, and surface composition.

The latter is one of the crucial factors in the modulation of interactions between MNPs with other entities, so that both pristine and aged MNPs can interact with different substances in the environment, including natural and anthropogenic pollutants, an interaction that can lead to the formation of composite pollutants, which may have a synergistic effect on toxicity and environmental impact (Al Harraq and Bharti 2021; Shukla et al. 2024; Xie et al. 2025). Therefore, MNPs can absorb, accumulate, transport, and desorb chemical, physical, and biological contaminants through environmental compartments, trophic levels, and tissues. This process increases the mobility of these contaminants in the environment and enhances the likelihood of their entry into organisms and natural systems at high concentrations, leading to problems such as soil structure alteration, bioaccumulation in aquatic and terrestrial organisms, and severe and chronic diseases in humans, such as tissue damage and carcinogenesis (Thompson et al. 2009; Shukla et al. 2024; Agboola and Benson 2021; Forero et al. 2022; Costigan et al. 2022; Porta 2021; Thomas 2022; Liang et al. 2024; Zhi et al. 2025; Beggel et al. 2024).

1.5 Advanced Functional Nanomaterials for Marine Anti-fouling Technology

Among the many anti-fouling technologies, the application of coatings (paints) is the most cost-effective and widely applicable method, accounting for about 76% of the market share. However, traditional anti-fouling coatings release harmful anti-fouling agents (such as organotin) that are difficult to decompose, polluting the ecological environment. Moreover, these substances can enter the human food chain through fish, shellfish, and other organisms, posing a serious threat to human health. Currently, most countries have restricted the use of traditional anti-fouling agents. The development of green anti-fouling coatings has become an important direction in marine anti-fouling. Due to the extremely small size of nanomaterials, they possess unique physical and chemical properties, and the resulting effects are commonly referred to as the nano-effect. By combining nanomaterials with polymers, nanocomposite materials can be produced. Nanocomposite coatings refer to the addition of nanomaterials (with at least one dimension in the three-dimensional space within the size range of 1 to 100 nm) to coatings (Nguyen-Tri et al. 2018), enhancing the physical and chemical properties of the coatings through their small size effects, surface effects, quantum size effects, and dielectric confinement effects. Currently, nanomaterials can be applied in fields such as antimicrobial coatings, anti-fouling coatings, waterproof coatings, weather-resistant coatings, stealth coatings, wear-resistant coatings, and anti-corrosion coatings (Necolau and Pandele 2020). For anti-fouling coatings, nanoscale particles form strong hydrogen bonding with the coating, improving the compactness and resistance to ion penetration of the coating. Nanoparticles can not only improve the rheological properties of the coating, enhance adhesion, hardness, and aging resistance, and reduce the surface roughness of the coating, but also enhance the surface roughness of the coating, forming a micro/nanocomposite structure, improving the sample's hydrophobicity, and achieving a self-cleaning effect. Additionally, some nanomaterials have catalytic functions and can induce the apoptosis of microorganisms or catalytically decompose adhered substances through the generation of reactive groups, effectively preventing biological fouling. The reactive groups produced during use can be rapidly degraded, ensuring good safety and not causing environmental pollution. In response to the basic requirements of environmental friendliness and stability of nano-anti-fouling coatings, the development of organic hybrid coatings containing nanofillers is an environmentally friendly alternative to traditional anti-fouling methods. By embedding nanoparticles into the polymer matrix, the mechanical properties, surface morphology, and active sites of the coating are fundamentally improved, with the aim of achieving better performance. At the same time, the doping of nanoparticles supports the preparation of ternary or quaternary nanocomposite materials, realizing the synergistic anti-fouling effect. This chapter focuses on the research progress of low surface energy nanocomposite anti-fouling coatings, superhydrophobic nanocomposite anti-fouling coatings, release-type nanocomposite anti-fouling coatings, catalytic nano-anti-fouling coatings, and multi-factor synergistic nano-anti-fouling coatings according to their different anti-fouling mechanisms.

1.6 Enhancing Bioremediation of Oil Spills Using Nanotechnology

A notable example is Corexit, a dispersant extensively used during the Deepwater Horizon spill cleanup. Studies on Corexit's efficacy and effects have produced mixed results. While it has been shown to effectively disperse oil, concerns have been raised about its potential to inhibit bioremediation processes. Additionally, adverse effects have been observed in human health studies, including reduced cell viability and DNA damage in cultured human bronchial cells after Corexit exposure. However, some findings have been challenged, citing inconsistencies in laboratory methods, high exposure thresholds, and reliance on self-reported data for human health impacts. Despite these concerns, dispersants remain a key tool for addressing large-scale oil spills, as they can enhance biodegradation by increasing oil droplet surface area (Dhaka and Chattopadhyay 2021).

Advantages of Chemical Methods

Chemical methods such as dispersants and oil herders are versatile and can be deployed in various environmental conditions, including those where traditional mechanical methods fail. Their ability to break down or contain oil slicks plays a critical role in minimizing environmental damage and facilitating subsequent cleanup efforts. However, concerns about long-term environmental persistence and potential health impacts highlight the importance of careful application and ongoing research. Bioremediation, the use of microorganisms to break down environmental contaminants, was first utilized six decades ago by George Robinson, who deployed microbes to address oil spills along the Santa Barbara coast in California (Upadhyay 2024). This process involves microorganisms and their metabolic byproducts to eliminate pollutants from aquatic systems and sediments. Indigenous microorganisms in sediments play a critical role in this process, as they can degrade organic compounds into simpler inorganic molecules or their elemental forms through a process known as mineralization. Microorganisms in bioremediation can neutralize, reduce, and transform oil contaminants into naturally occurring compounds in air, sediment, water, sludge, or soil. This process often occurs alongside detoxification, where microorganisms immobilize or remove waste materials through transformations and mineralization. For microbial cells to reproduce and remain viable, they require energy, typically derived from redox reactions such as respiration. Essential elements for this energy system include nutrients and electron donors/acceptors, such as oxygen, carbon dioxide, iron (III), and sulfate.

1.7 Marine Polysaccharides as Nano-drug Delivery Systems

Marine polysaccharides are primarily used in the food and cosmetic industries. Still, they are also widely present in pharmaceutical sciences, with an increasing interest in integrating them as materials for incorporating bioactive agents. Marine algae are the primary source of marine polysaccharides, but they can also be obtained from animal sources, such as the skeletons of crustaceans and cartilaginous fish tissue. Some polysaccharides can be extracted from marine microorganisms, like some prokaryotes. Marine polysaccharides are a large complex group of different macromolecules with different biological properties. Polysaccharides may exhibit different chemical structures and biological properties such as biocompatibility, biodegradability, adhesive properties, and the ability to form hydrogels. Among the many marine polysaccharides, one group stands out: sulfated polysaccharides (Cardoso et al. 2016) in comparison with the incorporation of bioactive agents (Laurienzo 2010). Marine algae are the primary source of marine polysaccharides, but they can also be obtained from animal sources, such as the skeletons of crustaceans and cartilaginous fish tissue. Some polysaccharides, like some prokaryotes, can be extracted from marine microorganisms (Senni et al. 2011). Marine polysaccharides are a large complex group of different macromolecules with different biological properties. Polysaccharides may exhibit different chemical structures and biological properties, such as biocompatibility, biodegradability, adhesive properties, and the ability to form hydrogels. Most conventional extraction processes (e.g., soxhlet, solid–liquid extraction, or liquid–liquid extraction) are characterized by long extraction times and comprise hazardous acidic and alkali treatments (Ibañez et al. 2011).

These conventional extraction methods are also associated with both low selectivity and extraction yields of bioactive compounds. To tackle these issues, the scientific community has been looking for eco-friendly and efficient approaches to designing new materials that follow the principles of Green Chemistry. Those techniques are mostly based on compressed fluids, including supercritical fluid extraction, pressurized hot water extraction, and pressurized liquid extraction, also known as accelerated solvent extraction (Sacramento et al. 2022). Aside from compressed fluids, extraction processes encompassing subcritical water extraction (SWE), ultrasound-assisted extraction (UAE), and microwave-assisted extraction (MAE) have also been used to surpass the bottlenecks of the standard extraction procedures since they could be applied at pilot scale, thus offering the possibility to be integrated into a process chain within a biorefinery; this summarizes each green extraction technique's main features, advantages, and significant drawbacks.

1.8 Applications of Bionanocomposites from Marine Waste

Biological methods have emerged as an eco-friendly approach for producing nanocomposites, utilizing bacteria, fungi, or enzymes instead of harsh chemicals. This makes them particularly suitable for applications in the medical and food sectors. For instance, enzymatic processes have been employed to synthesize nanocomposites from chitosan and alginate, which helps preserve the bioactivity of the source materials while minimizing toxicity. This is ideal for food packaging and biomedical uses (Niculescu and Grumezescu 2022; Khubiev et al. 2023). Additionally, Ag/chitosan nanocomposites have been synthesized through enzymatic methods and utilized as antibacterial films for food preservation, effectively preventing microbial growth and extending shelf life (Canama et al. 2023; Kyung and Ko 2014). Similarly, microbial processes had been used to prepare metal nanoparticles, such as silver or gold, combined with collagen or chitosan. These composites were produced using *Bacillus* or *Aspergillus* and demonstrate enhanced antibacterial effects, making them suitable for wound healing and medical device applications (El-Zawawy et al. 2023; Arif and Uddin 2021). Furthermore, green synthesis methods had been employed to create nanocomposites like collagen-gold and alginate-copper, which possess antioxidative properties. These composites were formulated using plant extracts and enzymes and are applied in cosmetics and dermatological products (Akturk et al. 2016; Anbalagan 2024).

Physical synthesis methods, such as electrospinning, freeze-drying, and mechanical milling, have been utilized to create nanocomposites with improved mechanical properties and specific structures, thereby expanding their applications in various fields. For instance, chitosan/HA nanofibers were produced using the electrospinning technique. These products demonstrated high mechanical strength and surface area, making them ideal for use as scaffolds in tissue engineering and bone regeneration (Frohbergh et al. 2012; Van Hong Thien et al. 2013). Additionally, the freeze-drying method helped maintain the porous structure of chitosan, alginate, or gelatin-based nanocomposites, which were well suited for wound dressings that required high biocompatibility and structural integrity. Mechanical milling and other physical techniques were reported to further enhance particle distribution and uniformity within these composites, broadening their applications in areas like environmental filtration and food packaging (Afjoul et al. 2020; Gorrasi and Sorrentino 2015). Typically, the characteristics of a composite are largely dependent on the properties of each component and their interactions. Given the valuable bioactive components, the applicability of marine-based nanocomposites spans a wide range of industries. These bionanocomposites are primarily divided into three groups: (i) nanocomposites consisting of metal nanoparticles and biopolymers, (ii) hybrids of metal oxides, hydroxides, carbon-based materials (graphene oxide (GO), graphene, carbon nanotubes (CNT)), or salt nanoparticles combined with biopolymers, and (iii) blends of two or more biopolymers at the nanoscale. The prepared marine-based nanocomposites demonstrated unique bioactive properties, including antioxidative, antimicrobial, biocompatible, and biodegradable activities, as well as effects related to their nanosize. Consequently, these nanocomposites were studied and applied in various fields, including biomedical and nutritional applications, functional food packaging, controlled drug delivery systems, cosmetics for antiaging products, food preservation, adjuncts in cancer therapies, tissue engineering scaffolds, wound dressings, and water pollution treatment.

1.9 Applications of Marine Nanoscience and Nanotechnology

Nanoscience and nanotechnology are innovative revolutionary fields focused on the study, manipulation, and application of materials at the nanoscale (typically between 1 and 100 nm) (Joudeh and Linke 2022; Mekuye and Abera 2023). At this scale, materials exhibit unique physical, chemical, and biological properties, opening doors to advanced and innovative technologies beyond traditional methods (Hassett 2021; Patra and Baek 2014; Munyalo and Zhang 2018). In the transportation sector, nanotechnology has contributed significantly to the progress made in reducing the weight of vehicles. This reduction translates into increased efficiency, higher speeds, improved safety, and lower energy consumption (Ahmed et al. 2019; Mathew et al. 2019). For example, nanoparticle-based steel alloys produce lighter yet stronger materials with higher strength-to-weight ratios than traditional metals. Nanoscience studies the sources of nanoparticles in ocean waters, which may originate from hydrothermal vents, volcanic activity, or chemical reactions. It also analyzes the chemical characteristics of these nanoscale structures and their interactions within marine ecosystems. Additionally, the transport of natural nanoparticles in the ocean, along with their functions in pollutant binding and nutrient recycling, is evaluated (Barhoum et al. 2022). Marine nanoscience is crucial for comprehending the influence of environmental and climate changes on marine life (Maltby 2022; Craig 2012). Research investigates how ocean acidification impacts nanoscale carbonate structures in microorganisms, specifically concerning fluctuations in pH. Moreover, the composition of nanoscale aerosols generated by ocean currents and their role in climate regulation is analyzed.

Nanotechnology is crucial in addressing biofouling, which happens when marine organisms like algae and barnacles accumulate on the hulls of ships and other submerged infrastructure (Sinha 2023; Krishna Mohan et al. 2024). Anti-fouling nanocoatings are notably effective in preventing the accumulation of these organisms, resulting in decreased maintenance expenses and enhanced operational efficiency. By reducing drag from biofouling, these coatings improve the speed and fuel efficiency of vessels, making them more environmentally friendly and economical. Nanomaterials also enhance the performance of tidal and wave energy conversion technologies (Rosa-Santos et al. 2022; Roy et al. 2023; Wang et al. 2019). Furthermore, nanotechnology supports the development of high-capacity, durable batteries for underwater drones and marine vessels while improving the durability of underwater energy harvesters and turbines via nanocoatings that lower maintenance expenses (Radhakrishnan 2022; Liu et al. 2023). Nanoparticles are employed specifically to target and manage invasive species while preserving native ones. These materials are also beneficial in the restoration of marine habitats, including aiding coral growth in reef restoration efforts. Nanoprobes and sensors are essential for examining marine microorganisms, plankton, and other minute organisms. Furthermore, they help evaluate the effects of pollution and climate change on these marine microorganisms (Plank and Law 2012; Fujii et al. 2020).

References

Abalansa S, Mahrad B, Vondolia G, Icely J, Newton A (2020) The marine plastic litter issue: a social-economic analysis. Sustainability 12(20):8677. https://doi.org/10.3390/su12208677

Afjoul H, Shamloo A, Kamali A (2020) Freeze-gelled alginate/gelatin scaffolds for wound healing applications: an in vitro, in vivo study. Mater Sci Eng C 113:110957

Agboola OD, Benson NU (2021) Physisorption and chemisorption mechanisms influencing micro (nano) plastics-organic chemical contaminants interactions: a review. Front Environ Sci 9:678574

Ahmed I et al (2019) Applications of nanotechnology in transportation engineering. Sustain Infrastruct Break Res Pract. https://doi.org/10.4018/978-1-7998-0948-7.ch044

Ajithkumar V, Philomina A, Meena K, Pothiaraj G, Dey D et al (2024) Insights of recent developments in microplastics pollution and its degradation in wastewater and other environment. J Taiwan Inst Chem Eng 166:105504

Akturk O et al (2016) Collagen/gold nanoparticle nanocomposites: a potential skin wound healing biomaterial. J Biomater Appl 31(2):283–301

Al Harraq A, Bharti B (2021) Microplastics through the lens of colloid science. ACS Environ Au 2(1):3–10

Anbalagan G et al (2024) Green synthesis of copper and copper oxide nanoparticles from brown algae turbinaria species' aqueous extract and its antibacterial properties. Cureus 16(4)

Antczak T (2012) Nanotechnology-methods of manufacturing cellulose nanofibres. Fibres Text East Eur 20(2):91

Arif R, Uddin R (2021) A review on recent developments in the biosynthesis of silver nanoparticles and its biomedical applications. Med Devices Sens 4(1):e10158

Banavath HN, Bhardwaj NK, Ray AK (2011) A comparative study of the effect of refining on charge of various pulps. Biores Technol 102(6):4544–4551

Barhoum A, Jeevanandam J, Danquah MK (2022) Fundamentals of bionanomaterials. https://doi.org/10.1016/C2020-0-00039-3

Beggel S, Kalis EJJ, Geist J (2024) Towards harmonized ecotoxicological effect assessment of micro- and nanoplastics in aquatic systems. Environ Pollut 366:125504

Belton PS, Tanner SF, Cartier N, Chanzy H (1989) High-Resolution solid-state 13C nuclear magnetic resonance spectroscopy of tunicin, an animal cellulose. Macromolecules 22(4):1615–1617

Bodus B, O'Malley K, Dieter G, Gunawardana C, McDonald W (2024) Review of emerging contaminants in green stormwater infrastructure: antibiotic resistance genes, microplastics, tire wear particles, PFAS, and temperature. Sci Total Environ 906:167195
Canama GJC et al (2023) Synthesis of chitosan-silver nanocomposite and its evaluation as an antibacterial coating for mobile phone glass protectors. ACS Omega 8(20):17699–17711
Canning-Clode J, Sepulveda P, Almeida S, Monteiro J (2020) Will covid-19 containment and treatment measures drive shifts in marine litter pollution? Front Mar Sci 7:00691. https://doi.org/10.3389/fmars.2020.00691
Cardoso MJ, Costa RR, Mano JF (2016) Marine origin polysaccharides in drug delivery systems. Mar Drugs 14(2):34
Corsi I, Cherr GN, Lenihan HS, Labille J, Hassellov M, Canesi L et al (2014) Common strategies and technologies for the ecosafety assessment and design of nanomaterials entering the marine environment
Costigan E, Collins A, Hatinoglu M, Bhagat K, MacRae J et al (2022) Adsorption of organic pollutants by microplastics: overview of a dissonant literature. J Hazard Mater Adv 6:100091
Craig RK (2012) Marine biodiversity, climate change, and governance of the oceans. Diversity 4(2):224–238
Dhaka A, Chattopadhyay P (2021) A review on physical remediation techniques for treatment of marine oil spills. J Environ Manage 288:112428
El-Zawawy NA et al (2023) Mycosynthesis of silver nanoparticles from endophytic Aspergillus flavipes AUMC 15772: ovat-statistical optimization, characterization and biological activities. Microb Cell Factories 22(1):228
Expósito N, Rovira J, Sierra J, Folch J, Schuhmacher M (2021) Microplastics levels, size, morphology and composition in marine water, sediments and sand beaches. Case study of Tarragona coast (western Mediterranean). Sci Total Environ 786:147453
Forero D, Fabiani M, Lassalle VL, Spetter CV, Fernandez D (2022) Critical review of the characteristics, interactions, and toxicity of micro/nanomaterials pollutants in aquatic environments. Mar Pollut Bull 174:113276
Frohbergh ME et al (2012) Electrospun hydroxyapatite-containing chitosan nanofibers crosslinked with genipin for bone tissue engineering, pp 9167–9178
Fujii T, Pondella DJ, Todd VL, Guerin A (2020) Seafloor heterogeneity: artificial structures and marine ecosystem dynamics. Front Mar Sci 7:378. https://doi.org/10.3389/fmars.2020.00378
Gorrasi G, Sorrentino A (2015) Mechanical milling as a technology to produce structural and functional bio-nanocomposites. Green Chem 17(5):2610–2625
Hassett KJ et al (2021) Impact of lipid nanoparticle size on mRNA vaccine immunogenicity. J Control Release 335:237–246
Hochella MF Jr (2002) Nanoscience and technology: the next revolution in the Earth sciences. Earth Planet Sci Lett 203(2):593–605
Ibañez E, Herrero M, Mendiola JA, Castro-Puyana M (2011) Extraction and characterization of bioactive compounds with health benefits from marine resources: macro and micro algae, cyanobacteria, and invertebrates. Marine bioactive compounds: sources, characterization and applications. Springer, Boston, MA, pp 55–98
Isogai A, Saito T, Fukuzumi H (2011) TEMPO-oxidized cellulose nanofibers. Nanoscale 3(1):71–85
Joudeh N, Linke D (2022) Nanoparticle classification, physicochemical properties, characterization, and applications: a comprehensive review for biologists. J Nanobiotechnology 20(1):262. https://doi.org/10.1186/s12951-022-01477-8
Khubiev OM et al (2023) Chitosan-based antibacterial films for biomedical and food applications. Int J Mol Sci 24(13):10738
Kim SW, Kim YT, Tsang YF, Lee J (2023) Sustainable ethylene production: recovery from plastic waste via thermochemical processes. Sci Total Environ 903:166789
Komorowska-Kaufman M, Marciniak W (2024) Removal of microplastic particles during municipal wastewater treatment: a current review. Desalin Water Treat 317:100006

Krishna Mohan MV, Srinivasa Reddy P, Bhanu Prakash TVK, Mukheerjee A (2024) Impact of titanium and silicon metal oxide nanoparticles on surface coatings for marine vehicles to mitigate biofouling. Part Sci Technol 42(6):915–928
Kyung GS, Ko SH (2014) Antimicrobial chitosan-silver nanocomposite film prepared by green synthesis for food packaging. Korean Journal of Food Science and Technology 46(3):347–351
Laurienzo P (2010) Marine polysaccharides in pharmaceutical applications: an overview. Mar Drugs 8(9):2435–2465
Lei K, Qiao F, Liu Q, Wei Z, Qi H, Cui S, Yue X, Deng Y, An L (2017) Microplastics releasing from personal care and cosmetic products in China. Mar Pollut Bull 123(1–2):122–126
Liang D, Shan R, Gu J, Cheng P, Wang S et al (2024) Analysis, risk assessment and treatment of aquatic micro/nanoplastics: a critical review. Sep Purif Technol 354:129418
Liu W, Zhang J, Liu H, Guo X, Zhang X et al (2021) A review of the removal of microplastics in global wastewater treatment plants: characteristics and mechanisms. Environ Int 146:106277
Liu J, Bao B, Chen J, Wu Y, Wang Q (2023) Marine energy harvesting from tidal currents and offshore winds: a 2-DOF system based on flow-induced vibrations. Nano Energy 114:108664
Maltby KM et al (2022) Marine climate change risks to biodiversity and society in the ROPME Sea Area. Clim Risk Manag 35:100411
Mathew J, Joy J, George SC (2019) Potential applications of nanotechnology in transportation: a review. J King Saud Univ Sci 31(4):586–594. https://doi.org/10.1016/j.jksus.2018.03.015
Mehmood A, Khan FSA, Mubarak NM, Mazari SA, Jatoi AS, Khalid M et al (2021) Carbon and polymer-based magnetic nanocomposites for oil-spill remediation—a comprehensive review. Environ Sci Pollut Res 28(39):54477–54496
Meijer L, Emmerik T, Ent R, Schmidt C, Lebreton L (2021) More than 1000 rivers account for 80% of global riverine plastic emissions into the ocean. Sci Adv 7(18):eaba5803. https://doi.org/10.1126/sciadv.aaz5803
Mekuye B, Abera B (2023) Nanomaterials: an overview of synthesis, classification, characterization, and applications. Nano Select 4(8):486–501
Munyalo JM, Zhang X (2018) Particle size effect on thermophysical properties of nanofluid and nanofluid based phase change materials: a review. J Mol Liq 265:77–87. https://doi.org/10.1016/j.molliq.2018.05.129
Necolau MI, Pandele AM (2020) Recent advances in graphene oxide-based anticorrosive coatings: an overview. Coatings 10(12):1149
Nguyen-Tri P, Nguyen TA, Carriere P et al (2018) Nanocomposite coatings: preparation, characterization, properties, and applications. Int J Corros 2018:1–19
Niculescu A, Grumezescu A (2022) Applications of chitosan-alginate-based nanoparticles—an up-to-date review. Nanomaterials 12(2):186
Palencia M, Berrio M, Melendrez M (2016) Nanostructured polymer composites with potential applications into the storage of blood and hemoderivates. J Sci Technol Appl 1:4–14
Patra JK, Baek KH (2014) Green nanobiotechnology: factors affecting synthesis and characterization techniques. J Nanomater. https://doi.org/10.1155/2014/417305
Plank MJ, Law R (2012) Ecological drivers of stability and instability in marine ecosystems. Theor Ecol 5(4):465–480
Porta R (2021) Anthropocene, the plastic age and future perspectives. FEBS Open Biol 11(4):948–953
Radhakrishnan S, Joseph S, Jelmy EJ, Saji KJ, Sanathanakrishnan T, John H (2022) Triboelectric nanogenerators for marine energy harvesting and sensing applications. Results Eng 15:100487
Rochman C (2015) The complex mixture, fate and toxicity of chemicals associated with plastic debris in the marine environment, 117–140. https://doi.org/10.1007/978-3-319-16510-3_5
Rosa-Santos P, Taveira-Pinto F, López M, Rodríguez CA (2022) Hybrid systems for marine energy harvesting. J Mar Sci Eng 10(5):633. https://doi.org/10.3390/jmse10050633
Roy RB, Alahakoon S, Arachchillag SJ, Rahman S (2023) Optimizing dynamic electric ferry loads with intelligent power management. Energy Rep 9:5952–5963

Sacramento MM, Borges J, Correia FJ, Calado R, Rodrigues JM, Patrício SG, Mano JF (2022) Green approaches for extraction, chemical modification, and processing of marine polysaccharides for biomedical applications. Front Bioeng Biotechnol 10:1041102

Sehaqui H, Mautner A, Perez de Larraya U, Pfenninger N, Tingaut P, Zimmermann T (2016) Cationic cellulose nanofibers from waste pulp residues and their nitrate, fluoride, sulphate and phosphate adsorption properties. Carbohydr Polym 135:334–340

Senni K, Pereira J, Gueniche F, Delbarre-Ladrat C, Sinquin C, Ratiskol J et al (2011) Marine polysaccharides: a source of bioactive molecules for cell therapy and tissue engineering. Mar Drugs 9(9):1664–1681

Shukla S, Khanna S, Khanna K (2024) Unveiling the toxicity of micro-nanoplastics: a systematic exploration of understanding environmental and health implications. Toxicol Rep 14:101844

Singh B, Kumar S, Kishore B, Narayanan TN (2020) Magnetic scaffolds in oil spill applications. Environ Sci Water Res Technol 6(3):436–463

Sinha S et al (2023) Nanotechnology-based solutions for antibiofouling applications: an overview. ACS Appl Nano Mater 6(14):12828–12848

Sudarsan JS, Dogra K, Kumar R, Raval NP, Leifels M et al (2024) Tricks and tracks of prevalence, occurrences, treatment technologies, and challenges of mixtures of emerging contaminants in the environment: with special emphasis on microplastic. J Contam Hydrol 265:104389

Thomas KV (2022) Understanding the plastics cycle to minimize exposure. Nat Sustain 5:282–284

Thompson RC, Swan SH, Moore CJ, von Saal FS (2009) Our plastic age. Phylosophical Trans R Soc B 364:1973–1976

Um M, Weerackody D, Gao L, Mohseni A, Evans B, Murdoch B, Schmidt J, Fan L (2023) Investigating the fate and transport of microplastics in a lagoon wastewater treatment system using a multimedia model approach. J Hazard Mater 446:130694

Upadhyay S (2024) Bioremediation: a sustainable solution

Van Hong Thien D, Hsiao SW, Ho MH, Li CH, Shih JL (2013) Electrospun chitosan/hydroxyapatite nanofibers for bone tissue engineering. J Mater Sci 48(4):1640–1645

Wang Z, Carriveau R, Ting DSK, Xiong W, Wang Z (2019) A review of marine renewable energy storage. Int J Energy Res 43(12):6108–6150. https://doi.org/10.1002/er.4444

Xie A, Chen S, Liang X, Li L, Song Y, Lv M et al (2025) Influence of microplastic aging on the adsorption and desorption behavior of Ni(II) under various aging conditions. Environ Geochem Health 47(3):87

Yap H, Shukor M, Yasid N (2022) The microplastics occurrence and toxic effects in marine environment. J Environ Microbiol Toxicol 10(2):1–6. https://doi.org/10.54987/jemat.v10i2.733

Zhi L, Li Z, Li R, Su Z, Chen F et al (2025) Micro/nanoplastics in aquaponic systems: sources, features and their effects on the immune system of fish. TrAC Trends Anal Chem 184:118158

Zhou J, Luo D (2024) The global governance of marine plastic pollution: rethinking the extended producer responsibility system. Front Mar Sci 11:1363269. https://doi.org/10.3389/fmars.2024.1363269

Chapter 2
Introduction to Marine Nanoscience and Nanotechnology

Abhay B. Fulke and Aishwarya P. Nair

Abstract Marine nanoscience and nanotechnology have become critical interdisciplinary fields that have the potential to revolutionize our understanding and management of marine ecosystems. This review provides a broad overview of the current advancements, applications, and future prospects of nanoscience and nanotechnology in marine environments. It explores how engineered nanomaterials, with their unique properties at the nanoscale, are being utilized to address pressing challenges in marine pollution, environmental monitoring, energy production, and marine biotechnology. The use of nanoparticles for the cleanup of marine pollutants such heavy metals, hydrocarbons, and plastics is one of the main applications covered. Another is the creation of sophisticated biosensors that can identify toxins and infections in real time. Additionally, it emphasizes how nanotechnology may improve marine renewable energy systems by increasing energy conversion device efficiency and providing corrosion-resistant materials for offshore structures. In addition to environmental and energy-related applications, marine nanoscience holds considerable promise for marine biotechnology, particularly in drug delivery systems, biosensors for disease detection, and marine bio-inspired materials. However, despite these promising advancements, the safe use of nanomaterials in marine environments is a major concern due to potential risks such as nanoparticle toxicity, bioaccumulation in marine organisms, and long-term ecological impacts. It is important to understand the interactions between nanoparticles and marine life to mitigate these risks and ensure the responsible development of nanotechnologies. It highlights key challenges and opportunities in marine nanoscience, including the need for better regulatory frameworks, standardized testing protocols, and further research into the environmental fate and behavior of nanomaterials. By addressing these issues, the field can continue to innovate while ensuring the protection and sustainability of marine ecosystems.

A. B. Fulke (✉) · A. P. Nair
Microbiology Division, Regional Centre, CSIR-National Institute of Oceanography (CSIR-NIO), Mumbai, Maharashtra, India
e-mail: afulke.nio@csir.res.in

A. B. Fulke
Academy of Scientific and Innovative Research (AcSIR), Ghaziabad, India

S. K. Panda and V. P.M. (eds.), *Bionanocomposites from Marine Origin*, Advanced Structured Materials 247, https://doi.org/10.1007/978-3-032-10749-7_2

Keywords Marine nanoscience · Marine nanotechnology · Nanomaterials · Pollution remediation · Biosensors · Environmental monitoring

2.1 Background of Nanoscience and Nanotechnology

The general public is familiar with the term "Nano," which is derived from the Greek word νάνος (pronounced "nanos"). The translation of literary means dwarf. In science and technology, the prefix nano is used in units of measurement. For example, one nanometer, or nm, is one billionth of a meter on a length scale. In Imperial unit conversions, one meter is equivalent to 3.28 feet, or 39.37 inches, and one inch is equivalent to 2.54 cm. In other words, an inch contains 25,400,000 nm. Another unit that is frequently used in many scientific and technological fields is the angstrom (Å), where 1 Å equals 0.1 nm (or 100 pm, where pm is the abbreviation for picometer). For a more concrete understanding of the nm-scale, consider that the typical diameter of a single human hair strand is 80,000 nm, or 80 μm (abbreviated as μm, which is one millionth of a meter) (Karanassios 2018). According to a commonly cited comparison, if one were to cut a single strand of this type of hair 80,000 times longitudinally that is, in the direction of its length and take just one slice from the result, it would be 1 nm thick. Put differently, depending on the atom used as a reference, 1 nm corresponds to a thickness of a few to about ten atoms. The nm-scale generally refers to the size of materials that can range from materials made of a few atoms or molecules made up of only a few atoms to materials that are one atom thick (such as a sheet of single layer graphene). However, some nanomaterials, like DNA and graphene, can be quite lengthy. One of the twenty-first century's most promising technologies is nanotechnology. It is the capacity to observe, measure, manipulate, assemble, regulate, and manufacture materials at the nanoscale scale in order to translate the theory of nanoscience into practical applications (Bayda et al 2019) Nanotechnology is "a science, engineering, and technology done at the nanoscale (1–100 nm), where unique phenomena permit revolutionary applications," according to the National Nanotechnology Initiative (NNI) of the USA. There are many different subsectors within the maritime sector, and each one has the potential to foster and support scientific and technological advancement. The following are the three broad types of marine industries: shipbuilding industries: producing many kinds of ships, including submarines, enormous oil tankers, and container ships; offshore industries: These comprise the building of offshore pipelines and platforms, both stationary and mobile, which are utilized in major oil and gas projects; port and coastal industries: This includes building oil terminals, piers, breakwaters, and other structures close to the coast. Determining the general requirements of a nation's maritime sectors can serve as both a growth engine and a reliable indicator for the use of nanotechnology products (Salmasi et al. 2022). This has led to the development of suitable coatings to protect against the impacts of the maritime environment, as well as novel materials for ship bodies and their constituent parts in an effort to strengthen them and lessen the noise and vibration they produce. The creation of novel materials also improves performance, such as fuel cells, new fuels, and batteries with extremely high energy storage.

Nanoscience-based solutions are especially intriguing because they can prolong the usable life of antimicrobial surfaces by providing distinctive surface modifications and making it possible to attach or insert active molecules to provide specific antibacterial properties. Either (i) bactericidal strategies that destroy organisms before or after contact with the surface, or (ii) antiadhesive properties that prevent germs from attaching, are the usual foundations of effective antimicrobial surface coatings. Active tactics (bactericidal activity), passive strategies (antiadhesive qualities), or, in more recent examples, a combination of both, are implemented in nanoscience-based antimicrobial surface designs to increase overall performance (Rigo et al. 2018). The rapid growth of nanoscience is evidence that nearly every field of science and technology will soon use nanotechnology production. The growing demand and effectiveness of nanotechnology is demonstrated by its current, contemporary applications in a range of industries, mainly in the food, pharmaceutical, cosmetic, automotive, oil and gas, chemical, biological, maritime, and mechanical sectors. Marine bio-nanotechnology is one area of research that is both intriguing and promising. Nanotechnology and nanoscience have great potential in the biologically diverse marine environment. Marine organisms produce incredible nanoparticles that range in size from 1 to 100 nm. These nanoparticles are used to create nanofabric structures such as pearls, seashells, and fish bones. The extraordinary architectures of coral reefs with calcium and diatoms and sponges are made of nanostructured silica cover. Because of the existence of nanoridges, the skin surface of dolphins and whales is rough. Biofoulers do not bind to these ridges because their pore size of 0.2 μm2 is smaller than that of marine fouling organisms. Despite its enormous promise, there is not much study being done on marine bio-nanotechnology.

2.2 Introduction

The twentieth century's scientific and technological advancements, which demonstrated the undeniable worth of humankind, established theoretical and practical underpinnings that made it feasible to examine the structure of matter and observe objects at the nanoscale (10–9 m).The size of made things, which must have at least one dimension no larger than 100 nm, is one of the criteria currently used to define nanotechnology (Antczak 2012). The physical and chemical characteristics of such objects should be controllable during creation, and they should be able to be used to construct larger objects. Because nanotechnology is interdisciplinary, it is frequently unnecessary and occasionally impossible to categorize it by prefixing the name of a certain scientific discipline with nano- (e.g., nanochemistry, nanobiotechnology). As a result, this field of study and technology ought to be regarded as a single nanotechnology. Nanotechnology serves as a platform that unites many fields to create profitable, easily marketable products.

One of the most active fields of research in modern scientific research is nanotechnology, which is developing daily and having an effect on every aspect of human existence. The engineering of atomic and molecular functional systems to create particles smaller than 100 nm by top-down and bottom-up methods is known as nanotechnology (Santhosh and Amarendra 2015). The creation of nanoparticles, nanowires, and nanotubes as a result of advances in nanotechnology has a wide range of uses in biotechnology, medical device manufacturing, and disease diagnosis and treatment. By precisely creating and controlling materials and devices at the atomic and molecular level, nanotechnology creates new qualities and functions that are not possible with traditional materials (Nasiri 2024). This translates into a wide range of applications in the maritime environment, ranging from materials science and environmental monitoring to pollution control and the generation of renewable energy. The oil spill has impacted the world for many decades, resulting in significant environmental problems and detrimental effects on human health. For effective oil spill cleanups, a variety of approaches and remediation methods are used; nevertheless, due to their limitations and environmental concerns, these processes were not entirely effective (Mishra et al. 2022). The term "nanotechnology" itself has captivated researchers and numerous industries alike, and there is a global competition to harness its potential and reap its benefits. Their remarkable qualities and compact size have demonstrated their capacity to offer technological answers to engineering challenges. With different biological entities being used to continuously produce nanoparticles, biological synthesis of nanoparticles has become a fast expanding field of study in nanotechnology worldwide, providing a viable substitute for traditional techniques. Simple prokaryotes and complex eukaryotic creatures, such as higher plants, can be converted into nanoparticles. One untapped potential is the utilization of marine microorganisms as nanofactories to produce nanoparticles. Marine microorganisms are known to interact with metal ions because marine ecosystems are constantly exposed to high quantities of metal salts. These microbes have the ability to quickly decrease metallic ions, forming nanoparticles with the appropriate size and shape.

Approximately 70% of the earth's surface is made up of the ecologically varied marine environment, which is yet mostly undiscovered. According to recent research, a number of marine plants can function as biofactories to produce nanoparticles. Several types of marine algae have the capacity to synthesize a wide range of nanoparticles, which has sparked interest in creating novel and strong medications from natural marine plant sources (Fawcett et al. 2017). Numerous types of these nanoparticles have been found to possess strong antibacterial properties. Microalgae and macroalgae are the two basic categories into which the diverse range of species found in marine algae can be separated. Microalgae species, such as phytoplankton, dwell suspended in the water column, but macroalgae, often known as seaweed, are plant-like organisms that can vary in size from a few centimeters to several meters.

The use of nanomaterials and nanoscale methods in marine environments is being investigated by the developing fields of marine nanoscience and nanotechnology. The main areas of innovation, as indicated by recent original research publications, include marine biotechnology, environmental monitoring, and the creation of innovative materials for energy harvesting and water treatment. The use of nanoparticles to identify contaminants in marine environments, such as heavy metals and oil spills, is one promising approach. Nanoparticles, due to their high surface area and reactivity, can selectively bind to contaminants, offering highly sensitive and rapid detection methods (Pete et al. 2021). For instance, a study published in Marine Pollution Bulletin demonstrated the potential of gold nanoparticles conjugated with specific ligands for the detection of toxic metals in seawater, highlighting the role of nanotechnology in environmental protection.

Marine biotechnology also heavily relies on nanotechnology, especially when it comes to creating new medicine delivery methods and bioactive substances from marine life. According to research published in Marine Drugs, nanostructured liposomes have the ability to improve the bioavailability of substances produced from marine environments, including anticancer drugs generated from marine sponges (Calcabrini et al. 2017). This illustrates how the pharmaceutical business can use nanotechnology to make it easier to extract and use bioactive ingredients. Nanomaterials are being investigated in the energy sector for their potential to increase the effectiveness of energy harvesting systems, such as those that use solar or ocean currents. A study in Renewable Energy detailed the use of nanostructured materials for enhancing the performance of marine-based solar cells and tidal energy devices, paving the way for more efficient, sustainable marine energy solutions (Pourhashem et al. 2022). However, challenges remain regarding the environmental impact of nanomaterials. Studies on the toxicity of nanoparticles to marine organisms are crucial for ensuring that the widespread use of nanotechnology does not inadvertently harm marine life. Research in Environmental Science and Technology has raised concerns about the bioaccumulation of nanoparticles in marine food chains, urging for more comprehensive ecological risk assessments (Fig. 2.1).

Marine nanoscience and nanotechnology represent an interdisciplinary field poised to address a wide range of challenges in marine environments. By leveraging the unique properties of nanomaterials, this field offers innovations in environmental monitoring, marine biotechnology, energy harvesting, and water purification. However, the potential applications of marine nanoscience are not without risks, particularly regarding the environmental impact of nanoparticles on marine ecosystems (Corsi et al. 2014). Nanotechnology, the manipulation of materials at the nanoscale (typically 1–100 nm), has revolutionized numerous scientific and industrial fields. In the marine environment, nanomaterials are being explored for applications that range from pollutant detection to energy production and drug delivery. Marine nanoscience, an emerging subdiscipline, takes advantage of the unique chemical, physical, and biological properties of nanomaterials to solve complex problems in marine ecosystems (Hochella Jr 2002). Despite its potential, concerns over the environmental impact and safety of nanomaterials in delicate maritime environments are also brought up by the field's quick development. The purpose of this review is to give a summary of the latest findings in the subject, describe the uses of marine nanoscience, and talk about nanotechnology in marine settings.

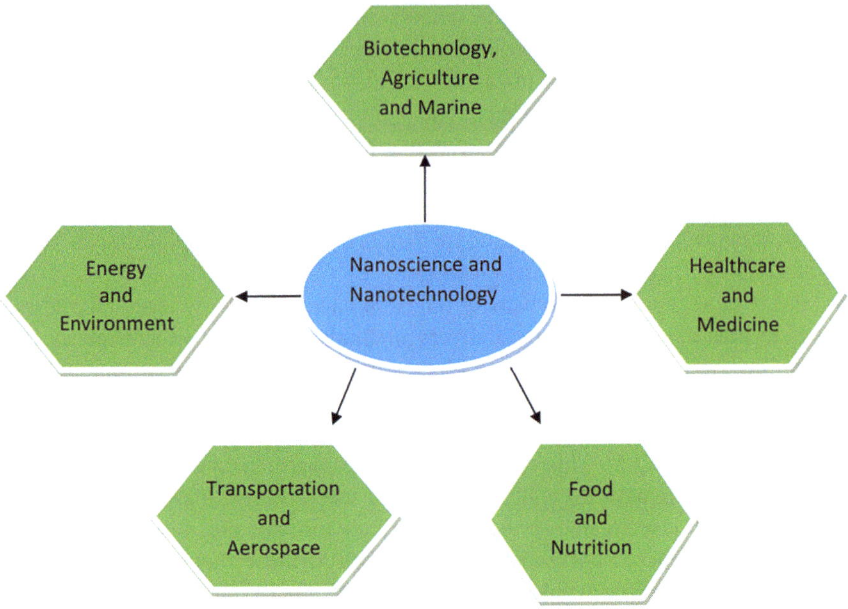

Fig. 2.1 Scopes of nanoscience and nanotechnology

2.3 Key Applications of Marine Nanoscience

2.3.1 Environmental Monitoring and Pollution Detection

The ability to detect and monitor pollutants in marine environments is crucial for protecting marine biodiversity and human health. Traditional monitoring techniques often struggle with sensitivity and speed, especially in detecting low levels of pollutants like heavy metals, pesticides, and oil spills. By using nanoparticles with high surface area-to-volume ratios that may concentrate and absorb pollutants, nanotechnology provides a possible remedy. Gold nanoparticles conjugated with particular ligands have been shown in studies to selectively bind harmful heavy metals like lead, cadmium, and mercury in saltwater. A study highlighted how these nanoparticles can be employed in sensors for real-time pollutant detection, enabling faster and more accurate assessments of water quality (Wang et al. 2023). Similarly, magnetic nanoparticles have been utilized to remove oil from seawater, showcasing their potential for environmental Cleanup (Mirshahghassemi and Lead 2015).

2.3.2 Marine Biotechnology and Drug Development

Marine organisms are a rich source of bioactive compounds, many of which have therapeutic potential. However, these compounds often have limited bioavailability due to their complex chemical structure or poor solubility. Nanotechnology offers solutions for improving the delivery and efficacy of marine-derived bioactive molecules, such as anticancer agents from marine sponges or antibacterial compounds from marine algae (Püsküllüoğlu and Michalak 2022). Nanostructured liposomes and micelles have been shown to encapsulate hydrophobic marine compounds, enhancing their stability and bioavailability. A recent study in Marine Drugs demonstrated that liposomal formulations of marine sponge-derived compounds increased their bioavailability in human cancer cell lines, suggesting that nanotechnology can facilitate the clinical application of marine bioactive substances (Kamat et al. 2022).

2.3.3 Energy Harvesting and Sustainable Technologies

Nanomaterials also show great promise in the field of renewable marine energy. The efficiency of marine-based renewable energy devices, such as tidal energy converters and solar cells, can be significantly enhanced by incorporating nanomaterials (Casini 2016). For example, nanostructured photovoltaic materials have been used to improve the performance of solar cells that are deployed in marine environments, where they must withstand harsh conditions. A study published in Renewable Energy explored the integration of quantum dots into marine solar panels, demonstrating an increase in efficiency by improving light absorption and reducing energy loss (Rodrigues et al. 2023). Similarly, nanocomposite materials are being developed for use in underwater turbines, offering greater durability and higher energy conversion efficiency compared to conventional materials. While the potential applications of marine nanoscience are vast, there are growing concerns about the environmental impact of nanoparticles once they are released into marine ecosystems. Nanomaterials, due to their small size and high reactivity, can interact with marine organisms in ways that larger particles cannot, potentially leading to toxicity, bioaccumulation, and disruption of food webs. A study published in Environmental Science and Technology examined the bioaccumulation of silver nanoparticles in marine organisms, including fish and mollusks. The study found that silver nanoparticles were absorbed by marine organisms and accumulated in their tissues, posing a risk to both marine life and human health through the food chain (Mat Lazim et al. 2023). Similarly, research on the toxicity of carbon-based nanoparticles has raised concerns about their potential to cause oxidative stress and impair the immune response in marine species. The challenge of mitigating these risks has led to the call for more stringent regulations and standardized testing protocols to assess the environmental fate and behavior of nanoparticles in marine environments. Additionally, research is needed to develop nanoparticles that are both effective in their applications and safe for marine ecosystems.

2.4 The Importance of Nanotechnology and Nanoscience in Marine Environments

The fields of nanoscience and nanotechnology have quickly become revolutionary and have the ability to solve difficult problems in marine environments. As the oceans deal with increasing environmental stresses including pollution, resource depletion, and climate change, the use of nanotechnology provides creative answers that were previously thought to be impossible. Because of their special qualities, which include high surface area, reactivity, and molecular-level interaction with biological systems, nanomaterials are especially well suited for a variety of marine applications (Corsi et al. 2014). The following are the main justifications for why nanoscience and nanotechnology are essential to marine research and conservation in future.

Monitoring the Environment and Identifying Pollution.

Whether from chemical runoff, plastics, heavy metals, or oil spills, pollution is one of the biggest problems facing marine ecosystems today. The development of more effective, sensitive, and real-time pollution monitoring techniques for the marine environment depends heavily on nanotechnology.

Elevated Selectivity and Sensitivity: It is possible to create nanoparticles, such as gold or carbon-based nanomaterials, to bind selectively with particular pollutants, including insecticides, organic contaminants, and hazardous metals (like lead and mercury). Monitoring efforts in marine environments are improved by their enhanced surface area and chemical reactivity, which enable highly sensitive detection at incredibly low concentrations. For example, research on gold nanoparticles conjugated with specific ligands for mercury detection in seawater has shown that such sensors can quickly identify contamination at trace levels, allowing for early intervention (Lim et al. 2021). Early Detection of Contaminants: Nanomaterial-based sensors and biosensors can detect pollutants before they reach harmful concentrations, offering a proactive approach to environmental protection. Early detection systems could help mitigate the damage from oil spills or chemical leaks, enabling quicker responses and reducing the impact on marine life.

2.4.1 Pollution Remediation and Cleanup

Innovative approaches to contaminant removal in maritime settings are provided by nanotechnology. The cost, effectiveness, and scalability of traditional remediation techniques are frequently constrained, particularly in large-scale operations like oil spill response. Oil may be efficiently captured and removed from the water's surface using nanomaterials like magnetic nanoparticles and nanostructured sponges. Cleaning can be made simpler and more effective by engineering these materials to be extremely absorbent and magnetically separable (Singh et al. 2020). Removal of Heavy Metals: By removing dangerous heavy metals from seawater, nanomaterials can lower the possibility that they will bioaccumulate in marine food chains. Studies on nanocomposites and nanofiltration systems have demonstrated the potential to extract toxic metals from water with higher efficiency than conventional methods. For instance, nanoparticles with magnetic properties have been employed to clean oil and chemical contaminants from water surfaces, offering a cleaner and faster alternative to traditional methods like dispersants and skimmers (Mehmood et al. 2021).

2.4.2 *Marine Biotechnology and Drug Development*

Numerous bioactive substances, such as antibiotics, antiviral drugs, and anticancer agents, are abundant in marine species and may find application in medicine. However, because of their weak solubility or molecular characteristics, these substances frequently have restricted bioavailability. By encapsulating marine-derived bioactive compounds in nanocarriers as liposomes, micelles, or dendrimers, nanotechnology can enhance their delivery (Gorain et al. 2023). These carriers improve bioactive chemicals' solubility and stability, improve their capacity to pass through biological membranes, and enable targeted distribution to particular tissues or cells. Because of their distinct biochemical characteristics, nanomaterials made from marine organisms like algae or marine sponges can be used. Nanostructures from these organisms may be used to create drugs or therapeutic devices that have advantages over traditional methods, including increased efficacy, lower toxicity, and reduced side effects. For example, liposomal formulations of marine sponge-derived anticancer compounds have shown promise in improving drug efficacy and bioavailability in cancer therapies (Mahaddalkar and Lopus 2017).

2.4.3 *Sustainable Energy Solutions*

The oceans represent an enormous potential resource for renewable energy, particularly through tidal, wave, and solar energy. Nanotechnology can enhance the efficiency and durability of energy devices used in marine environments. Nanomaterials, such as quantum dots and nanostructured thin films, can significantly improve the efficiency of solar cells deployed in marine environments. These materials are highly efficient in absorbing light and converting it into energy, even under low-light conditions common in coastal areas. Nanomaterials can also be integrated into the construction of underwater turbines or wave energy converters, making these devices more durable and resistant to corrosion. Nanocomposite materials, for instance, can be used to create lighter, stronger, and more corrosion-resistant components for energy harvesting devices. Research into nanostructured photovoltaic cells has led to promising results in enhancing solar energy conversion, while nanocomposites are being applied to improve the performance and lifespan of marine energy devices (Rakesh Tej Kumar et al. 2018).

2.4.4 Advancing Marine Science and Monitoring Marine Ecosystems

Nanotechnology offers advanced tools for exploring and understanding marine ecosystems at the molecular and cellular levels. Nanoparticles can be used as tracers in environmental studies to track the movement of pollutants, nutrients, and other substances through marine food webs. By embedding nanoparticles into the food web, researchers can track how materials move through marine ecosystems, providing insights into nutrient cycling, pollution dispersion, and bioaccumulation (Petersen et al. 2019). These techniques offer a more precise understanding of ecosystem dynamics than traditional methods. Nanotechnology-based sensors can be used to monitor the health of marine species and ecosystems. By detecting biomarkers of stress or disease at the molecular level, these sensors provide a more comprehensive picture of ecosystem health and help guide conservation efforts.

2.4.5 Addressing Climate Change Impacts on Marine Ecosystems

Additionally, by improving carbon capture and sequestration technologies, nanotechnology can help lessen the impact of climate change on marine ecosystems. In order to mitigate the wider consequences of global warming and lessen ocean acidification, nanomaterials are being created to absorb carbon dioxide from saltwater (Creamer and Gao 2015). The capacity of nanomaterials like graphene oxide and carbon nanotubes to absorb and store carbon dioxide in ocean water is being investigated in hopes of lowering atmospheric CO_2 levels and delaying the pace of climate change. Nanotechnology and nanoscience have revolutionized many scientific fields, including microbiology and biotechnology, by offering new tools, techniques, and solutions that were previously unattainable. The unique properties of nanomaterials—such as their small size, large surface area, high reactivity, and the ability to interact with biological systems—make them particularly well suited for advancing research and applications in these disciplines (Khan et al. 2022). Below, we explore the importance of nanotechnology and nanoscience in microbiology and biotechnology, highlighting key areas where these fields intersect and provide transformative possibilities.

2.4.6 *Advanced Diagnostic Tools and Sensors*

The sensitivity and specificity of diagnostic instruments used in microbiology and biotechnology have been greatly enhanced by nanotechnology. Traditional diagnostic methods, such as culturing bacteria or PCR (polymerase chain reaction), can be time-consuming and costly, especially in detecting low-abundance pathogens or rare genetic mutations. Nanomaterials offer more rapid, efficient, and sensitive alternatives. Nanomaterials, such as gold nanoparticles, carbon nanotubes, and quantum dots, have been integrated into diagnostic biosensors to detect a wide range of microorganisms (bacteria, viruses, fungi) and their biomarkers at early stages (Alba-Patio et al. 2022). Nanoparticles' enormous surface area and small size allow them to interact with target molecules at the molecular level, increasing sensitivity. The development of portable, inexpensive, and extremely sensitive diagnostic tools for the quick identification of infections even in environments with limited resources is made possible by nanotechnology. These devices can detect bacterial infections, viral diseases (like COVID-19), and even antimicrobial resistance (AMR) markers with high accuracy and minimal sample requirements.

2.4.7 *Antimicrobial Nanomaterials and Resistance Management*

With the growing threat of antimicrobial resistance (AMR), there is an urgent need for new antimicrobial agents that can overcome resistant bacteria and fungi. Nanotechnology offers novel solutions by enabling the development of nano-antimicrobials—nanoparticles that can directly interact with microbial cells in ways that conventional antibiotics cannot (Kalashgarani and Babapoor 2022).

Nanostructured Antimicrobial Agents: When treating conditions like cancer, where customized administration can minimize adverse effects and enhance therapeutic results, this strategy is very helpful. Gene Transfer: The transfer of genetic material for gene therapy is another important application of nanotechnology. According to Jiang et al. (2022), nanocarriers have the ability to shield genetic payloads—like DNA, RNA, or CRISPR-Cas9 systems—from nuclease degradation and facilitate their delivery to target cells. This makes it possible to precisely modify genes, providing therapy for cancer, genetic abnormalities, and other illnesses that call for gene-based therapeutics. This approach, called nano-enhanced antibiotics, has been studied for its ability to restore the activity of antibiotics against resistant bacterial strains, offering a potential solution to the AMR crisis.

Fig. 2.2 Application on cancer care using marine nanotechnology

2.4.8 Targeted Drug Delivery and Gene Therapy

Nanotechnology has enabled remarkable advances in targeted drug delivery and gene therapy fields that are at the core of modern biotechnology. Traditional drug delivery methods often fail to deliver therapeutic agents to the desired location in the body, or they may have off-target effects, leading to toxicity. Nanocarriers can overcome these limitations by providing controlled, site-specific drug release. Nanocarriers for Drug Delivery: Liposomes, polymeric nanoparticles, dendrimers, and micelles are examples of nanomaterials that can encapsulate drugs, protect them from degradation, and release them at a specific site in the body (Chamundeeswari et al. 2019). When treating conditions like cancer, where customized administration can minimize adverse effects and enhance therapeutic results, this strategy is very helpful. Gene Transfer: The transfer of genetic material for gene therapy is another important application of nanotechnology. According to Jiang et al. (2022), nanocarriers have the ability to shield genetic payloads—like DNA, RNA, or CRISPR-Cas9 systems—from nuclease degradation and facilitate their delivery to target cells. This makes it possible to precisely modify genes, providing therapy for cancer, genetic abnormalities, and other illnesses that call for gene-based therapeutics. Research into lipid nanoparticles (LNPs) has been particularly successful in developing mRNA vaccines, as demonstrated by the rapid deployment of COVID-19 vaccines (Moderna, Pfizer-BioNTech). LNPs efficiently deliver mRNA into human cells, which then produce the spike protein needed to stimulate an immune response (Wilson and Geetha 2022) (Fig. 2.2).

2.4.9 Biomaterials for Tissue Engineering and Regenerative Medicine

The development of biomaterials, especially in tissue engineering and regenerative medicine, is significantly influenced by nanotechnology. More efficient scaffolds for cell growth, tissue healing, and regeneration are made possible by the development of nanomaterials that resemble the extracellular matrix (ECM) of tissues.

Nanostructured Scaffolds: Nanofibers and nanocomposites are being used to create scaffolds that provide structural support and promote the growth of new tissues (Barnes et al. 2007). These scaffolds can be made from biodegradable materials, ensuring that they are safely absorbed by the body once the tissue is regenerated. Regenerative Medicine: Nanomaterials are also being incorporated into stem cell therapies to improve cell adhesion, differentiation, and proliferation (He et al. 2021). For example, nanodiamonds and carbon nanotubes have been shown to enhance the effectiveness of stem cell-based therapies for the regeneration of bone, cartilage, and nerve tissues.

Enzyme and Biocatalyst Enhancement

Enzymes are essential to biotechnology activities including bioremediation, the synthesis of valuable chemicals, and the creation of biofuels. Nanotechnology offers the ability to improve the activity, stability, and selectivity of enzymes and biocatalysts by modifying their structure at the nanoscale. Enzyme Immobilization: Nanomaterials, such as mesoporous silica, magnetic nanoparticles, and carbon nanotubes, are used to immobilize enzymes (Razzaghi et al. 2022). This enhances enzyme stability, reusability, and overall catalytic efficiency in industrial applications, such as biocatalysis and wastewater treatment. Enhanced Biocatalysis: Nanomaterials can also be used to enhance the efficiency of biocatalytic reactions by providing more efficient substrates and promoting better enzyme-substrate interactions (Anwar et al. 2023). The sustainable manufacture of medicines, biofuels, and other fine chemicals depends on this.

Biosensing and Environmental Monitoring

Nanotechnology offers powerful biosensing capabilities that are particularly valuable in environmental microbiology and biotechnology applications. Nanomaterial-based sensors can be used to detect specific microbial communities, monitor the presence of pollutants, and even assess the health of ecosystems. Biosensors for Pathogen Detection: Nanomaterials can be engineered to detect specific microbial pathogens in environmental samples. These sensors are highly sensitive and capable of detecting pathogens at low concentrations, which is critical for public health monitoring, food safety, and environmental protection (Nnachi et al. 2022). Bioremediation Monitoring: Nanotechnology can also be used to monitor the effectiveness of bioremediation processes, where microorganisms are employed to clean up pollutants (Benjamin et al. 2019). Nanoparticles can be used as tracers to track the degradation of contaminants and assess the success of remediation efforts.

Microbiome Modulation and Health Applications

Immunity, metabolism, and even mental health are all impacted by the human microbiome, which is vital to both health and illness. Nanotechnology offers new ways to manipulate the microbiome for therapeutic purposes. Targeting Microbiota: Nanoparticles can be designed to interact selectively with specific microbes in the gut, allowing for targeted modulation of the microbiome (Rajoka et al. 2021). This opens up new possibilities for treating conditions like irritable bowel syndrome, inflammatory bowel disease, and obesity by restoring balance to the microbiome. Microbial Biotherapeutics: Nanotechnology can also be used to deliver live microbes (probiotics) to the gut or other body sites, improving their survival and efficacy (Trivedi et al. 2022) (Fig. 2.3).

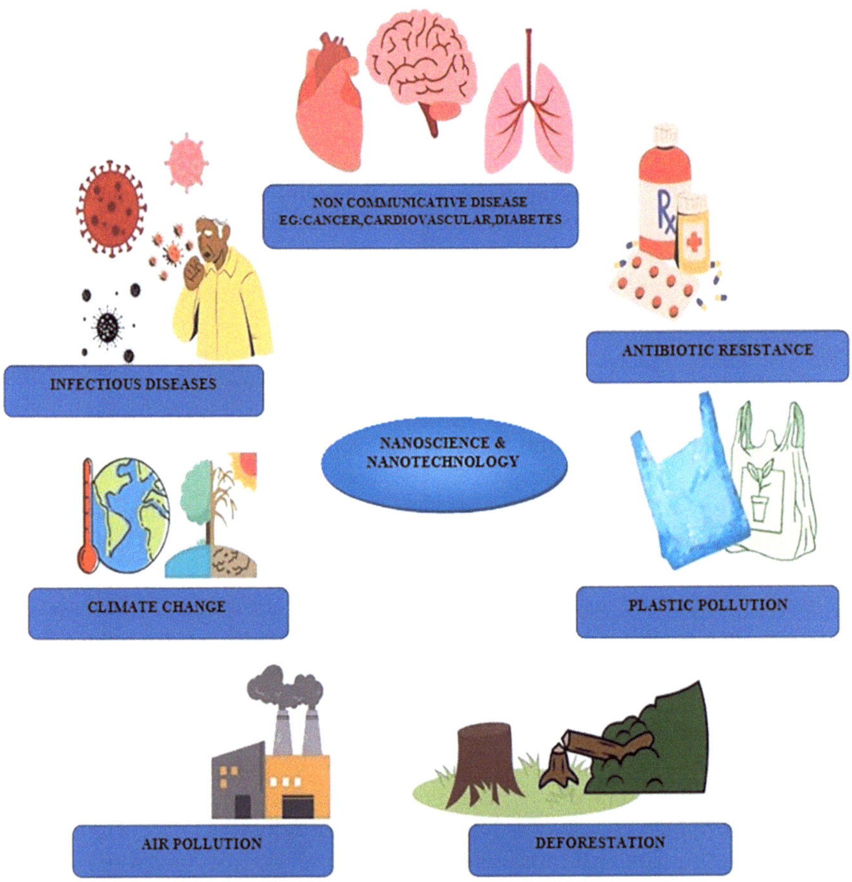

Fig. 2.3 Nanoscience and nanotechnology applications across the world

2.5 Marine Nanoscience and Nanotechnology: Traditional and Advancements

Marine nanoscience and nanotechnology is an interdisciplinary field that combines principles of nanotechnology with marine science to explore the potential applications of nanomaterials in marine environments. This rapidly growing field has profound implications for environmental sustainability, marine biology, medicine, and the development of new materials and technologies that can address the challenges faced by marine ecosystems. The field integrates innovations in materials science, chemistry, physics, biology, and engineering, offering exciting possibilities for advancements in marine research and industry.

2.5.1 Tradition of Marine Nanoscience and Nanotechnology

The field of marine nanoscience has its origins in the late twentieth-century nanotechnology revolution. In disciplines like electronics, health, and materials science, nanotechnology, the manipulation and use of materials at the nanoscale (1–100 nm) started to gain traction. But, the use of nanoscience in marine ecosystems is a relatively new development, having gained popularity in the last 20 years. The study of natural nanomaterials has long been a primary focus of marine nanoscience. Numerous creatures, including coral, mollusks, and marine diatoms, have developed nanoscale structures that have specialized functions like chemical defense, mechanical strength, and light manipulation (Green et al. 2014). The creation of synthetic nanomaterials with comparable qualities has been influenced by these natural nanostructures. The potential of nanotechnology in marine applications was initially explored in areas such as:

1. Water Purification and Desalination: The development of nanomaterials that could efficiently filter and desalinate seawater, reducing the environmental impact of conventional methods.
2. Marine Pollution Detection: The application of nanomaterials in maritime environments to identify contaminants such as pesticides, heavy metals, and oil spills.
3. Antifouling Coatings: Nanotechnology has been used to create coatings for marine vessels that prevent the growth of barnacles and algae, which can cause corrosion and drag on ships.

2.5.2 Advancements in Marine Nanoscience and Nanotechnology

In recent years, the field has seen significant advancements in both fundamental research and practical applications. Nanomedicine in Marine Organisms: Advances in nanomedicine have begun to show promise in marine biology and conservation. Nanoparticles are being used to deliver drugs to marine organisms suffering from diseases, and they can bc uscd in monitoring marine animal health (Vaseashta et al. 2007). These sensors, made from nanoparticles or nanowires, have high sensitivity and can detect minute concentrations of pollutants such as microplastics, heavy metals, and oil in seawater. This provides a powerful tool for early detection and monitoring of environmental changes that can have devastating effects on marine life. Environmental Remediation: Nanotechnology offers significant promise in the area of marine pollution cleanup. Nanomaterials, such as nanostructured carbon-based materials, magnetic nanoparticles, and nanoscale zero-valent iron, are being investigated for their ability to adsorb, neutralize, or degrade contaminants in the ocean, including oils, heavy metals, and even radioactive materials. For instance, magnetic nanoparticles can be used to collect and remove oil spills, while nanomaterials are being developed to break down plastic waste. Marine Biotechnology: Nanotechnology is increasingly being used in marine biotechnology for the production of high-value compounds, such as biofuels, pharmaceuticals, and chemicals. Marine organisms such as algae and bacteria, when combined with nanomaterials, have the potential to generate bio-based alternatives to petrochemicals and other synthetic materials. Additionally, engineered nanoparticles can be used to enhance the extraction of valuable compounds from marine organisms. Antifouling and Corrosion Resistance: New advancements in nanotechnology have resulted in the creation of environmentally friendly antifouling coatings that prevent the buildup of marine organisms on ships and offshore platforms (Thomas et al. 2024). These coatings use nanoparticles to disrupt the attachment of biofilms, which reduces the need for toxic chemical-based antifouling paints. Similarly, nanomaterials are being used to improve the corrosion resistance of metal surfaces exposed to seawater, thus enhancing the longevity of marine infrastructure. Nano-enabled Materials for Marine Structures: The development of nanostructured composite materials for marine infrastructure is another significant area of advancement. These materials combine the strength and durability of nanomaterials with the flexibility and corrosion resistance required for underwater structures, such as pipelines, oil rigs, and seawalls. For example, shown in Table 2.1, Nanotechnology also offers solutions to improve the performance and reduce the environmental impact of materials used in marine construction.

Table 2.1 Widespread utilization of nanoscience and nanotechnology

Application area	Nanomaterial type	Key purpose	Scientific findings	References
Marine pollution detection	Nanoparticles, nanowires	To identify contaminants in seawater, such as oil, microplastics, and heavy metals	Nanomaterials are highly sensitive and selective, capable of detecting trace amounts of pollutants in real-time	Liang et al. (2021), sensors and actuators B: chemical
Water purification and desalination	Carbon nanotubes (CNTs), graphene, membranes	To filter seawater and eliminate impurities such as organic pollutants, heavy metals, and salts	CNTs and graphene-based membranes show high efficiency in desalination and removal of pollutants with low energy consumption	Cioffi et al. (2021), environmental science and technology
Antifouling coatings	Silver nanoparticles, nanostructured polymers	To prevent biofouling on ship hulls, offshore platforms, and other marine equipment	Silver nanoparticles disrupt biofilm formation, while nanostructured coatings create smooth surfaces, reducing organism attachment	Zhang et al. (2023), marine pollution bulletin
Nanomedicine for marine organisms	Gold nanoparticles, lipid nanoparticles	To deliver drugs and monitor the health of marine organisms	In marine species, nanoparticles can be utilized to target particular cells, tissues, or organs in order to cure conditions like cancer or infections	Della Porta et al. (2023), marine drugs
Environmental remediation	Magnetic nanoparticles, zero-valent iron (nZVI)	To get rid marine habitats of harmful pollutants such as heavy metals and oil	nZVI and magnetic nanoparticles are effective in adsorbing and removing pollutants from the ocean with minimal environmental impact	Rajendran et al. (2022), environmental science and technology letters

(continued)

Table 2.1 (continued)

Application area	Nanomaterial type	Key purpose	Scientific findings	References
Marine biotechnology	Nanocellulose, biodegradable nanoparticles	To increase the production of biofuel and the extraction of bioactive substances from marine organism	Nanomaterials can enhance enzymatic processes and increase yields of biofuels, pharmaceuticals, and other marine-based compounds	Sun et al. (2021), biotechnology for biofuels
Corrosion resistance	Zinc oxide nanoparticles, carbon nanotubes	To improve the durability and longevity of marine structures exposed to seawater	Zinc oxide nanoparticles are used for corrosion protection, while carbon nanotubes enhance the mechanical properties of coatings	Zhao et al. (2023)
Nanostructured Marine Sensors	Quantum Dots, Nanowires	To monitor environmental changes and detect pollutants in real time in marine environments	Quantum dots offer high sensitivity and multiplexing capability, enabling real-time detection of various marine pollutants	Zhao et al. (2022), nature nanotechnology
Nano-enabled materials for marine structures	Nanocomposites, nanoclay reinforced polymers	To decrease the weight and enhance the mechanical qualities of materials utilized in marine infrastructure	Nanocomposite materials show enhanced resistance to corrosion, wear, and impact, making them ideal for underwater structures	Yang et al. (2021), composites science and technology
Biofilm inhibition	Titanium dioxide (TiO_2), graphene oxide	To stop bacteria and biofilms from growing on surfaces in maritime environments	TiO_2 nanoparticles are effective against microbial growth, and graphene oxide disrupts bacterial biofilm formation	Lee et al. (2020), colloids and surfaces B: biointerfaces

2.6 Challenges and Future Directions

There are still a number of obstacles in the field of marine nanoscience and nanotechnology, despite the quick advancements. The possible effects of manmade nanomaterials on the environment are among the main worries. Comprehensive research on the toxicity of nanoparticles in marine environments is necessary since their extensive use may put marine life at unexpected risk. The behavior, bioaccumulation, and long-term effects of nanoparticles in the food chain are areas of ongoing research. Furthermore, scaling up nanotechnology applications from the laboratory to real-world scenarios poses practical challenges. Ensuring that nanomaterials are cost-effective, safe, and environmentally sustainable is key to the widespread adoption of marine nanotechnology. Marine nanoscience and nanotechnology is an exciting and rapidly evolving field with vast potential to transform marine conservation, resource management, and industrial practices. The advancements in nanomaterials, sensors, pollution control, and marine biotechnology are pushing the boundaries of what is possible in preserving marine ecosystems and supporting sustainable practices in marine industries. However, there is a need for continued research, especially on the environmental impact of nanomaterials, to ensure that the benefits of these technologies outweigh the potential risks. As the field continues to grow, it promises to play a crucial role in addressing some of the most pressing challenges facing the oceans and marine life today (Ray et al. 2009).

Nanomaterials, particularly nanoparticles and nanowires, are highly sensitive to small concentrations of pollutants in seawater. These materials can be designed to possess particular chemical characteristics that allow them to attach to organic contaminants like pesticides and oil, heavy metals (including lead and mercury), and microplastics in a selective manner. Real-time monitoring and detection of pollutants in maritime environments is the main use case. These sensors have major advantages over conventional techniques that could necessitate more time-consuming analysis since they can quickly alert users to changes in the environment. High surface-area-to-volume ratios in nanoparticles have been found to improve their capacity to interact with pollutants. For example, illustrated in Table 2.1, because of their special optical characteristics, which alter as they attach to particular pollutants, gold nanoparticles can be utilized to detect heavy metals (Liang et al. 2021). Additionally, because nanowires' electrical conductivity varies when they come into touch with contaminants, they have been utilized to detect microplastics. Nanomaterials Used: Carbon Nanotubes (CNTs), Graphene, and Membranes. Nanomaterials like carbon nanotubes and graphene have exceptional filtration properties, making them ideal for applications in water purification and desalination. These materials can filter out salts, heavy metals, and organic pollutants at the nanoscale level, offering a highly efficient way to purify seawater. To enhance desalination processes and create more efficient water filtration systems, thus addressing the growing need for freshwater in regions dependent on seawater desalination, CNTs and graphene-based membranes are known for their high permeability and selective filtration properties. Graphene oxide membranes, for example, have been demonstrated to reject salts while allowing water molecules to pass through, significantly reducing energy consumption compared to traditional desalination methods (Cioffi et al. 2020), and this is shown in Table 2.1. This makes them highly suitable for large-scale applications in marine environments.

Nanomaterials Used: Silver Nanoparticles and Nanostructured Polymers. Marine biofouling where organisms like barnacles and algae attach to the surfaces of ships and offshore structures can lead to corrosion, increased drag, and efficiency losses. Antifouling coatings are a novel approach made possible by nanotechnology. To lessen the environmental effect and maintenance expenses by stopping the formation of marine organisms and biofilms on ship hulls, offshore platforms, and undersea pipelines, silver nanoparticles are used as they have antibacterial qualities. As shown in Table 2.1, for example, they prevent microbial growth by releasing silver ions, which interfere with microbial cell functions (Zhang et al. 2023). Additionally, nanostructured polymer coatings provide smooth surfaces that discourage organism attachment. These coatings are also more environmentally friendly compared to traditional toxic chemical-based antifouling treatments. Nanomaterials Used: Gold Nanoparticles and Lipid Nanoparticles. In marine biology, nanomedicine refers to the use of nanoparticles to administer medications or treatments to fish, corals, and other marine life in order to cure illnesses or stop infections. This includes the use of nanoparticles for health monitoring and detection in marine organisms. By targeting particular cells in marine species with nanoparticles, therapeutic medicines can be delivered directly to the site of infection or sickness, effectively treating marine wildlife with conditions like tumors or bacterial infections.

Gold nanoparticles are being explored for their ability to cross cell membranes and target specific tissues, which can be useful in treating diseases in marine organisms (Della Porta et al. 2020) as illustrated in Table 2.1. Similarly, lipid nanoparticles are being researched for their potential to deliver antiviral or antimicrobial agents to marine species. Nanomaterials Used: Magnetic Nanoparticles and Zero-Valent Iron (nZVI). Nanotechnology is also being used to clean up marine environments affected by oil spills, heavy metals, and other pollutants. Magnetic nanoparticles and zero-valent iron are particularly effective in adsorbing contaminants and helping to remove them from the ocean, offering economical and effective ways to remove organic contaminants, hazardous heavy metals, and oil spills from the water. Magnetic nanoparticles can be easily manipulated using magnetic fields to separate contaminants from water, while nZVI nanoparticles are able to reduce the toxicity of pollutants such as heavy metals and oils through chemical reactions (Rajendran et al. 2022). These technologies are effective at removing a wide range of pollutants and have been shown to have minimal environmental impact.

Nanomaterials Used: Nanocellulose and Biodegradable Nanoparticles. Marine biotechnology involves using marine organisms or by-products for biotechnological applications, including the production of biofuels, pharmaceuticals, and other chemicals. The extraction of useful substances from marine organisms, including bacteria and algae, can be improved using nanotechnology to boost marine biotechnology processes' productivity and efficiency, particularly in the manufacturing of biofuel and medicinal applications. Nanocellulose and biodegradable nanoparticles have been used to enhance the extraction of valuable compounds from marine species like algae and corals. These nanoparticles can improve the catalytic efficiency of enzymes used in biofuel production, while also enabling the extraction of bioactive compounds from marine organisms (Sun and Jin 2021). Nanomaterials Used: Zinc Oxide Nanoparticles and Carbon Nanotubes. Marine structures, such as oil rigs, ships, and underwater pipelines, are highly susceptible to corrosion due to constant exposure to seawater. Nanotechnology is being used to develop more effective and durable materials for preventing corrosion in marine infrastructure. Carbon nanotubes and zinc oxide nanoparticles have demonstrated significant promise in enhancing the mechanical characteristics and resistance to corrosion of materials exposed to seawater. These substances create shields that keep marine structures from deteriorating and increase their longevity (Yano et al. 2023).

2.7 Advantages of Marine Nanoscience and Nanotechnology

2.7.1 Environmental Remediation

Pollution Control and Oil Spill Cleanup

Addressing environmental contamination in seas and coastal waterways is one of the most promising uses of marine nanotechnology. Because of their huge surface areas, nanomaterials—like iron, silica, or carbon nanoparticles—are able to absorb or react with contaminants more effectively than traditional materials. For example, heavy metals like lead, arsenic, and mercury as well as organic pollutants like oil have been cleaned up in laboratory settings using nanoscale iron particles. Because of their greater reactivity, these substances can remove pollutants from marine environments more quickly (Das et al. 2018). By creating nanostructured materials that can absorb or degrade oil more effectively than conventional techniques, nanotechnology can be applied to the cleanup of oil spills. Nanomaterials have more surface area to react with pollutants and can absorb more oil. Additionally, they may be designed to specifically target certain oil kinds, which would make them extremely useful in a variety of maritime situations.

2.7.2 Water Filtration and Desalination

Improved Water Purification

Nanotechnology offers significant advancements in water purification, especially for desalination and purification of seawater. Nanomaterials such as carbon nanotubes, graphene oxide membranes, and nanocomposites can be used to filter out salts, heavy metals, and microorganisms from seawater (Al-Maliki et al. 2022). These nanomaterials possess highly selective filtration properties, making them ideal for removing pollutants or pathogens from water with a high degree of efficiency. The process of desalination, which converts seawater into potable water, often requires a great deal of energy. Nanotechnology can improve the efficiency of desalination plants by using advanced filtration technologies that reduce energy consumption and increase water recovery rates. Nanomaterials can also improve the durability and lifespan of desalination membranes, making them more cost-effective and sustainable.

2.7.3 Marine Biotechnology and Healthcare

Drug Delivery and Marine-Based Pharmaceuticals

In marine biotechnology, nanotechnology is creating new opportunities. Because of their small size and large surface area, nanoparticles can be used to create more efficient drug delivery systems, which could lead to better treatments for both human populations in coastal areas and marine wildlife. Furthermore, it has been demonstrated that marine animals like corals, sponges, and algae contain substances with bioactive qualities. These molecules may be extracted and delivered with the aid of nanotechnology, which could result in the creation of innovative therapeutic solutions for both human and marine health. Nanotechnology can have a big impact on aquaculture, which is the rearing of fish and other marine life (Tayel et al. 2019). Fish feed can benefit from the addition of nanomaterials to improve nutritional absorption, encourage development, and lessen disease outbreaks. In order to maximize fish health and water quality, nanoparticles can also be utilized to create more intelligent sensors and aquaculture monitoring systems.

2.7.4 Renewable Energy from Marine Sources

Marine Energy Harvesting and Enhanced Solar Energy Storage

Nanotechnology can be employed in renewable energy systems like tidal, wave, and ocean current energy. Nanomaterials could be integrated into energy conversion systems to improve their efficiency by enhancing the performance of components such as turbines, generators, and energy storage systems. For example, the use of nanocoatings and nanomaterials in marine turbines could improve energy conversion efficiency and prevent corrosion from the harsh marine environment (Salmasi et al. 2022). Nanomaterials are being investigated for their ability to improve the efficiency of solar cells, making them an excellent fit for energy generation in coastal regions. Additionally, advances in nanotechnology may lead to better energy storage systems, which are essential for managing intermittent renewable energy sources like solar and tidal energy.

2.7.5 *Marine Sensors and Monitoring Systems*

Enhanced Sensors for Marine Ecosystems

Nanotechnology has enabled the development of ultra-sensitive sensors that can detect pollutants, changes in water temperature, salinity, and even the presence of pathogens in marine environments. These nanosensors are smaller, more cost-effective, and faster than traditional sensors. They can be used for real-time monitoring of marine ecosystems, allowing for the early detection of environmental changes or contamination (Martins et al. 2022). The ability to monitor water quality at a high level of detail will help manage marine resources more effectively.

2.8 Disadvantages of Marine Nanoscience and Nanotechnology

2.8.1 *Environmental and Health Risks*

Toxicity to Marine Life and Bioaccumulation

One of the most significant concerns regarding marine nanotechnology is the potential toxicity of nanoparticles to marine organisms. Some nanoparticles, depending on their composition and size, may be toxic to marine life. For instance, some metal nanoparticles may build up in marine species' tissues and impair their ability to reproduce or maintain good health. Additionally, these materials may also impact the overall health of marine ecosystems, disrupting food webs. Nanomaterials can be absorbed by marine organisms, leading to the potential for bioaccumulation, where the concentration of nanoparticles increases as they move up the food chain (Mona et al. 2023). This raises concerns about the long-term impact of nanomaterials on marine food safety and human health, particularly if contaminated seafood is consumed by humans.

One of the most significant concerns in marine nanoscience is the potential environmental impact of nanomaterials. Although nanoparticles are designed to perform specific tasks such as water purification or sensing, their behavior in marine environments is still not fully understood. Their small size, high surface area, and reactivity may lead to unintended interactions with marine organisms, potentially causing toxicity and disrupting ecosystems. The long-term fate of nanomaterials in marine environments whether they degrade, accumulate, or migrate remains an open question.

The release of nanoparticles into the ocean could affect not only microorganisms but also larger marine species, with possible consequences for food chains. There is an urgent need for more studies to assess the biocompatibility and biodegradability of these materials in real-world marine conditions.

2.8.2 *Unpredictable Behavior in Marine Environments*

Behavior of Nanomaterials in Complex Marine Systems

It is yet unclear how nanomaterials behave in the diverse and dynamic maritime environment. Unexpected outcomes could result from elements including temperature fluctuations, ocean currents, and interactions with other substances. Unpredictable aggregation or degradation of nanoparticles could result in novel types of pollution or environmental risks. Although nanomaterials can be designed for certain uses, they may have negative effects if they accidentally interact with other materials in marine environments. The long-term behavior of nanoparticles in marine ecosystems is still a crucial research topic because these impacts are not always predicted. While nanotechnology has shown promise in applications such as water desalination, pollutant removal, and marine sensors, its real-world implementation is still limited by several technological challenges. The synthesis of nanomaterials that are effective in marine environments, cost-effective to produce, and scalable remains difficult.

Moreover, the stability of nanoparticles when exposed to the harsh marine environment, with its constant saltwater, UV radiation, and varying temperatures, is a significant issue. Nanomaterials must be designed to withstand these conditions without losing their functionality or breaking down into potentially harmful by-products. In the field of marine sensors, for example, while nanomaterial-based sensors can offer superior sensitivity and selectivity, their integration into marine monitoring systems requires advanced, durable technologies. The miniaturization of devices for oceanic deployment is still a barrier, and the need for low-power, high-performance devices remains a challenge.

2.8.3 *High Costs and Technological Barriers*

Expensive Development and Technological Limitations

The development of nanotechnology for marine applications requires significant investment in research, development, and infrastructure. The high cost of producing nanoparticles and nano-devices can limit their widespread adoption, particularly in low-income or developing regions where marine pollution or water scarcity may be most acute. Furthermore, the integration of nanotechnology into existing marine systems (e.g., desalination plants, pollution cleanup efforts) can be technologically challenging and expensive. The manufacturing of nanomaterials at large scales suitable for marine applications poses technical challenges (Al-Maliki et al. 2022). Issues such as scalability, quality control, and the development of appropriate materials for specific marine conditions need to be addressed for nanotechnology to achieve its full potential. Despite the considerable advances in nanoscience, the production of nanomaterials at scale remains prohibitively expensive. In marine applications, especially those involving large-scale environmental remediation or monitoring, the high cost of producing and applying nanomaterials could limit their widespread use. Scalability is particularly important for projects like large-scale water purification or the deployment of nanosensors across vast marine areas. The lack of cost-effective methods to mass-produce and apply nanotechnology-based solutions remains a critical hurdle.

2.8.4 Regulatory and Ethical Issues

Lack of Regulation and Standards

There is a general lack of globally accepted regulatory frameworks or safety standards for the use of nanomaterials in marine environments. The absence of comprehensive guidelines for their manufacture, usage, and disposal presents challenges for governance. Moreover, the potential for unintended environmental consequences of nanoparticle release into marine environments further complicates the development of appropriate regulations (Lai et al. 2018). Deployment of nanotechnology in marine environments raises several ethical issues, especially related to the safety of marine life and the potential for harm to coastal communities that depend on marine resources. The integration of nanotechnology in marine environments also faces regulatory challenges. Currently, there is no uniform global framework for regulating the use of nanomaterials, and marine-specific regulations are sparse. Different countries may have varying standards regarding the safety and deployment of nanotechnology, which can complicate the development and commercialization of new technologies.

Ethically, the introduction of nanomaterials into marine ecosystems raises questions about unintended consequences. The long-term effects of nanomaterial pollution are still not fully understood, and it is challenging to predict how these materials might interact with marine organisms and ecosystems over time. This uncertainty necessitates the development of comprehensive risk assessment methodologies and ethical guidelines for marine nanotechnology applications. The concerns about the safety of nanomaterials in the marine food chain and the potential for disproportionate benefits to wealthy regions also contribute to ethical debates.

2.8.5 Public Perception and Trust

Fear of Nanomaterials

Public perceptions of nanotechnology are often influenced by concerns about its safety and environmental impact. Misunderstandings or misinformation about the risks associated with nanomaterials can lead to resistance or opposition to their use. Public trust in the technology will require transparent communication, rigorous testing, and strict regulatory measures to ensure that nanomaterials used in marine environments do not pose unacceptable risks. Another limitation in the development of marine nanoscience is the public perception and acceptance of nanotechnology. As with other environmental technologies, the deployment of nanomaterials in marine environments faces resistance from communities and environmental groups who fear the potential dangers of nanomaterial exposure. Building public trust through education, transparent research, and demonstrable benefits is key to gaining wider acceptance of marine nanotechnology applications (Fig. 2.4).

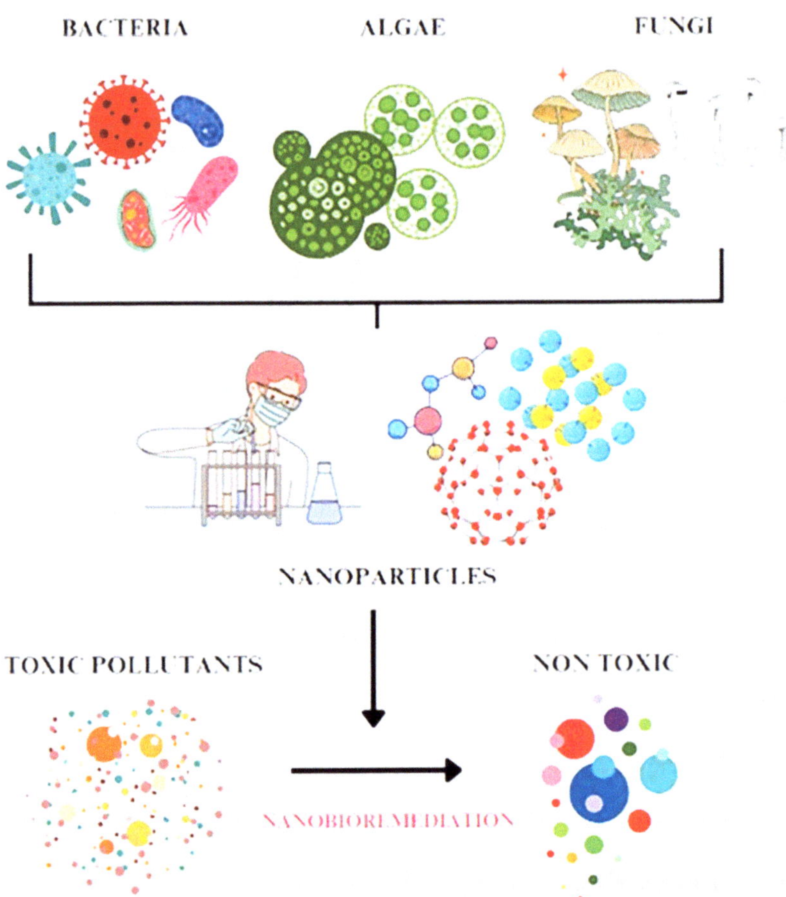

Fig. 2.4 Use of marine microorganism for nanobioremediation

2.9 Future Prospects for Marine Nanoscience and Nanotechnology

Marine nanoscience and nanotechnology represent an evolving frontier of research that is poised to provide groundbreaking solutions to global challenges. As scientific understanding deepens and technological advancements continue, the future of this field offers considerable promise in multiple domains, from environmental remediation to sustainable energy production. Marine nanoscience is inherently interdisciplinary, requiring expertise from fields like nanotechnology, marine biology, chemistry, and environmental science. One of the limitations of the field is the gap between these disciplines. Researchers in nanoscience may not always have a deep understanding of marine ecosystems, and marine biologists may not be well versed in the latest developments in nanotechnology. Bridging this knowledge gap is essential for the effective application of nanotechnology in marine contexts.

2.9.1 Environmental Remediation and Pollution Control

Marine pollution, including oil spills, plastic waste, and toxic metals, continues to be a major environmental issue. Nanotechnology offers the potential for highly efficient, targeted, and sustainable solutions for mitigating marine pollution. The future holds exciting prospects for the development of advanced nanomaterials capable of cleaning up various types of contaminants more effectively than conventional methods.

- Advanced Nanoparticle Adsorption Systems: In future, nanoparticles such as carbon nanotubes, graphene, and magnetic nanoparticles could be fine-tuned to absorb or capture a wider range of pollutants, including heavy metals, oil, and plastics. Researchers are likely to develop nanoparticle-based systems that can be deployed in situ, making them easier to use in large-scale cleanups.
- Self-Healing Materials: In the next decade, we may see the advent of self-healing nanomaterials, which can automatically repair themselves after being used for cleaning up contaminants. These materials could enhance the sustainability and longevity of nanomaterials deployed in marine environments.
- Biodegradable Nanomaterials: As concerns about the environmental impact of synthetic nanomaterials grow, there will likely be a shift toward biodegradable nanomaterials that can break down safely after they have performed their cleanup function (Bhavya et al. 2021). This will be crucial for ensuring that these technologies do not inadvertently contribute to the pollution they are designed to mitigate.

2.9.2 Protection and Durability of Marine Infrastructure

The increasing severity of corrosion and biofouling in marine environments places a strain on the longevity and maintenance costs of marine infrastructure. Nanotechnology could be pivotal in creating coatings and materials that provide longer-lasting, more effective protection against these challenges.

- Smart Coatings with Self-Cleaning Abilities: Future marine coatings may incorporate nanomaterials capable of self-cleaning, self-repairing, and preventing biofouling. These coatings would use nanoparticles to break down organic materials, preventing the accumulation of harmful microorganisms. Because of their photocatalytic qualities, nanomaterials like zinc oxide and titanium dioxide may be used to successfully stop the production of biofilms.
- Enhanced Corrosion Resistance: Research into nanostructured coatings will likely result in the development of more durable materials that can resist the aggressive conditions of saltwater. Nanomaterials like silica, zinc oxide, and nanocomposites are expected to offer superior corrosion resistance, extending the lifespan of critical infrastructure such as ships, oil rigs, and underwater pipelines (Zhao et al. 2023).
- Antimicrobial Nanomaterials: Utilizing antimicrobial nanoparticles like copper, silver, and copper oxide may result in coatings that actively prevent bacterial and fungal growth, lowering biofouling and the related hazards of disease transmission in marine environments.

2.9.3 Marine Monitoring and Biosensing Technologies

With growing concerns over ocean health, the need for advanced monitoring systems to track pollution, temperature changes, and other environmental variables is increasing. Nanosensors and nanomaterial-based monitoring devices hold the key to providing real-time, high-precision data for these applications.

- Nano-Enabled Sensor Networks: Future developments may involve creating large-scale sensor networks that could be deployed across oceans to provide continuous monitoring of marine environments (Martins et al. 2022). These nanosensors will be able to detect a wide range of pollutants and environmental variables (e.g., pH, salinity, and temperature) at very low concentrations, allowing for better predictive modeling of oceanic changes.
- Miniaturized and Low-Cost Devices: As nanotechnology advances, we can expect the development of miniaturized biosensors and sensors that are low cost and easy to deploy, offering real-time monitoring for oceanographers, environmental scientists, and regulatory bodies. This will democratize access to monitoring technology, enabling more widespread environmental stewardship.
- Wearable Marine Biosensors: Another exciting future prospect is the development of wearable or portable marine biosensors that could be used by researchers,

fishermen, and even tourists to monitor water quality and the presence of harmful microorganisms, enabling on-site decision-making.

2.9.4 Marine Biotechnology and Drug Delivery

Bioactive substances with possible medical uses have long been abundant in marine species. Nanotechnology could dramatically enhance the extraction, delivery, and effectiveness of these marine-derived drugs, with significant implications for medicine and aquaculture.

- Targeted Drug Delivery Systems: In order to create targeted medicine delivery systems for marine organism, nanotechnology will remain essential (Chamundeeswari et al. 2019). Nanocarriers, such as lipid-based nanoparticles, dendrimers, and solid lipid nanoparticles, will be used to improve the bioavailability and effectiveness of marine-derived compounds. These systems will enable more controlled and efficient delivery of drugs, minimizing side effects and improving therapeutic outcomes.
- Nanotechnologies for Aquaculture Health: Nanotechnology could also improve the health and productivity of farmed marine species. Future applications may include the development of nanoscale vaccines, antibiotic alternatives, or nutrient delivery systems that enhance the immune response and growth of marine organisms (Tayel et al. 2019). Such applications could revolutionize aquaculture by increasing production while reducing the environmental impact of farming operations.
- Marine-Derived Nanomaterials for Human Medicine: The development of nanomaterials derived from marine organisms, such as sponges, corals, and algae, will be crucial for creating novel medical treatments. These nanomaterials may be used in cancer therapies, anti-inflammatory treatments, or drug delivery platforms, opening up new possibilities in medical biotechnology.

2.9.5 Sustainable Marine Energy Solutions

Marine renewable energy, such as tidal, wave, and ocean thermal energy, holds enormous promise as a clean energy source. Nanotechnology could play a pivotal role in making these energy sources more efficient, cost-effective, and reliable.

- Enhanced Energy Conversion Efficiency: It is anticipated that nanomaterials will be employed in the creation of more effective energy conversion devices, such as enhanced tidal and wave turbines. In challenging marine conditions, nanostructured materials with improved strength-to-weight ratios and corrosion resistance will enable more robust and effective energy capture devices.

- Nano-Enabled Energy Storage Systems: To fully harness marine energy, effective storage systems are essential. Nanotechnology can contribute to the development of next-generation batteries and supercapacitors that store energy generated from tidal and wave energy systems. Nanomaterials such as graphene, carbon nanotubes, and advanced lithium-ion systems could significantly improve energy density, charge/discharge rates, and lifespan, making renewable marine energy more viable.
- Organic Photovoltaic Cells for Marine Environments: Nanotechnology holds great promise for enhancing solar energy capture, particularly in the form of organic photovoltaics (Rodrigues et al. 2023). These flexible and lightweight materials could be integrated into marine-based energy systems, where traditional solar panels are often less efficient due to space constraints and challenging conditions.

2.9.6 Marine Nanotoxicology and Safety Concerns

The growing use of nanomaterials in marine environments necessitates a deeper understanding of their potential risks to marine organisms and ecosystems. Future research in marine nanotoxicology will be critical to ensuring the safe deployment of nanotechnology.

- Risk Assessment Models: Advanced risk assessment models will be created in future to forecast how nanoparticles will affect the environment based on their size, shape, surface characteristics, and chemical composition. The fate of nanoparticles in marine ecosystems and their possible toxicity to marine species will be evaluated with the use of these models.
- Safe-by-Design Nanomaterials: The development of safe-by-design nanomaterials will be a priority, with an emphasis on minimizing any potential harm to the environment. Researchers will focus on designing nanomaterials that degrade safely in marine environments and do not accumulate in marine food chains.
- Long-Term Ecological Monitoring: The long-term effects of nanoparticles on marine ecosystems, including possible bioaccumulation and trophic transmission, will also be the subject of future research (Mehmood et al. 2021). Standards and regulations for the safe application of nanotechnology in marine applications will be developed with the help of this research.

2.10 Conclusion

Marine nanoscience and nanotechnology represent a dynamic and rapidly evolving intersection of scientific disciplines, offering promising solutions to a wide array of challenges faced by the marine environment. The integration of nanomaterials into marine science has ushered in a new era of possibilities for environmental protection, infrastructure enhancement, biotechnology, energy, and more. By manipulating materials at the nanoscale, these technologies allow for the creation of innovative materials and systems that can achieve previously unattainable levels of efficiency, specificity, and sustainability. Traditionally, marine science relied on mechanical, chemical, and biological methods to address challenges such as pollution, biofouling, and infrastructure degradation. While these approaches laid the foundation for addressing marine issues, they often fell short in terms of effectiveness, scalability, and long-term environmental sustainability. For example, traditional methods for cleaning up oil spills, controlling biofouling, and maintaining infrastructure involved techniques that were not always efficient, required continuous maintenance, or introduced additional pollutants into marine ecosystems.

Developments in nanotechnology and marine nanoscience have greatly enhanced these procedures. Because of their exceptional efficacy at adsorbing pollutants, nanomaterials like carbon nanotubes, graphene, and magnetic nanoparticles have made it possible to remediate marine habitats more quickly and thoroughly. Likewise, the development of advanced nanocoatings has greatly improved the durability and corrosion resistance of marine infrastructure, reducing the need for toxic chemicals and frequent maintenance. In the realm of marine monitoring, nanosensors have provided unprecedented levels of sensitivity, enabling real-time detection of pollutants, pathogens, and environmental changes at a scale and precision that traditional techniques could not achieve. These advancements have the potential to radically transform how we monitor and manage marine ecosystems. In future, marine nanoscience and nanotechnology are expected to continue advancing and expanding into new areas. The development of "smart" materials and self-healing nanoparticles holds promise for improving the sustainability and effectiveness of infrastructure protection and marine pollution maintenance. Furthermore, nanomaterials that improve energy conversion efficiency and storage capabilities will supplement the capacity to capture renewable energy from the ocean, such as through tidal and wave energy systems. In biotechnology, the potential for marine-derived drugs and therapies, facilitated by nanotechnology, could lead to breakthroughs in disease treatment while reducing the environmental impact of resource extraction. However, as these technologies advance, there are significant challenges that must be addressed. The potential ecological risks posed by the introduction of nanomaterials into marine environments are not yet fully understood, and research into nanotoxicology will be essential to mitigate any harmful effects. The safe-by-design principle, which advocates for the development of nanomaterials that are environmentally benign or easily degradable, will be crucial in ensuring that the benefits of marine nanoscience are not outweighed by unintended environmental consequences.

Marine nanoscience and nanotechnology hold tremendous potential to revolutionize the way we manage marine ecosystems, address pollution, and harness renewable energy. The advantages are clear: They offer solutions for environmental cleanup, sustainable water purification, improved aquaculture, and enhanced renewable energy systems. However, these benefits must be carefully weighed against the risks, particularly in terms of environmental toxicity, unintended consequences in marine ecosystems, high costs, and regulatory challenges Marine nanoscience and nanotechnology can only reach their full potential for sustainable development and environmental protection through a balanced approach that takes into account both the benefits and drawbacks. To achieve this, clear regulatory frameworks governing the safe development, use, and disposal of nanomaterials must be established, along with ongoing research to better understand the long-term effects on marine life and ecosystems.

Moreover, the scalability and commercialization of these technologies will require continued interdisciplinary collaboration and innovation. Regulatory frameworks will need to evolve in parallel with these advancements to ensure that nanomaterials are used safely and responsibly, particularly in sensitive marine ecosystems. In conclusion, marine nanoscience and nanotechnology hold immense potential for transforming how we approach and solve challenges related to the oceans. These technologies provide more economical, ecologically friendly, and efficient solutions by overcoming the shortcomings of conventional techniques. The future of marine nanoscience is promising, but it must be approached with a mindful consideration of safety, regulation, and ecological impact to ensure that the benefits can be fully realized without compromising the health of marine ecosystems. As research and innovation continue, marine nanoscience and nanotechnology will undoubtedly play an increasingly critical role in promoting the sustainability and resilience of the world's oceans.

Acknowledgements Authors are grateful to Director, CSIR-National Institute of Oceanography (CSIR-NIO), Goa, India; Scientist-in-Charge, CSIR-NIO, Regional Centre, Mumbai; and Academy of Scientific and Innovative Research (AcSIR) for their encouragement and support.

References

Alba-Patio A, Vaquer A, Barón E, Russell SM, Borges M, de la Rica R (2022) Micro-and nanosensors for detecting blood pathogens and biomarkers at different points of sepsis care. Microchim Acta 189(2):74

Al-Maliki RM, Alsalhy QF, Al-Jubouri S, Salih IK, AbdulRazak AA, Shehab MA et al (2022) Classification of nanomaterials and the effect of graphene oxide (GO) and recently developed nanoparticles on the ultrafiltration membrane and their applications: a review. Membranes 12(11):1043

Antczak T (2012) Nanotechnology-methods of manufacturing cellulose nanofibres. Fibres Text East Eur 20(2):91

Anwar A, Imran M, Iqbal HM (2023) Smart chemistry and applied perceptions of enzyme-coupled nano-engineered assemblies to meet future biocatalytic challenges. Coord Chem Rev 493:215329

Barnes CP, Sell SA, Boland ED, Simpson DG, Bowlin GL (2007) Nanofiber technology: designing the next generation of tissue engineering scaffolds. Adv Drug Deliv Rev 59(14):1413–1433

Bayda S, Adeel M, Tuccinardi T, Cordani M, Rizzolio F (2019) The history of nanoscience and nanotechnology: from chemical-physical applications to nanomedicine. Molecules 25(1):112. https://doi.org/10.3390/molecules25010112

Benjamin SR, Lima FD, Florean EOPT, Guedes MIF (2019) Current trends in nanotechnology for bioremediation. Int J Environ Pollut 66(1–3):19–40

Bhavya G, Belorkar SA, Mythili R, Geetha N, Shetty HS, Udikeri SS, Jogaiah S (2021) Remediation of emerging environmental pollutants: a review based on advances in the uses of eco-friendly biofabricated nanomaterials. Chemosphere 275:129975

Calcabrini C, Catanzaro E, Bishayee A, Turrini E, Fimognari C (2017) Marine sponge natural products with anticancer potential: an updated review. Mar Drugs 15(10):310

Casini M (2016) Smart buildings: advanced materials and nanotechnology to improve energy-efficiency and environmental performance. Woodhead Publishing

Chamundeeswari M, Jeslin J, Verma ML (2019) Nanocarriers for drug delivery applications. Environ Chem Lett 17:849–865

Cioffi R, Travaglioni M, Piscitelli G, Petrillo A, Parmentola A (2020) Smart manufacturing systems and applied industrial technologies for a sustainable industry: a systematic literature review. Appl Sci 10(8):2897

Corsi I, Cherr GN, Lenihan HS, Labille J, Hassellov M, Canesi L et al (2014) Common strategies and technologies for the ecosafety assessment and design of nanomaterials entering the marine environment

Creamer AE, Gao B (2015) Carbon dioxide capture: an effective way to combat global warming, vol 62. Springer, Cham

Das S, Chakraborty J, Chatterjee S, Kumar H (2018) Prospects of biosynthesized nanomaterials for the remediation of organic and inorganic environmental contaminants. Environ Sci Nano 5(12):2784–2808

Fawcett D, Verduin JJ, Shah M, Sharma SB, Poinern GEJ (2017) A review of current research into the biogenic synthesis of metal and metal oxide nanoparticles via marine algae and seagrasses. J Nanosci 2017(1):8013850

Gorain B, Karmakar V, Sarkar B, Dwivedi M, Leong JT, Toh JH, Seah E, Ling KY, Chen KY, Choudhury H, Pandey M (2023) Biomacromolecule-based nanocarrier strategies to deliver plant-derived bioactive components for cancer treatment: A recent review. Int J Biol Macromolecules 253:126623

Green DW, Lai WF, Jung HS (2014) Evolving marine biomimetics for regenerative dentistry. Mar Drugs 12(5):2877–2912

He F, Cao J, Qi J, Liu Z, Liu G, Deng W (2021) Regulation of stem cell differentiation by inorganic nanomaterials: recent advances in regenerative medicine. Front Bioeng Biotechnol 9:721581

Hochella MF Jr (2002) Nanoscience and technology: the next revolution in the earth sciences. Earth Planet Sci Lett 203(2):593–605

Jiang Q, Xie Y, Peng M, Wang Z, Li T, Yin M, Shen J, Yan S (2022) A nanocarrier pesticide delivery system with promising benefits in the case of dinotefuran: Strikingly enhanced bioactivity and reduced pesticide residue. Environ Sci Nano 9(3):988–999

Kalashgarani MY, Babapoor A (2022) Application of nano-antibiotics in the diagnosis and treatment of infectious diseases. Adv Appl Nano Bio Technol 3(1):22–35

Kamat S, Sureesh D, Modi S, Kumari M, Jayabaskaran C (2024) Marine fungi as a bioresource of medicinal entities. In: Fungi Bioactive Metabolites: Integration of Pharmaceutical Applications, pp 207–239. Singapore: Springer Nature Singapore

Karanassios V (2018) Brief introduction to nanoscience and nanotechnology. Nanosci J 1(1):1–6

Khan Y, Sadia H, Ali Shah SZ, Khan MN, Shah AA, Ullah N et al (2022) Classification, synthetic, and characterization approaches to nanoparticles, and their applications in various fields of nanotechnology: a review. Catalysts 12(11):1386

Lee DH, Kim BS, Kang SS (2020) Bacteriocin of Pediococcus acidilactici HW01 inhibits biofilm formation and virulence factor production by Pseudomonas aeruginosa. Probiotics Antimicrob Proteins 12(1):73–81

Liang X, Chen M, Bhattarai P, Hameed S, Tang Y, Dai Z (2021) Complementing cancer photodynamic therapy with ferroptosis through iron oxide loaded porphyrin-grafted lipid nanoparticles. ACS nano 15(12):20164–20180

Lim JW, Kim TY, Woo MA (2021) Trends in sensor development toward next-generation point-of-care testing for mercury. Biosens Bioelectron 183:113228

Mahaddalkar T, Lopus M (2017) From natural products to designer drugs: development and molecular mechanisms action of novel anti-microtubule breast cancer therapeutics. Curr Top Med Chem 17(22):2559–2568
Martins R, Figueiredo J, Sushkova A, Wilhelm M, Tedim J, Loureiro S (2022) "Smart" nanosensors for early detection of corrosion: Environmental behavior and effects on marine organisms. Environ Pollut 302:118973
Mat Lazim Z, Salmiati S, Marpongahtun M, Arman NZ, Mohd Haniffah MR, Azman S et al (2023) Distribution of silver (Ag) and silver nanoparticles (AgNPs) in aquatic environment. Water 15(7):1349
Mchmood A, Khan FSA, Mubarak NM, Mazari SA, Jatoi AS, Khalid M et al (2021) Carbon and polymer-based magnetic nanocomposites for oil-spill remediation—a comprehensive review. Environ Sci Pollut Res 28(39):54477–54496
Mirshahghassemi S, Lead JR (2015) Oil recovery from water under environmentally relevant conditions using magnetic nanoparticles. Environ Sci Technol 49(19):11729–11736
Mishra S, Chauhan G, Verma S, Singh U (2022) The emergence of nanotechnology in mitigating petroleum oil spills. Mar Pollut Bull 178:113609
Mona C, Salomé MM, Judit K, José-María N, Eric B, María-Luisa FC (2023) Considerations for bioaccumulation studies in fish with nanomaterials. Chemosphere 312:137299
Nasiri H (2024) Nanotechnology in marine industries: innovations for sustainability
Nnachi RC, Sui N, Ke B, Luo Z, Bhalla N, He D, Yang Z (2022) Biosensors for rapid detection of bacterial pathogens in water, food and environment. Environ Int 166:107357
Pete AJ, Bharti B, Benton MG (2021) Nano-enhanced bioremediation for oil spills: a review. ACS ES&T Eng 1(6):928–946
Petersen EJ, Mortimer M, Burgess RM, Handy R, Hanna S, Ho KT et al (2019) Strategies for robust and accurate experimental approaches to quantify nanomaterial bioaccumulation across a broad range of organisms. Environ Sci Nano 6(6):1619–1656
Pourhashem S, Seif A, Saba F, Nezhad EG, Ji X, Zhou Z et al (2022) Antifouling nanocomposite polymer coatings for marine applications: a review on experiments, mechanisms, and theoretical studies. J Mater Sci Technol 118:73–113
Püsküllüoğlu M, Michalak I (2022) An ocean of possibilities: a review of marine organisms as sources of nanoparticles for cancer care. Nanomedicine 17(22):1695–1719
Rajendran S, Priya TA, Khoo KS, Hoang TK, Ng HS, Munawaroh HS, Karaman C, Orooji Y, Show PL (2022) A critical review on various remediation approaches for heavy metal contaminants removal from contaminated soils. Chemosphere 287:132369
Rajoka MSR, Mehwish HM, Xiong Y, Song X, Hussain N, Zhu Q, He Z (2021) Gut microbiota targeted nanomedicine for cancer therapy: challenges and future considerations. Trends Food Sci Technol 107:240–251
Rakesh Tej Kumar K, Ramakrishna M, Durga Sukumar G (2018) A review on PV cells and nanocomposite-coated PV systems. Int J Energy Res 42(7):2305–2319
Ray PC, Yu H, Fu PP (2009) Toxicity and environmental risks of nanomaterials: challenges and future needs. J Environ Sci Health C 27(1):1–35
Razzaghi M, Homaei A, Vianello F, Azad T, Sharma T, Nadda AK et al (2022) Industrial applications of immobilized nano-biocatalysts. Bioprocess Biosyst Eng 45:1–20
Rigo S, Cai C, Gunkel-Grabole G, Maurizi L, Zhang X, Xu J, Palivan CG (2018) Nanoscience-based strategies to engineer antimicrobial surfaces. Adv Sci 5(5):1700892
Rodrigues AV, da Costa Romeiro F, Rodrigues CV (2023) The impact of nanomaterials on the environment and perspectives for sustainable energy generation. In: Implications of nanoecotoxicology on environmental sustainability. IGI Global, pp 156–177
Salmasi A, Abraham J, Salmasi F (2022) Prospects for application of nanotechnology in marine industries: a brief review
Santhosh RS, Amarendra V (2015) Nanotechnology–from a marine discovery perspective. In: Springer handbook of marine biotechnology, pp 1113–1129

Singh B, Kumar S, Kishore B, Narayanan TN (2020) Magnetic scaffolds in oil spill applications. Environ Sci Water Res Technol 6(3):436–463

Sun W, Dou F, Li C, Ma X, Ma LQ (2021) Impacts of metallic nanoparticles and transformed products on soil health. Crit Rev Environ Sci Technol 51(10):973–1002

Sun L, Jin YS (2021) Xylose assimilation for the efficient production of biofuels and chemicals by engineered Saccharomyces cerevisiae. Biotechnol J 16(4):2000142

Tayel AA, Elsaied BE, Diab AM (2019) Nanotechnology for aquaculture. In: Nanoscience for sustainable agriculture, pp 479–544

Thomas P, Sahoo BN, Thomas PJ, Greve MM (2024) Recent advances in emerging integrated anticorrosion and antifouling nanomaterial-based coating solutions. Environ Sci Pollut Res 31:1–27

Trivedi R, Upadhyay TK, Kausar MA, Saeed A, Sharangi AB, Almatroudi A et al (2022) Nanotechnological interventions of the microbiome as a next-generation antimicrobial therapy. Sci Total Environ 833:155085

Vaseashta A, Vaclavikova M, Vaseashta S, Gallios G, Roy P, Pummakarnchana O (2007) Nanostructures in environmental pollution detection, monitoring, and remediation. Sci Technol Adv Mater 8(1–2):47

Wang C, Sun S, Wang P, Zhao H, Li W (2023) Nanotechnology-based analytical techniques for the detection of contaminants in aquatic products. Talanta 269:125462

Wilson B, Geetha KM (2022) Lipid nanoparticles in the development of mRNA vaccines for COVID-19. J Drug Deliv Sci Technol 74:103553

Yang W, Wang L, Mettenbrink EM, DeAngelis PL, Wilhelm S (2021) Nanoparticle toxicology. Annu Rev Pharmacol Toxicol 61(1):269–89

Yano H, Chitsazi R, Lucaj C, Tran P, Hoffman AF, Baumann MH, Lupica CR, Shi L (2023) Subtle structural modification of a synthetic cannabinoid receptor agonist drastically increases its efficacy at the CB1 receptor. ACS Chem Neurosci 14(21):3928–3940

Zhang L, Li H, Zhang X, Li Q, Zhu G, Liu FQ (2023) A marine coating: self-healing, stable release of Cu^{2+}, anti-biofouling. Mar Pollut Bull 195:115524

Zhao Z, Liu Z, Zhang A, Yan X, Xue W, Peng B et al (2022) Graphene-nanopocket-encaged PtCo nanocatalysts for highly durable fuel cell operation under demanding ultralow-Pt-loading conditions. Nat Nanotechnol 17(9):968–975

Zhao Y, Peng L, Xie H, Zhang W, Huang S, Yang Z et al (2023) Enhancing the erosion–corrosion resistance of cupronickel alloy through grain boundary engineering. Corros Sci 219:111228

Chapter 3
Nanocellulose for Sustainable Water Purification

Ran Chen, Panqi Xu, Liang Wang, Feng Gao, and Qinghua Zhang

Keywords Nanocellulose · Water purification · Material development

3.1 Introduction

Water resources, as one of the natural resources essential for human survival, are facing a severe crisis. With the acceleration of global industrialization and urbanization, the discharge of industrial wastewater, agricultural runoff, and domestic sewage has led to increasingly serious water pollution. These pollutants include heavy metal ions (such as Pb, Cd, and Hg), organic pollutants (such as dyes, drug residues, and pesticides), and microplastics, which seriously affect the ecological environment, animal and plant health, and human quality of life. In addition, water scarcity and deterioration of water quality have further exacerbated the global water crisis. According to the United Nations' 2021 Water Resources Report, at least one-third of the global population is facing varying degrees of water scarcity, with the issue of safe and clean drinking water supply being particularly prominent. At present, traditional water treatment technologies include adsorption, membrane filtration, chemical precipitation, photocatalytic degradation, etc. Although these technologies have to some extent solved the problem of water pollution, they still face issues such as low efficiency, high costs, secondary pollution, susceptibility to pollution, and poor durability.

Nanocellulose (NC), as a renewable and environmentally friendly biomaterial, not only has unique physical and chemical properties such as high specific surface area, excellent mechanical properties, and rich surface functional groups (such as hydroxyl

R. Chen · P. Xu · L. Wang · F. Gao · Q. Zhang (✉)
College of Chemical and Biological Engineering, Zhejiang University, Hangzhou, China
e-mail: qhzhang@zju.edu.cn

S. K. Panda and V. P.M. (eds.), *Bionanocomposites from Marine Origin*, Advanced Structured Materials 247, https://doi.org/10.1007/978-3-032-10749-7_3

and carboxyl groups), but also facilitates chemical modification to introduce adsorption functions. It also has efficient removal ability, such as the excellent adsorption and filtration effects of nanocellulose-based materials on heavy metal ions, organic pollutants, and small particles, demonstrating excellent selectivity and efficiency. Additionally, it can also be functionalized and composite designed: through chemical modification or composite with other materials such as metal oxides and carbon materials, nanocellulose can achieve multifunctionality and further improve water treatment performance.

This chapter provides theoretical support for material development by systematically summarizing the sources, structures, and functionalization methods of nanocellulose. This study presents an in-depth analysis of the specific applications of nanocellulose-based materials in water treatment, including the removal of heavy metals, the degradation of organic pollutants, membrane filtration technology, and microplastic treatment. Explore performance optimization and challenges, sort out the adsorption mechanism, filtration efficiency, and regeneration capacity of nanocellulose water treatment materials, and analyze their application bottlenecks in complex water environments. Looking ahead to future development directions, propose a sustainable development path and research focus for nanocellulose-based water treatment technology. Based on the above content, this review aims to provide a systematic theoretical summary and practical guidance for the research of nanocellulose in the field of water treatment and promote the development of green and sustainable water treatment technologies.

3.2 Classification and Properties

Nanocellulose (NC) is a nanomaterial characterized by a diameter measuring less than 100 nm, obtained from natural cellulose through specialized processing techniques. Its specific surface area can attain approximately 500 m^2/g, in stark contrast to soft cellulose pastes, which typically exhibit a surface area ranging from 1 to 4 m^2/g (Banavath et al. 2011). This substantial increase in surface area is associated with a greater availability of hydroxyl groups on the surface of nanocellulose, facilitating the grafting of functional groups or molecules via various methods, including carboxylation, etherification, silicification, sulfonation, esterification, 2,2,6,6-Tetramethylpiperidin-1-yloxy radical (TEMPO)-mediated oxidation, phosphorylation, and amidation (Isogai et al. 2011). The field of biomaterials science currently leverages the structural and functional properties of nanocellulose to develop innovative materials with multifunctional characteristics, addressing contemporary demands across diverse sectors such as biomedicine, environmental science, energy, pharmaceuticals, agriculture, and food technology. Notably, it has been established that the incorporation of carboxyl, sulfonate, and phosphonate functional groups onto nanocellulose can effectively improve the selective absorption of contaminants in water remediation processes (Sehaqui et al. 2016). Furthermore, nanocellulose-based membranes and filters have proven effective in the removal of metal ions, dyes, and microorganisms, while modified nanocellulose matrices with hydrophobic or lipophilic functional groups can also eliminate organic contaminants, including oil and cyclohexene.

The primary sources of nanocellulose are renewable plant materials, such as wood, bamboo, and agricultural residues. Wood, originating from trees, exhibits a stratified composition that includes cells or fibers, filaments, cell walls, and cellulose molecules (Gibson 2012). Cellulose fibers predominantly reside within the secondary cell walls of plants. Nanocellulose can be obtained from ultrafine fiber bundles utilizing mechanical and physical extraction techniques, typically following biochemical pretreatments. The structural composition of nanocellulose is characterized by numerous parallel chains of cellulose molecules, each comprising three hydroxyl groups (–OH) linked to a glucose ring. The close proximity of these hydroxyl groups promotes the formation of hydrogen bonds, leading to a highly organized hydrogen bonding network within the nanocellulose. These hydrogen bonds play a significant role in establishing the crystalline regions of nanocellulose, thereby endowing it with the distinctive properties associated with cellulose crystals (Liu et al. 2024a). Although the chemical compositions of various types of nanocellulose are relatively similar, significant variations exist in terms of particle dimensions, crystallinity, and morphological properties (Noremylia et al. 2022). The morphology and properties of nanocellulose can be categorized based on raw materials, preparation methods, and morphology, as these factors are closely interconnected, as illustrated in Fig. 3.1.

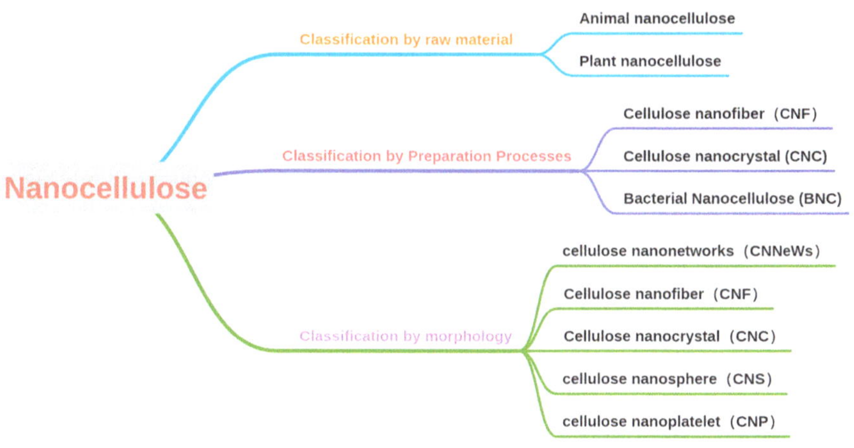

Fig. 3.1 Classification of nanocellulose

3.2.1 Classification by Source of Raw Materials

NC can be categorized based on its sources into two primary types: animal nanocellulose and plant nanocellulose. The former primarily encompasses tunicate cellulose, which is sourced from organisms such as sea squirts, whereas the latter includes cellulose derived from various plant materials, including wood pulp, cotton, and bamboo.

Animal Nanocellulose

Tunicates, commonly referred to as sea squirts, are a group of marine invertebrates characterized by their unique biological structures. Within these organisms, sea squirt nanocellulose is present, forming a protective outer layer in conjunction with proteins. Notably, cellulose has been identified exclusively in tunicates among the animal kingdom. The outer shell of sea squirts is a complex, stratified structure composed predominantly of cellulose and protein, with approximately 60% cellulose and 40% protein, alongside trace amounts of lipid components. This shell features a slender, compact external layer accompanied by a more substantial internal layer composed of cellulose (Matthysse et al. 2004). Sea squirt nanocellulose is characterized by its distinct structural characteristics, which include higher levels of crystallinity and a greater aspect ratio, setting it apart from cellulose derived from plants and microorganisms. The dimensions of the microfibers vary across different species, with nanofibers derived from periphyton measuring between 100 and 2000 nm in length and approximately 10–30 nm in cross-sectional diameter, resulting in an aspect ratio ranging from 70 to 100. The cross-sectional morphology of these fibers is typically parallelogram-shaped (Helbert et al. 1998). Furthermore, the lifestyle of the encapsulated gastropods exerts a considerable influence on the quality of cellulose generated; nanofibrillar cellulose from solitary gastropods exhibits higher crystallinity, larger crystal size, and enhanced mechanical properties compared to that derived from group-living gastropods (Chanthathamrongsiri et al. 2021).

The high crystallinity of sea squirt nanocellulose presents challenges for its solubility in water, as it tends to swell rather than dissolve, thereby limiting its potential applications. Consequently, extracted sea squirt nanocellulose is often processed into a more functional form. This tunicate-derived nanocellulose is characterized by exceptional properties, including high strength, purity, and Young's modulus, as well as notable optical, electrical, magnetic, and rheological attributes. The high concentration of hydroxyl groups present on the surface of nanocellulose enables various potential functional modifications. The applications of nanocellulose extracted from tunicates span various fields, including biomedicine, polymer-reinforced materials, papermaking, food packaging, and environmental engineering, which have attracted considerable scholarly interest. Nonetheless, the extraction and processing of cellulose from tunicates pose significant challenges that hinder its wider application. The methodology employed for extraction has a significant impact on the quality of the cellulose produced. A variety of research efforts have effectively isolated nanocellulose from the outer integument of tunicates. As it reported, Belton et al. (1989) utilized Dyson's enzyme to eliminate non-cellulosic constituents from the external layer of periphyton, subsequently subjecting the material to reflux with sodium hydroxide to obtain crude cellulose, albeit with residual colored components. At present, the primary techniques for cellulose extraction from tunicates consist of alkali treatment and the kraft process.

Plant Nanocellulose

Plants represent the most abundant source of cellulose, commonly comprising materials such as wood, cotton, sisal, flax, rice straw, and bagasse. The synthesis of cellulose takes place via hexagonal configurations known as rosettes, which contribute to the formation of nanofibrous structures within the cell wall. This process is enhanced by the presence of hemicellulose and lignin, which are non-cellulosic constituents that are essential for providing structural support and distributing mechanical stresses encountered by the plant. The interaction between neighboring nanofibers is facilitated by hydrogen bonding and van der Waals forces, leading to a stratified architecture of cellulose fibers (Gómez et al. 2016). Wood presents numerous advantages as a source material for nanocellulose, including a high yield, widespread availability, and cost-effectiveness. The microfibrils derived from wood-based nanocellulose typically have diameters ranging from approximately 3~4 nm. However, there are notable drawbacks; for example, nanofibers produced via mechanical methods often retain lignin, hemicellulose, and other constituents. The preparation of pulp from wood necessitates high-pressure alkaline cooking, bleaching, and various other treatments, which may pose environmental concerns. Nonetheless, the cost of pulp as a precursor for nanofiber production is considerably lower than that of cellulose derived from animal sources and other plant-based celluloses. In contrast, sisal fiber boasts higher cellulose content, ranging from 67 to 87%, surpassing that of wood fiber, which typically contains 35% to 50% cellulose. Additionally, sisal fiber demonstrates enhanced mechanical properties, including elevated tensile strength and modulus. Moreover, nanocellulose can be derived from the cell walls of different algal species, including both red and green algae, typically through bleaching methods.

3.2.2 Classification by Preparation Process

Nanocellulose can be categorized into three distinct types based on the method of preparation: cellulose nanofiber (CNF), cellulose nanocrystal (CNC), and bacterial nanocellulose (BNC), whose microscopic morphology is shown in Fig. 3.2.

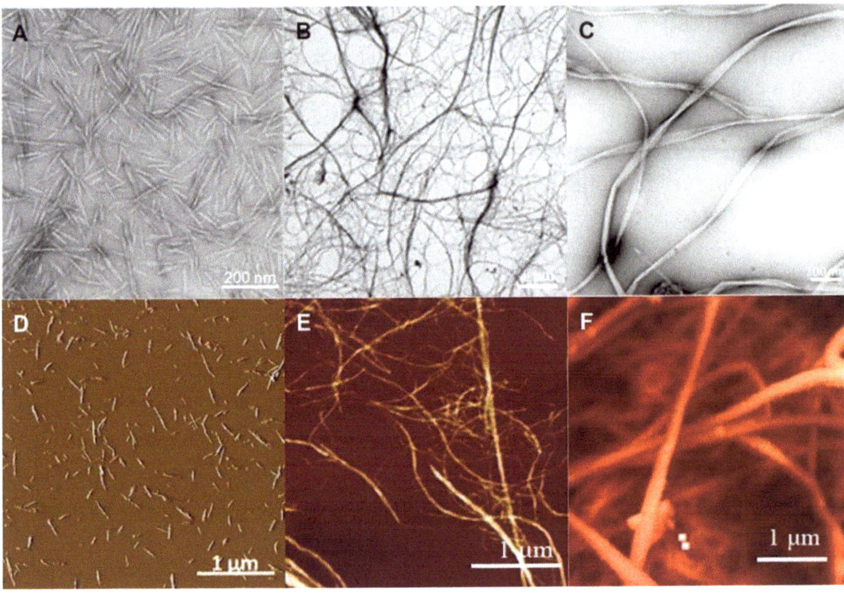

Fig. 3.2 TEM and AFM images of CNCs (**a**, **d**), CNFs (**b**, **e**), and BCs (**c**, **f**) (Chu et al. 2020): reproduced with permission (Castro et al. 2012; Chen et al. 2015; George et al. 2011; Hoeng et al. 2015; Pelissari et al. 2014)

Cellulose Nanofiber

Cellulose nanofiber (CNF), which is alternatively referred to as nanofibril cellulose or cellulose nanofibril, constitutes a collection of elongated cellulose structures characterized by their intricate entanglement and high specific surface area. These fibers exhibit long, flexible chains with diameters that range from 10 nm to several hundred nanometers and lengths that extend to the micrometer scale. The isolation of CNF is typically achieved through a variety of mechanical, chemical, and biological methods. Notably, mechanical techniques such as high-pressure homogenization, cryogenic crushing, and milling are commonly employed for the extraction of cellulose nanofibers (Pääkkö et al. 2007). Chemical treatments typically involve alkali processes, while biological methods generally employ enzymatic treatments (Tonoli et al. 2012). Numerous researchers employ a combination of techniques to attain the desired characteristics of CNF. The inherent characteristics of CNF can demonstrate significant variability, contingent upon the origin of the raw materials and the particular extraction techniques utilized. Consequently, the attributes of the generated cellulose nanofibers (CNF) may vary significantly with respect to their shape, degree of fibrillation, morphology, and overall characteristics.

CNF is composed of two distinct regions: crystalline and amorphous, with a crystallinity that typically ranges from 50 to 70%. It is characterized by a very high surface area, consistent dispersion, significant chemical tunability, and excellent hydrophilicity, among other attributes. Nevertheless, CNF exhibits two significant limitations related to their fundamental physical characteristics: (1) the high concentration of hydroxyl groups, which promotes robust hydrogen bonding interactions between the two varieties of nanofibrils, resulting in the development of a gel-like structure during the manufacturing process; and (2) its pronounced hydrophilicity, which constrains its applicability in certain domains, such as paper coating (due to exacerbated dehydration effects) and composites (due to a propensity for agglomeration in petrochemical polymers). Consequently, a promising approach to mitigate these limitations is through chemical surface modification, which can diminish the quantity of hydroxyl interactions and enhance compatibility with diverse matrices (Abol-Fotouh et al. 2020).

CNF exhibits a variety of distinctive advantages that not only support its production at an industrial scale but also promote applications across diverse industrial sectors. This is largely attributable to the presence of numerous functional groups and the availability of several appealing industrial processing methods (Naderi 2017). It is recognized as superior nanoscale building block owing to its remarkable mechanical properties, which include high strength comparable to that of steel and Kevlar, as well as a significant stiffness characterized by a Young's modulus of 138 GPa. Additionally, CNF possesses a low coefficient of thermal expansion (1×10^{-7} K^{-1}), high levels of crystallinity, hydrophilic properties, and surfaces that are amenable to facile modification. Furthermore, magnetic CNF-based nanomaterials demonstrate a variety of outstanding characteristics, such as thermal stability and extractability, which position them as viable candidates for application as adsorbents in the context of the remediation of hazardous metals and organic contaminants (Sarafraz-Yazdi et al. 2015).

Cellulose Nanocrystal

Cellulose nanocrystal (CNC) is characterized by its elongated, crystalline rod-like morphology and exhibits greater rigidity in comparison with CNF. The crystallinity of CNC, which ranges from 54 to 88%, can be ascribed to the significant elimination of amorphous regions throughout the processing stages. Although enzymatic hydrolysis methods are available for the production of CNC, the predominant technique employed is strong acid hydrolysis, particularly utilizing sulfuric acid (Filson et al. 2009). CNC is derived from pretreated fibers through a process of acid hydrolysis. This procedure involves the application of controlled strong acid hydrolysis to the pretreated fibers, which facilitates the dissolution of residual amorphous and aromatic structural domains, thereby promoting the longitudinal cleavage of microfibrils. In the process of strong acid hydrolysis, hydrated hydrogen ions penetrate both the aromatic and amorphous domains of the cellulose chain. This infiltration results in the hydrolytic cleavage of glycosidic bonds, ultimately yielding individual cellulose microcrystals, a process that is frequently augmented by sonication. The resultant nanoparticles, which are typically suspended in aqueous solutions, are commonly designated as CNC (Nagarajan et al. 2021). Subsequent processing techniques, such as lyophilization (freeze-drying) or spray-drying, may be employed as required. The condition under which the reaction occurs, along with the cellulose source, play a critical role in determining various characteristics of the resultant CNC, including their crystallinity, aspect ratio, size distribution, and morphology. Typically, CNC manifests as rod-like or whisker-like structures, with widths ranging from 3 to 50 nm and lengths from 50 to 500 nm. These nanocrystals are characterized by high axial stiffness (105−168 GPa), a substantial Young's modulus (20−50 GPa), a low coefficient of thermal expansion (0.1 ppm/K), impressive tensile strength (9 GPa), a high aspect ratio (10–70), significant thermal stability (up to 260 °C), low density (1.5−1.6 g/cm^3), the ability to exhibit soluble liquid crystal behavior, and shear-thinning rheological properties (Thomas et al. 2018).

In comparison with CNF, CNC demonstrates diminished flexibility due to the absence of amorphous regions. Furthermore, the diameter of CNC is heterogeneous, which results from the random cleavage of cellulose chains that occurs during the hydrolysis process (Thomas et al. 2020). Furthermore, the surface of nanocellulose is characterized by a high concentration of reactive hydroxyl groups, which facilitates a range of chemical modifications aimed at improving its surface properties. Nonetheless, the abundance of these hydroxyl groups may result in agglomeration, thereby hindering effective dispersion. As a result, chemical modifications are essential to enhance the miscibility and compatibility of solvents or matrices utilized in the production of polymer nanocomposites. Consequently, a substantial amount of research has concentrated on modifying the surface characteristics of CNC to diminish the density of hydroxyl groups in targeted areas. This modification subsequently leads to a reduction in the strength of interparticle interactions and enhances compatibility with hydrophobic polymer matrices. Additionally, the low contrast of nanocellulose in transmission electron microscopy (TEM) images presents a notable challenge for morphological characterization and sizing. Moreover, the processes involved in the fabrication of CNC and CNF can inflict structural damage, making it imperative to implement strategies that minimize such damage during the extraction process. Rafieian et al. (2019) research indicated that polyethersulfone (PES) was developed as a substrate, and amine-functionalized CNC is utilized as innovative membrane for the extraction of copper ions and Red-16 nanofillers from aqueous solutions. The surface modification of CNC was conducted utilizing (3-aminopropyl) triethoxysilane (APTES), with verification achieved through Fourier transform infrared (FTIR) spectroscopy and energy dispersive X-ray spectroscopy. Following this, PES membranes were synthesized by incorporating modified CNC (MCNC) at various concentrations. The results indicated that the maximum adsorption capacity for copper ions was 90% in membranes containing 1 wt% MCNC. Regarding dye removal, pure PES membranes demonstrated an efficiency of 89%, which improved to 99% with the addition of 1 wt% MCNC. These findings imply that the incorporation of modified CNC into PES membranes significantly enhances the effectiveness of contaminant removal from water, thereby offering a promising strategy for water filtration applications.

It is noteworthy that hairy cellulose nanocrystals (HCNCs) have been recognized as a novel class of cellulose-derived nanomaterials following their identification in

2012 (Yang et al. 2012). In comparison with traditional CNC production methods, the preparation of HCNCs is characterized by a significantly milder process that does not require the use of strong acids. Instead, the synthesis involves either periodate oxidation or chlorite oxidation of cellulose. HCNCs are characterized by their distinctive morphological and chemical attributes. They are composed of central rod-shaped nanocrystals from which hair-like molecular chains protrude, and they exhibit a greater diversity and abundance of functional groups compared to conventional CNC. This improvement can be ascribed to the incorporation of rigid crystalline nanorods, which increase the reactivity of the amorphous regions within HCNCs toward chemical modifications, thereby increasing the availability of functional groups for subsequent modification and functionalization. To date, five primary types of HCNCs have been successfully developed: electrosterically stabilized nanocrystalline cellulose (ENCC), which has a high carboxyl group content; cationic nanocrystalline cellulose (CNCC), characterized by a carboxyl group content of up to 6.6 mmol/g and a quaternary ammonium salt content of 1.68 mmol/g; aminated nanocrystalline cellulose (ANCC), which contains a high primary amine content of up to 5.5 mmol/g; sterically stabilized nanocrystalline cellulose (SNCC), with an aldehyde content of up to 8.2 mmol/g; and bifunctional nanocrystalline cellulose (BNCC), which possesses both aldehyde and carboxyl groups, as demonstrated in Fig. 3.3.

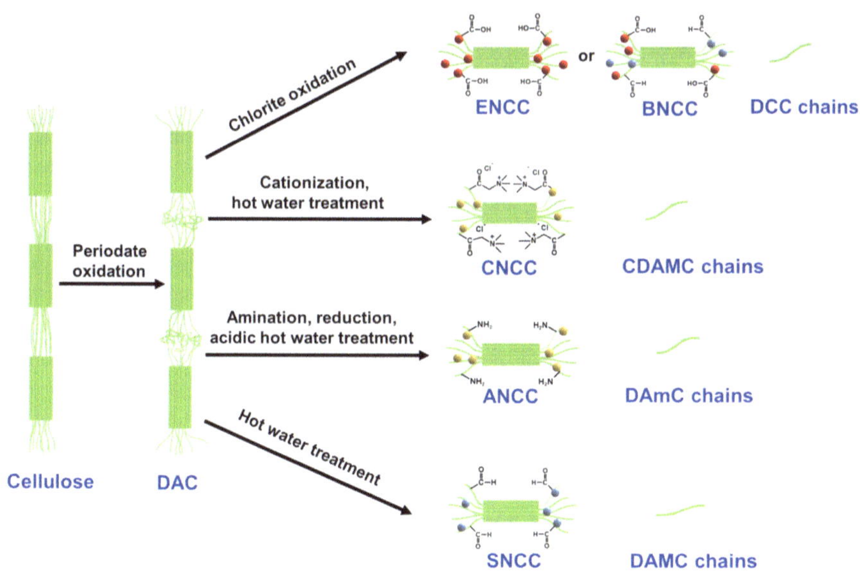

Fig. 3.3 Schematic illustration depicting the methodologies employed in the synthesis of HCNCs (DAC: dialdehyde cellulose; DCC: dicarboxylated cellulose; CDAMC: cationic dialdehyde cellulose; DAMC: dissolved diaminated cellulose; DAMC: dialdehyde modified cellulose): reproduced with permission (Ding et al. 2024)

Bacterial Nanocellulose

Bioprocessing technologies frequently utilize metabolically engineered microorganisms, such as *Escherichia coli*, to transform biomass into valuable biomaterials. These biomaterials encompass biofuels, bioplastics, platform chemicals, and extracellular polymers (Gilmore et al. 2015). Researchers often employ a variety of biomass sources for numerous environmental applications. Due to the limitations inherent in both synthetic and natural materials, there has been an increasing focus among scientists on the creation of more efficient and safer materials that can be utilized across a broad spectrum of applications. Bacterial nanocellulose (BNC) emerges as a promising biopolymer, produced as an extracellular macromolecule composed of β-D-glucopyranose by certain bacterial species, including *Agrobacterium tumefaciens*, *Achromobacter*, *nitrogen-fixing Bacillus*, *Rhizobium*, *Salmonella*, and *Escherichia*, among others. BNC exhibits a unique combination of cellulose properties and those of nanomaterials (Sharma and Bhardwaj 2019). The biosynthesis of BNC is a meticulously structured and rigorously controlled process that unfolds in two distinct phases: the initial establishment of 1, 4-glucan linkages, succeeded by the assembly and crystallization of cellulose. This process initiates with the uptake of carbon substrates, including glucose and fructose, into the bacterial cell, which subsequently facilitates the synthesis of the cellulose precursor, uridine diphosphate glucose (UDPG). Subsequently, cellulose synthase (Bcs) catalyzes the polymerization of glucose from UDPG to generate the 1, 4-glucan chain. In conclusion, cellulose chains are excreted as sub-fibrils via pores in the cell membrane and are then arranged into ribbons within a three-dimensional network of hydrogen-bonded nanofibers. The synthesis of BNC takes place through an oxidative fermentation process that employs sugars and organic acids in either synthetic or non-synthetic sugar and nitrogen-rich media, generally maintained at temperatures between 25 and 30 °C and pH levels ranging from 4.5 to 7.5 (Jozala et al. 2016).

BNC is characterized by its outstanding chemical purity, lacking impurities such as hemicellulose, lignin, or pectin. It features a highly porous nanofibrous architecture, with pore dimensions generally less than 10 μm. This structural configuration contributes to its remarkable permeability to gases and liquids, in addition to an exceptionally high water-binding capacity, with moisture levels surpassing 90% (Portela et al. 2019). Furthermore, BNC exhibits a high crystallinity index, generally ranging from 80 to 90%, which contributes to its improved thermal stability—an important attribute for applications that necessitate heat sterilization. The polymerization degree of BNC is considerably elevated, generally falling within the range of 4000 to 10,000 anhydrous glucose units. Its crystalline nanofiber architecture also bestows remarkable mechanical properties. Although some variations are noted in the existing literature, it is widely recognized that BNC possesses a tensile strength ranging from 200 to 300 MPa, along with a Young's modulus that can reach values of 15 to 35 GPa (Cacicedo et al. 2016). Beyond its remarkable physicochemical and mechanical characteristics, bacterial nanocellulose (BNC) demonstrates considerable in vivo dermal biocompatibility, as evidenced by studies involving human subjects. This attribute renders BNC particularly advantageous for use in dermatological applications (Almeida et al. 2014).

The supramolecular architecture of BNC, as well as its physical and mechanical characteristics, can be meticulously tailored during the biosynthetic process through the modulation of different production parameters. These conditions encompass production methodologies, carbon sources, bacterial strains, and the inclusion of additives in the culture medium, which provides a notable advantage over plant-derived cellulose. Consequently, it is imperative to meticulously select and regulate these parameters in accordance with the desired characteristics and specific applications. Furthermore, akin to any cellulosic substrate, BNC can be readily modified post-purification through chemical treatments (such as acetylation, alkylation, and carboxymethylation) or by integrating additional materials into the BNC matrix (Hu et al. 2014). The elevated surface area of BNC nanofibrils facilitates interactions between the abundant hydroxyl groups present on the BNC surface and the functional groups of various other components, including reactive low molecular weight compounds, polymers, metal oxides, and metal nanoparticles. The integration of BNC with these materials imparts novel or enhanced properties to the native BNC, such as bioactivity (including antioxidant, anti-inflammatory, and antimicrobial effects), electrical conductivity, as well as optical and magnetic characteristics. Additionally, it can lead to improved mechanical properties. These advancements significantly broaden the applicability of BNC, enabling the development of high-value, functional BNC-based materials suitable for a diverse array of applications.

3.2.3 Classification by Ration Process

Nanocellulose can be categorized based on its morphological characteristics into several distinct types: cellulose nanonetworks (CNNeWs), cellulose nanofiber (CNF), cellulose nanocrystal (CNC), cellulose nanosphere (CNS), and cellulose nanoplatelet (CNP).

Cellulose Nanonetworks

Cellulose nanonetworks (CNNeWs) are produced from plant cellulosic fibers. In contrast to disposable dried plant cellulosic fibers, such as cotton shorts, which possess a cellulose content greater than 95%, CNNeWs can be efficiently derived from these fibers without the necessity for any pretreatment (Abe et al. 2007). The CNNeWs/aqueous dispersions display a white, cloudy, or translucent appearance, a phenomenon that can be ascribed to the existence of substantial protofibril bundles that exceed the wavelength of visible light. The solid consistency of these dispersions can be increased to approximately 20 wt% through the application of pressure. Additionally, these dispersions can be converted into dry nanopaper through methods such as filtration, papermaking processes, or continuous drying, without experiencing

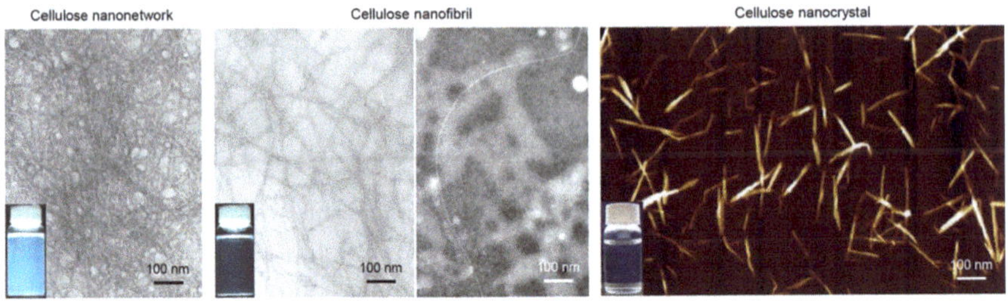

Fig. 3.4 Morphology of cellulose nanonetworks (CNNeWs), cellulose nanofibril (CNF), and cellulose nanocrystal (CNC) (Isogai 2021): reproduced with permission (Saito et al. 2006, 2007, 2013; Iwamoto et al. 2009)

significant challenges such as clogging or extended filtration times (Varanasi and Batchelor 2013).

The aqueous dispersion is depicted in Fig. 3.4. CNNeWs consist of protofibrils that range from 5 to 100 nm in width, representing aggregates of cellulose microfibrils. These networks display a heterogeneous structure characterized by branching. CNF that is dispersed exclusively in water does not aggregate into fibril bundles, whereas CNC generally adopts a fusiform morphology.

Cellulose Nanosphere

In contrast to nanocellulose materials characterized by high aspect ratios, cellulose nanosphere (CNS), which is spherical in shape and exists within the nanometer scale, represents an innovative class of materials. The morphology of these nanoparticles plays a crucial role in determining their efficacy across a variety of applications. Spherical polymer nanoparticles are associated with numerous advantages and find extensive use in domains such as pharmaceutical delivery systems, disease detection, and diagnostics within the biomedical sector.

It is widely recognized that the substantial inter- and intramolecular hydrogen bonding inherent in cellulose imparts exceptional mechanical properties. However, this same characteristic presents difficulties concerning the insolubility of cellulose in both aqueous and common organic solvents. Consequently, CNS has been derived from natural cellulose, which primarily exists in the cellulose I crystalline form, utilizing top-down approaches such as chemical hydrolysis, mechanical disintegration, and enzymatic hydrolysis. These methodologies are similar to those employed in the production of CNF and CNC. Alternatively, a bottom-up strategy has been utilized to regenerate cellulose via a solvation process, leading to the formation of the cellulose II crystalline structure. This approach provides a multifunctional nanocellulose substance, as the crystalline structure and crystallinity of CNS can be modified over a considerably wider range than those of CNF and CNC. In the crystalline region, cellulose II is distinguished by antiparallel, closely packed molecular chains, in contrast to the parallel configuration of molecular chains characteristic of cellulose I.

CNS can be synthesized from a varied assortment of raw materials, encompassing sources of high-purity cellulose such as cotton, wood pulp, CNC, and CNF. Furthermore, low-value agro-industrial byproducts, including used diapers and plant residues, may also be employed as raw materials. It is particularly noteworthy that agro-industrial wastes with a cellulose content of at least 40% can be directly utilized for the preparation of CNS (Chen et al. 2019). To produce CNS, highly polymerized cotton cellulose fibers undergo a pretreatment process involving a 24-h treatment in a ball mill to diminish their degree of crystallinity. Alternatively, the fibers may be treated in a concentrated sodium hydroxide solution, followed by exposure to dimethylsulfoxide (DMSO) for several hours to facilitate the expansion of the crystalline regions (Meng et al. 2020). The treatments facilitated the development of nanorods due to hydrolysis occurring in both the surface and internal regions of the fibers, as the interior of the treated cellulose fibers exhibited enhanced accessibility to the acid solution. A comparable effect was observed when the cellulose was immersed in water overnight. Likewise, the two-step pretreatment process, which involved enzymatic process succeeded by high-pressure homogenization (HPH), led to the exposure of a greater number of amorphous regions within the bamboo fibers. This increased hydrolysis efficiency enabled the preparation of CNS using lower concentrations of sulfuric acid.

Cellulose Nanoplatelet

In recent years, scholars have documented a novel morphology characterized by a two-dimensional (2D) nanostructure referred to as cellulose nanosheets. The acquisition of these nanosheets is facilitated when they are deposited onto silicon wafers, which enables their analysis using optical microscopy. Cellulose nanofilms sourced from banana pseudostem nanoplatelets demonstrate a transparency level of approximately 83% (Chávez-Guerrero et al. 2019). In recent years, scholars have identified a novel morphology characterized by a 2D nanostructure referred to as cellulose nanosheets. These nanosheets can be more readily obtained when deposited on silicon wafers, facilitating their examination through optical microscopy. Cellulose nanofilms derived from banana pseudostem nanoplatelets demonstrate a transparency of approximately 83%, rendering them promising candidates for various applications within the packaging and food industries. Unlike BC, CNC, or CNF, CNP consists of randomly entangled nanofibrils that are integrated to form sheets with a thickness of approximately 80 nm, as opposed to aligned nanofibrils. This structural configuration offers several advantages in terms of both chemical composition and physical structure, thereby enhancing the potential for enzymatic hydrolysis (e.g., via endoglucanases). Consequently, this unique structure presents a valuable opportunity for investigating enzyme-promoted interactions. Chávez-Guerrero et al. (2021) conducted a study in which CNP and cellulose fiber (CF) were extracted from banana pseudostems through a process of acid hydrolysis following alkalization, resulting in the production of self-supporting transparent films. The films, composed of cellulose nanosheets and cellulose fibers, exhibited a crystallinity index of 78.05, demonstrating the ability to withstand a tensile stress of approximately 180 MPa and an average strain of 6%. The tensile testing of these films revealed favorable mechanical properties, indicating their significant potential as substitutes for traditional polymers in various applications, particularly within the food and packaging sectors. The transition of fiber dimensions from the micron scale to the nanoscale leads to a decrease in defects within the nanofibrils, thereby improving their mechanical strength. This enhancement is primarily due to the augmented number of junctions within the cellulose matrix.

3.2.4 Summary

Nanocellulose is a biomass-derived material with extensive potential applications, which can be classified according to its source of raw materials, the techniques employed in its preparation, and its morphological properties. Presently, it is primarily classified into three types: cellulose nanofiber (CNF), cellulose nanocrystal (CNC), and bacterial nanocellulose (BNC). Although these types share similar chemical compositions, they exhibit significant variations in crystallinity, particle size, and other properties. The relationship among these conformations encompasses various factors, including physical structure, chemical characteristics, and macroscopic properties. These attributes can be optimized by manipulating microstructural and macroscopic preparation conditions to fulfill the requirements of diverse applications.

3.3 Structure Property Relationship

3.3.1 Structural Characteristics and Classification

Nanocellulose (NC) consists of several main types, such as cellulose nanocrystal (CNC), bacterial nanocellulose (BNC), and cellulose nanofiber (CNF). Each type differs slightly in structure and properties (Fig. 3.5).

Molecular Structure and Crystallinity

Cellulose is a polysaccharide composed of *D*-glucose units connected by β-1, 4-glycosidic bonds, forming a linear chain. Each glucose residue has three hydroxyl groups, enabling the formation of an extensive hydrogen bond network within the

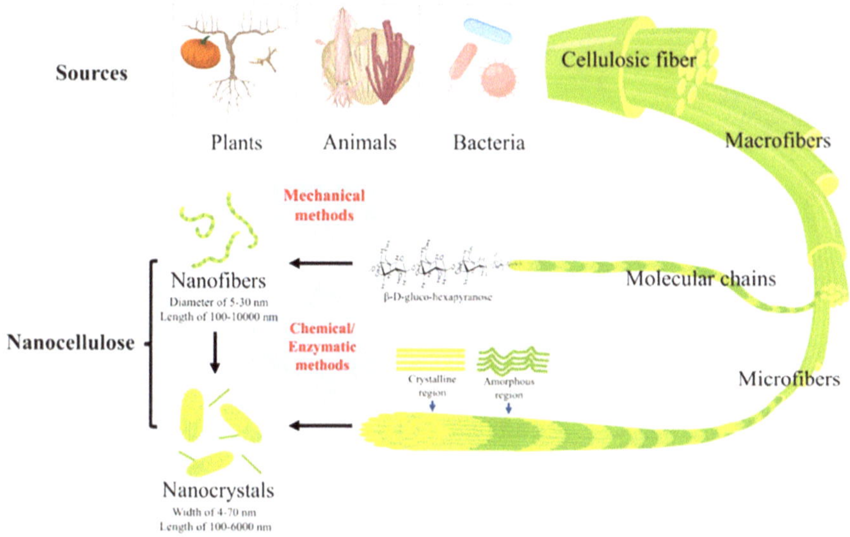

Fig. 3.5 Sources and structures of different types of cellulose: reproduced with permission (Zhang et al. 2023)

structure. Hydrogen bonding not only stabilizes the molecular chain configuration, but also increases the rigidity and strength of cellulose.

The molecular structure of NC consists of crystalline and amorphous regions. In the crystalline region of cellulose, molecular chains are arranged in a highly organized and tightly packed manner, resulting in significant structural order. In contrast, the amorphous regions exhibit a lack of such organization, leading to a more disordered molecular structure. The crystalline region gives cellulose high mechanical strength and thermal stability, while the amorphous region is softer, making cellulose more susceptible to mechanical dissociation and chemical modification.

The crystalline structure of natural cellulose predominantly exists as cellulose I, which can be further categorized into two distinct subcrystalline polymorphs: I_α and I_β. I_α subcrystalline forms are mainly found in algal and bacterial cellulose, while I_β subcrystalline forms are commonly found in higher plants. I_β crystalline forms are more stable in hydrogen bonding structure, and therefore, they are more heat-resistant and chemically stable. In alkaline environments or at high temperatures, cellulose I can be transformed into cellulose II. Cellulose II has a crystalline structure distinct from that of cellulose I. It is known for its enhanced chemical stability and higher absorbency, making it advantageous for certain applications. However, its mechanical strength is typically slightly inferior compared to cellulose I. This structural variation arises from differences in molecular arrangement, contributing to the unique properties of cellulose II (Fig. 3.6).

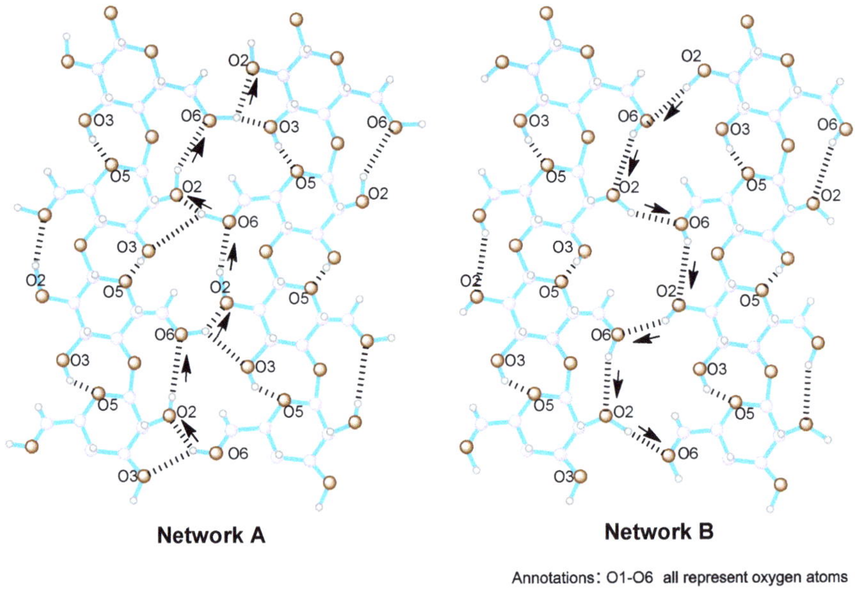

Fig. 3.6 Illustration of a cellulose chain showing the cooperative network of hydrogen bonds. Thick dashed lines represent the hydrogen bonds, while arrows indicate the donor–acceptor–donor sequences within networks A and B. Additionally, thinner dashed lines denote specific hydrogen bonds, such as O3-H···O5 and O6-H···O3 interactions

Nanofiber Structure and Morphology

Cellulose nanocrystal (CNC) is produced by acid hydrolysis of natural cellulose, which removes most of the amorphous regions and leaves only highly crystalline nanofiber fragments. Typical CNC is rod-like or needle-like, with diameters ranging from 5 to 20 nm and lengths of about 100–500 nm, and its high crystallinity imparts excellent mechanical strength and thermal stability. CNC typically has low aspect ratios and is suitable for use in reinforced composites and liquid crystal displays.

Cellulose nanofiber (CNF), prepared by mechanical dissociation or enzymatic treatment, retains a certain percentage of amorphous regions, exhibits high aspect ratios (3–30 nm in diameter and several micrometers in length), and is flexible and has excellent network formation capabilities. The surface of CNF contains a large number of hydroxyl and other functional groups, which make it easy to achieve a wide range of functions through chemical modification. CNF is suitable for use as a thickener, reinforcing material, filtration material, and can even be used to mimic cellular matrices in biomedical materials.

Bacterial nanocellulose (BNC) is synthesized through the fermentation of specific bacteria in a bioreactor and has a highly connected three-dimensional network structure. BNC is extremely pure and does not require further purification and has a high aspect ratio and excellent biocompatibility. BNC is highly suitable for a wide range

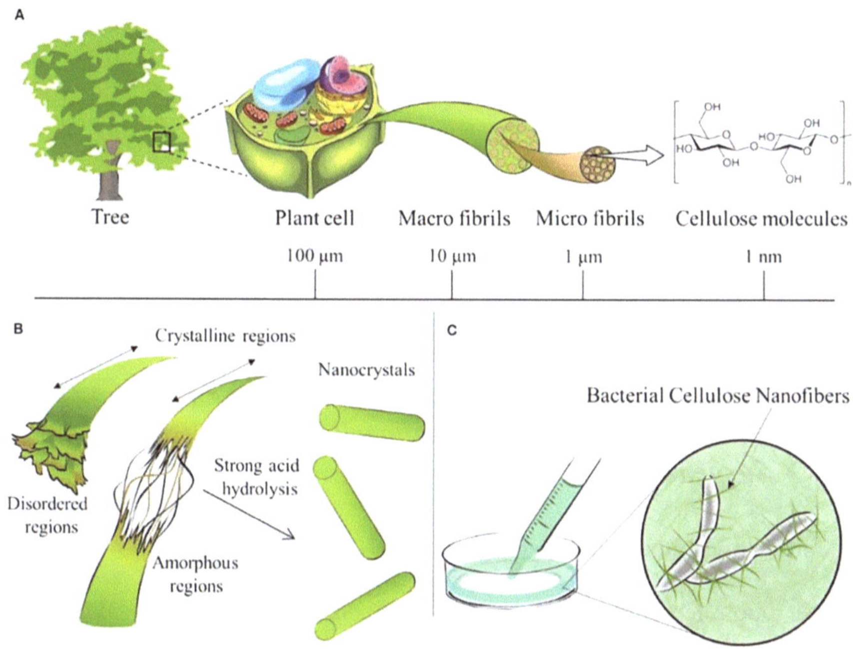

Fig. 3.7 Cellulose found in plants and trees exhibits a hierarchical structure that spans from the meter scale down to the nanometer scale, as illustrated in panel (**a**). Panel (**b**) presents a schematic diagram showing the reaction between cellulose and strong acid to produce nanocellulose. Meanwhile, panel (**c**) depicts bionanocellulose cultivated from cellulose-synthesizing bacteria. Reproduced with permission: reproduced with permission (Miyashiro et al. 2020)

of biomedical applications, including as wound dressings, scaffolds for tissue engineering, and in drug delivery systems. Its versatility makes BNC an ideal material for addressing various medical needs, particularly in areas requiring biocompatibility and structural support (Fig. 3.7).

Surface Chemistry and Functionalization

The hydroxyl, carboxyl, and other functional groups on the surface of NC can be modified by chemical methods to enhance its application performance.

(i) Carboxylation and sulfonation

Introducing charged groups on the surface of CNF through TEMPO oxidation, sulfonation and other methods can improve its dispersibility and stability in water. These functionalized groups also give NC the potential to be compounded with other materials, especially in metal ion adsorption and water treatment applications.

(ii) Hydrophobic modification

The hydrophobicity of NC can be enhanced by esterification, silylation, etc., rendering it appropriate for applications in oil–water separation and for the creation of more hydrophobic polymer composites.

(iii) Multiple modification

Some studies have also attempted to introduce multiple functional groups into NC to achieve specific adsorption, conductivity, catalysis, and other multifunctional applications, such as flexible electronics, catalytic carriers, and water purification materials.

3.3.2 Relationship Between Aspect Ratio, Volume Fraction, and Phase Formation of Nanocellulose and Its Suspensions

Definition and Impact of Aspect Ratio

Aspect ratio refers to the ratio of length to the diameter of nanofibers. The aspect ratio of NC largely determines its performance, including its mechanical properties, network formation ability, rheological properties, and self-assembly behavior. The following is an analysis of effect of aspect ratio on NC characteristics (Fig. 3.8):

(i) Mechanical properties

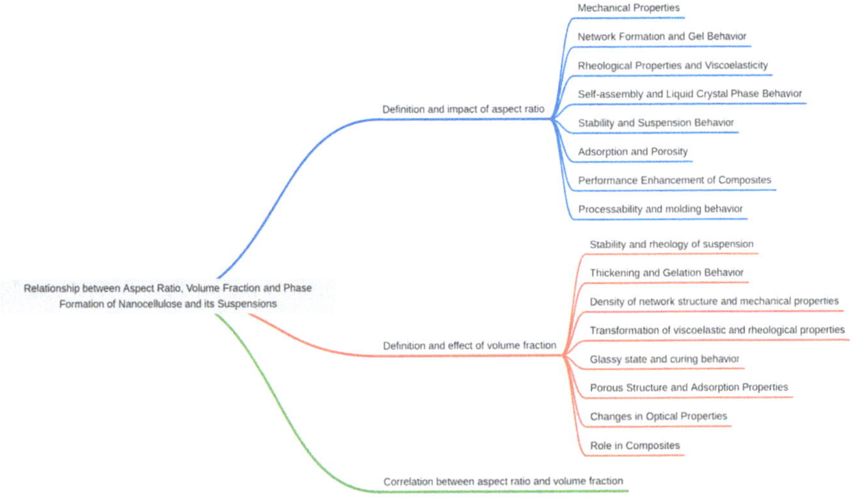

Fig. 3.8 Guide map of relationship between aspect ratio, volume fraction, and phase formation of nanocellulose

The higher NC's aspect ratio, the better the long chains of individual fibers can transfer the stresses, thus improving the tensile and compressive characteristics of the material. This high aspect ratio allows the fibers to form a stronger network in the composite material, which significantly increases the material's elasticity modulus and tensile strength. High aspect ratios allow the fibers to entangle and cross-link with each other, enhancing the toughness and flexibility of the material. Especially in cellulose nanofiber (CNF), high aspect ratios help to form soft and flexible networks, which are suitable for biocompatible materials and toughened composites.

(ii) Network formation and gel behavior

NC that has a high aspect ratio is more likely to form stable network structures at low volume fractions. This is due to the fact that longer fibers are more likely to come into contact with and physically entangle in the solution, forming a stable gel network. This behavior makes NC with a high aspect ratio an ideal thickener that provides good viscoelasticity, especially at low concentrations. The network formed by NC with a high aspect ratio has a higher density and strength and is able to withstand greater external forces and deformations. The more tangles and intersections there are between fibers with high aspect ratios, the denser the network is and the more rigid and structurally stable the material is.

(iii) Rheological properties and viscoelasticity

NC suspensions with high aspect ratios usually exhibit shear-thinning behavior. This is due to the rearrangement of fibers under shear, which reduces interactions. Higher aspect ratio fibers flow better under these conditions and are suitable for applications where both flow and stability are required, such as coatings, inks, and food additives. In addition, NC suspensions and gels with high aspect ratios exhibit high viscoelastic moduli. The entangled network formed between long fibers can be deformed by external forces, but will return to its initial shape when the forces are removed, a property that is particularly important in soft materials and medical hydrogels.

(iv) Self-assembly and liquid crystal phase behavior

When the concentration of fibers with high aspect ratio increases to a certain level, especially cellulose nanocrystal (CNC), a liquid crystal phase is formed spontaneously. This liquid crystal phase has a highly ordered molecular arrangement and can be oriented in response to external electric or magnetic fields, making it suitable for optical displays, sensors, and the construction of conductive materials. NC with high aspect ratios can form highly ordered arrangements and structures through interactions in solution. This self-assembly behavior makes it advantageous for nanocomposites and functional membrane materials, where the fiber arrangement can be tuned to achieve conductivity, optical or mechanical properties in specific directions.

(v) Stability and suspension behavior

NC with high aspect ratio shows good dispersion stability in water and other solvents. Longer fibers are difficult to agglomerate through electrostatic repulsion and hydrogen bonding interactions, which facilitates the formation of stable suspensions. This stabilization is particularly important in applications such as coatings, inks, and cosmetics where delamination and sedimentation are avoided. Since fibers with high aspect ratios can intertwine at lower concentrations, thickening can be achieved at lower volume fractions. This property makes it widely used in thickeners and emulsion stabilizers, where a similar thickening effect can be achieved with a reduced addition.

(vi) Adsorption and porosity

NC that has a high aspect ratio usually has a high specific surface area, conducive to the absorption of contaminants, metallic ions, or biological molecules. In particular, high-aspect-ratio fibers with functionalized surfaces can be used in water treatment, air filtration, and biosensing. Moreover, NC with high aspect ratio is very likely to form a porous network during gelation and drying, enhancing the material's porosity and specific surface area. This porous structure is advantageous in aerogels, hydrogels, and filtration materials for applications such as adsorption, thermal, and acoustic insulation.

(vii) Performance enhancement of composites

High-aspect-ratio NC provides greater reinforcement in composites. Longer fibers form more effective interfaces with the polymer matrix and enhance stress transfer between the interfaces, thereby improving the tensile strength, impact resistance, and heat resistance of the composite material. Due to the low volume fraction of high-aspect-ratio fibers required to form a network in the composite, reinforcement can be achieved with a low filler content. This low filler content and high performance is particularly important in lightweight structural materials, such as automotive components and packaging materials.

(viii) Processability and molding behavior

NC with its high aspect ratio has good processing properties and is suitable for the preparation of coatings and films. The fibers are uniformly distributed by shear forces and form a stable film structure during drying. This property gives it great potential for applications in paper coatings, food packaging films, and antimicrobial coatings. When high-aspect-ratio NC is compounded with conductive fillers or graphene oxides, the electrical and thermal conductivity of the material is significantly improved. The network structure with the high-aspect-ratio fibers provides a continuous path for electrical or thermal conductivity, making them ideal for flexible electronics and thermal management materials.

The aspect ratio of NC has a wide and far-reaching effect on its properties. High aspect ratios can enhance the material's mechanical properties and can also show significant advantages in rheology, stability, network formation, self-assembly, and multifunctional composites. By adjusting NC's aspect ratio, it is possible to optimize its performance in a diverse applications, including thickeners, composites, adsorbent materials, and biomedical materials.

Definition and Effect of Volume Fraction

Volume fraction is the percentage of volume of NC in the dispersion medium. An increase in volume fraction affects the phase transition behavior of NC, such as the transition from suspension to gel to glass. At low volume fractions, NC is usually in suspension, and the fibers are independent of each other, forming a stable dispersion. As the volume fraction increases, the interactions between the fibers increase, and a gel network is formed, which is viscoelastic and somewhat solid and can support external stresses. As the volume fraction is increased further, the NC network becomes denser and changes to a glassy state, which is characterized as a solid state; i.e., the material becomes solidified and immobile. These different phase states exhibit unique properties in a variety of applications.

The effect of volume fraction on NC properties is as follows:

(i) Stability and rheology of suspension

At low volume fraction, the NC dispersion is in liquid suspension, and the fibers are mostly dispersed and independent of each other. At this time, the viscosity of the solution is low, and the fluidity is strong. At low volume fractions, NC dispersions generally exhibit Newtonian fluid behavior (viscosity does not change with shear rate) due to low fiber interactions and the fact that the fibers are freely suspended and relatively stable. Low volume fraction suspensions are suitable for fluid applications such as coatings and thickeners.

(ii) Thickening and gelation behavior

With the increased of volume fraction of NC, the distance between the NC decreases, and the interaction among the fibers increases, leading to the formation of tangles and preliminary network structures. At this point, the solution begins to show thickening effect, and the rheology changes to a shear-thinning non-Newtonian fluid. This thickening effect is very useful in food thickeners and coatings. When the volume fraction reaches a critical point, a stable network structure is formed among the fibers, and the dispersion transitions to the gel state. This gelation process is known as the "sol-gel transition", and in this state the NC dispersion exhibits solid elastic properties rather than fluid properties. NC in the gel state is commonly used for thickening, fixing, and supporting structural applications such as drug release vehicles and soft matter materials.

(iii) Density of network structure and mechanical properties

With a further increase in the volume fraction, the density of the NC network structure increases, the crosslinking points between fibers increase, and the network structure becomes denser and more compact. The formation of this high-density network increases the stiffness and compression resistance of the material, resulting in a good elastic modulus. The dense network formed at high volume fractions significantly increases the tensile and shear strength of NC, resulting in good mechanical properties. This property gives NC material potential for load-supporting applications (e.g., structural reinforcements), such as reinforcements for composite matrices and protective coatings.

(iv) Transformation of viscoelastic and rheological properties

At low volume fractions, NC dispersions behave mainly as liquid suspensions with viscous behavior, whereas when the volume fraction is increased to a certain level, NC dispersions transform from a fluid state to a viscoelastic gel with solid properties. In the gel state, the shear modulus (G′) of NC is bigger than the loss modulus (G″), showing that the material has a more elastic component. This viscoelasticity increases with increasing volume fraction, and the viscoelastic modulus of the material increases significantly. Such viscoelastic gels have important applications in biomedical materials, drug delivery systems, and food engineering.

(v) Glassy state and curing behavior

When the volume fraction is further increased, the NC dispersion gradually enters the glassy state. At this time, the NC network becomes very dense, and with restricted molecular movement, the material exhibits rigid and solid state properties and is essentially no longer mobile. In the glassy state, the mechanical strength and abrasion resistance of NC materials are considerably increased, which makes them suitable for use in hard coatings, protective materials, and structurally reinforced composites. NC materials in the glassy state show great potential for applications such as load support and mechanical protection.

(vi) Porous structure and adsorption properties

In gelatinized or glassy NC networks, fibers form a porous structure characterized by a high specific surface area, which is critical for both adsorption and filtration applications. For example, by controlling the drying method (e.g., freeze-drying) and the volume fraction, the NC's pore size and distribution can be adjusted to control its adsorption capacity (Table 3.1).

NC gels and aerogels with porous structure can show excellent adsorption properties at high volume fraction, which is suitable for adsorbing heavy metal ions in water treatment, oil–water separation, and gas filtration. This porosity also makes them valuable for thermal and acoustic insulation and lightweight materials.

(i) Changes in optical properties

At low volume fractions, NC dispersions are usually transparent because of the large fiber spacing and small fiber size, which results in less light scattering. This transparency is promising for applications in optical coatings and transparent films.

As the volume fraction increases, the fiber–fiber interactions increase, and the network structure becomes denser, leading to an increase in light scattering and a gradual loss of transparency and a cloudy appearance. In some packaging materials, this property can be used to adjust the visual effect, and in applications where transparency is required, a low volume fraction is more advantageous.

(ii) Role in composites

Table 3.1 Morphology of NC obtained by different drying methods and their advantages and disadvantages

Drying method	Nanocellulose morphology after drying	Advantages	Disadvantages
Oven drying	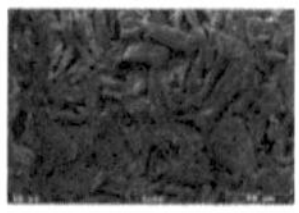	Low cost; fast drying process	Fibers produce irreversible agglomeration during drying; Lose nanoscale
Freeze drying	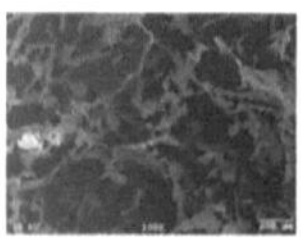	Can keep some fibers at nanoscale; the process is also relatively stable; applicable to both CNC and CNF	Long drying time; high energy consumption; fibers may clump, affecting performance
Spray drying		Continuous production, low cost; mainly used for CNC drying	Dried fibers are granular; fibers will have slight agglomeration; low drying efficiency
Supercritical extraction drying	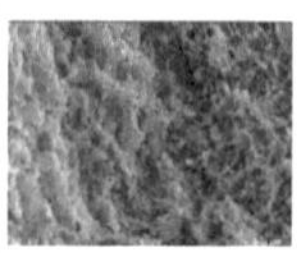	High drying efficiency; keep fibers at nanoscale; mainly used for CNF drying	High cost; Impractical for actual production; complex process, using a lot of solvents; not applicable to CNC drying

NC is used as filler in composites, and an increase in its volume fraction can significantly enhance the material's mechanical properties. With a high volume fraction, NC can form a continuous network structure in the composite matrix, effectively transferring and dispersing stresses, and improving the tensile, flexural, and impact resistance of the material. However, too high a volume fraction may make the composite more difficult to process and reduce the fluidity of the material. Therefore, a balance of NC volume fractions is needed in practical applications to achieve the best combination of properties and processability. The volume fraction of NC significantly affects its suspension, gelation, mechanical, rheological, and optical properties in solution. By adjusting the volume fraction, the structural morphology of NC can be controlled to meet different application requirements. Low volume fractions are suitable for transparent coatings and fluid applications, medium volume fractions for thickeners and gels, and high volume fractions for reinforced composites and porous adsorbents. This structure–property modulation opens up multiple possibilities for the use of NC in biomedical materials, composites, and filtration materials.

Correlation Between Aspect Ratio and Volume Fraction

NC that has a high aspect ratio is able to form a stable gel network at lower volume fractions because the longer fibers are more likely to form contact points and entangle with each other. Fibers with higher aspect ratios transition to the gel or glassy state more easily than shorter fibers at the same volume fraction. Phase diagram studies have shown that volume fraction and the aspect ratio together determine the phase state of NC materials: high aspect ratios and high volume fractions are more likely to lead to the formation of gels or glassy states, while low aspect ratios and low volume fractions usually remain in suspension. By varying the aspect ratio and volume fraction, the rheological and mechanical properties of NC can be designed and tuned to suit different applications such as filtration materials, structural composites, and biomedical materials (Fig. 3.9).

3.3.3 Influence of Chemical Functionalization

The surface chemical properties and functionalization of NC have a vital influence on its properties and applications. The surface of NC is high in hydroxyl groups that can be chemically modified to introduce different functional groups to change the dispersibility, adsorption, mechanical properties, and compatibility of NC. The following is an analysis of the effects of surface chemistry and functionalization on the performance of NC (Fig. 3.10).

(i) Surface hydroxyl groups and hydrogen bonding

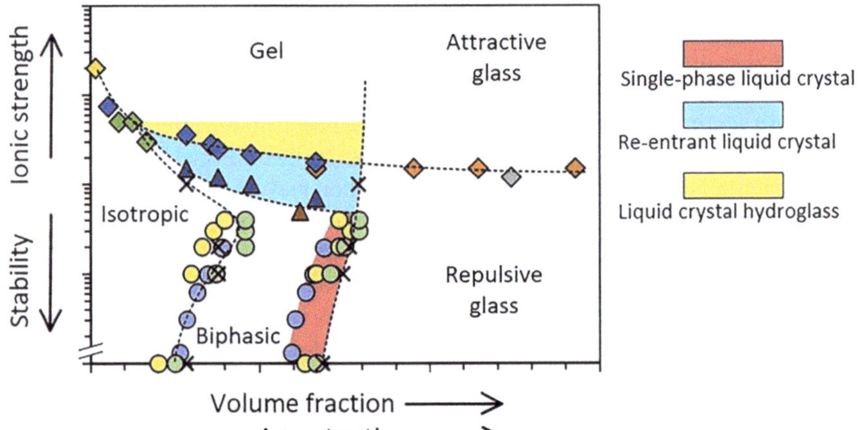

Fig. 3.9 Aspect ratio plotted against volume fraction to identify the boundaries of distinct phases such as NC suspensions, gels, and glasses: reproduced with permission (Xu et al. 2020)

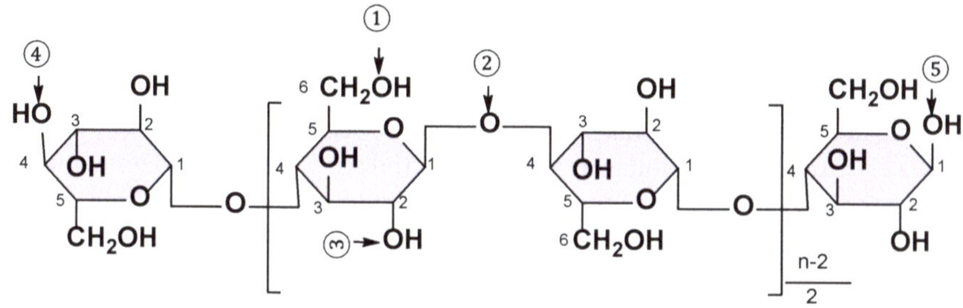

Fig. 3.10 Reactivity groups of cellulose fibers ①-primary hydroxyl, ②-glycosidic bond, ③-secondary hydroxyl, ④-terminal hydroxyl, ⑤-terminal hydroxyl (with hemiacetal properties)

Numerous hydroxyl groups are present on the NC surface, capable of establishing hydrogen bonding networks. These hydrogen bonds give NC good mechanical properties and viscoelasticity. Hydrogen bonding also increases the interactions between fibers, leading to a gel structure at low concentrations. Surface hydroxyl groups offer sites for modifying the chemical structure of NC. Through these hydroxyl groups, various functional groups can be introduced to impart specific functions to NC, such as hydrophilicity, hydrophobicity, and electrical conductivity, thus expanding its applications in different environments (Fig. 3.11).

(ii) Carboxylation and sulfonation functionalization

TEMPO Oxidation: TEMPO oxidation is a common functionalization method that oxidizes the hydroxyl group on the surface of NC to a carboxyl group, giving it

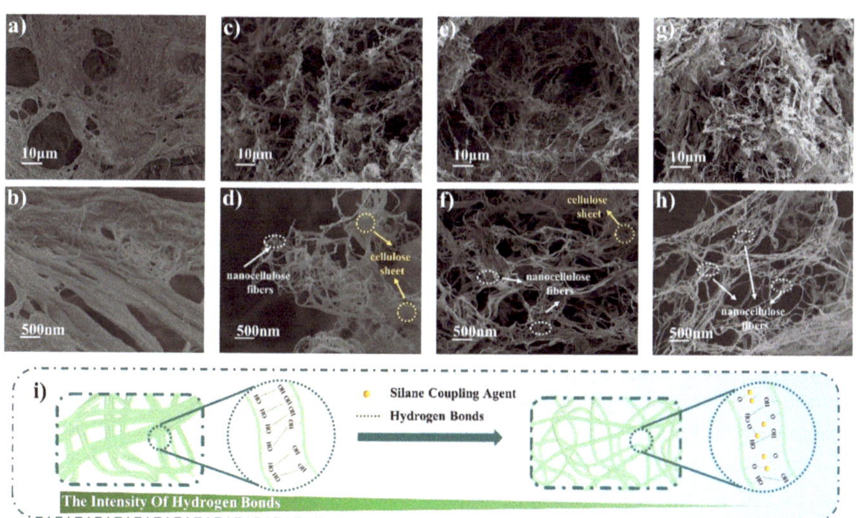

Fig. 3.11 Schematic of the SEM images and microstructural changes of NC with the addition of silane coupling agent: reproduced with permission (Cao et al. 2023)

a negative charge. This carboxylated NC shows better dispersion and stability in water and is suitable for use in aqueous systems. Carboxylated NC exhibits a robust capacity for adsorbing heavy metal ions (e.g., lead and copper) because the carboxyl groups can establish ligand bonds with the metal ions. Carboxylated NC has potential for water treatment and pollutant removal. By introducing sulfonic acid groups to the NC surface, the surface charge density can be further increased, making it suitable for ion exchange, adsorption, and catalytic applications. This treatment can significantly improve the adsorption performance and chemical stabilization of NC.

(iii) Hydrophobicity modification

The surface of NC can be modified to be hydrophobic through esterification, silylation, and other methods. The hydrophobicity modification enables NC to be stably dispersed in nonpolar solvents and shows good performance in oil–water separation and hydrophobic composites. The hydrophobically modified NC is more compatible with hydrophobic polymer matrices (e.g., polypropylene and polyethylene). The NC thus treated has higher dispersion uniformity in composite materials, which can substantially enhance the mechanical strength and water repellency of composite materials. Moreover, hydrophobic NC can be used as water-resistant agent in coating materials to improve the waterproof and dirt-proof performance of the materials, which is suitable for packaging materials and construction materials.

(iv) Multifunctional modification

The introduction of amino or thiol groups that are on NC's surface can enhance its activity in special adsorption and chemical reactions. The aminated NC has strong trapping ability and is suitable for carbon dioxide adsorption, dye adsorption, and metal ion removal.

By combining different functional groups (e.g., phosphate groups, carboxyl, and amino), NC can exhibit multifunctional properties. For example, the coexistence of amino and carboxyl groups enables NC to have both positive and negative charge adsorption ability, which is conducive to improving its stability and adsorption efficiency under different pH environments.

(v) Surface charge regulation

The dispersion stability of NC is influenced by its surface charge. The negatively charged NC in water, due to the electrostatic repulsion phase, can improve its stability and adsorption efficiency under different pH conditions, and show good suspension performance. The dispersion and gelation behavior of NC in solution can be controlled by adjusting the charge density. In water treatment applications, the introduction of diverse charged groups enables the achievement of effective adsorption of heavy metal ions and various pollutants. For example, NC surfaces with sulfonic acid groups have a high charge density and are highly capable of adsorbing positively charged metal ions. Under certain conditions, the charge on the NC surface can be reversed to meet different adsorption needs.

(vi) Enhancement in composite materials

The interfacial compatibility of NC with the matrix polymer can be improved through surface chemical modification. For example, NC with carboxyl or sulfonic acid groups can form hydrogen or covalent bonds with polymer matrices (e.g., polyurethanes) containing amino or hydroxyl groups, thus enhancing the interfacial bond strength of composites. Functionalized NC can be loaded with different types of functional materials such as graphene oxide, quantum dots, and metal nanoparticles. These composites find applications in high value-added applications like catalyst carriers, conductive films, and sensors. These composites can be used in high value-added applications such as catalyst carriers, conductive films, and sensors. For example, NC loaded with silver nanoparticles exhibits good antimicrobial properties and is suitable for use in medical devices and hygiene materials.

(vii) Catalytic and electrochemical applications

NC after surface functionalization can be used as an ideal carrier for metal nanoparticles (e.g., silver, palladium, and gold) for catalytic and electrochemical applications. Carboxy and amino groups can firmly fix the metal nanoparticles on the fiber surface to improve the stability and efficiency of the catalytic reaction. By loading conductive materials (e.g., carbon nanotubes and graphene) on the surface of NC or by modification with conductive polymers, NC can exhibit good electrical conductivity for use in electrode materials and flexible electronic devices. Surface chemical modification allows NC to be stably dispersed in the electrode structure and provides a conductive network, which contributes to the electrochemical performance of batteries and supercapacitors.

(viii) Biocompatibility and biomedical applications

Surface chemical modifications such as loading of antimicrobial agents (e.g., silver ions and copper ions) can make NC exhibit good antimicrobial properties, which makes it suitable for use in medical dressings and antimicrobial packaging materials. The surface chemical modifications of NC can selectively kill bacteria without causing toxicity to human cells, which increases its biocompatibility. The adsorption and release behavior of NC can be controlled by surface modification. NC loaded with carboxyl or phosphate groups can achieve targeted drug release at specific pH environments, making it appropriate for use in controlled-release drug delivery systems and enteric-coated tablets. NC materials with biocompatible modifications can also be used as tissue engineering scaffolds to support cell adhesion and growth.

The surface chemistry and functionalization of nanocellulose provides a wide range of properties that can significantly enhance its performance in different applications. Chemical modification of the NC surface enhances its dispersion stability, adsorption capacity, mechanical properties, and compatibility, giving it unique advantages in water treatment, biomedical materials, composites, catalysis, and electrochemical applications. This versatility in functionalization gives NC significant potential for use in the advancement of sustainable materials.

3.3.4 *Aggregation Mechanisms and Interactions*

The aggregation mechanisms and interactions of NC determine its dispersion behavior, structure formation, and functionality in solution, gel, and composite materials, etc. The mechanisms of NC aggregation include hydrogen bonding, electrostatic interactions, van der Waals forces, and the regulation of aggregation behavior by external chemical modifications, and the aggregation behavior of NC in water is influenced by the surface charge, pH, and the added electrolytes:

Electrostatic Interactions

The surface of NC features hydroxyl groups, which can be chemically modified (e.g., TEMPO oxidation, and sulfonation) to introduce negative charges (e.g., carboxyl and sulfonate groups) and form negatively charged cellulose chains. In aqueous solution, electrostatic repulsion between these negative charges keeps the fibers away from each other and enhances the stability and dispersion of NC.

The addition of electrolytes (e.g., salt) can affect the charge shielding effect on the NC surface. When low concentrations of electrolytes are added, the electrolyte ions partially neutralize the charge on the NC surface, leading to a weakening of the electrostatic repulsion, which promotes slight aggregation or gelation of the NC, whereas high concentrations of electrolytes may induce stronger charge shielding, leading to aggregation or settling of the NC (Fig. 3.12).

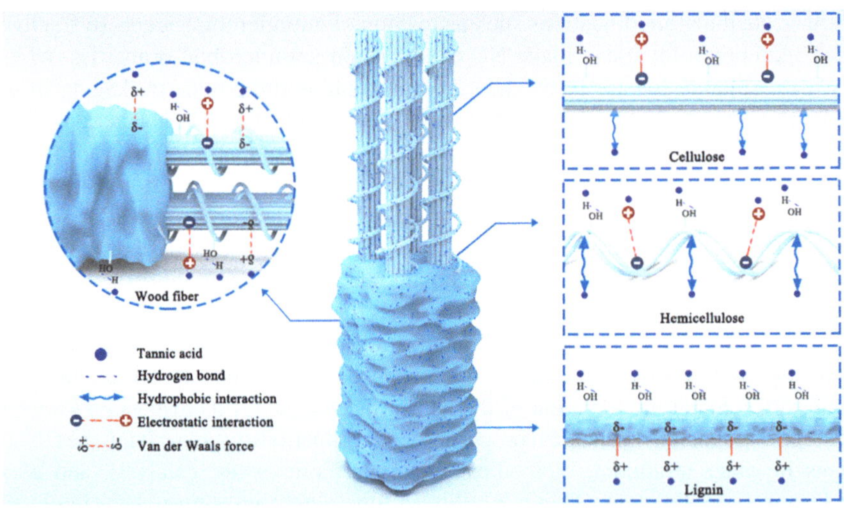

Fig. 3.12 Schematic diagram of hydrogen bonding, van der Waals force interactions, and electrostatic interactions between wood fibers and TA: reproduced with permission (Liu et al. 2024b)

Hydrogen Bonding Interactions

NC molecular chains are abundant in hydroxyl groups which facilitate the formation of hydrogen bonds. These bonds occur both between fibers and between fibers and water molecules. This enhanced bonding significantly strengthens the interactions within NC, leading to increased viscosity and improved mechanical properties in solution.

At the right concentration, hydrogen bonding causes NC to develop a robust three-dimensional network structure, resulting in a gel state. The inter-fiber hydrogen bonding network increases the elasticity and strength of the gel, making it suitable for use in thickeners, soft materials, and biomedical materials. The dynamic reversibility of this hydrogen bonding allows the gels to be deformed without breaking under external forces and to exhibit viscoelasticity (Figs. 3.13 and 3.14).

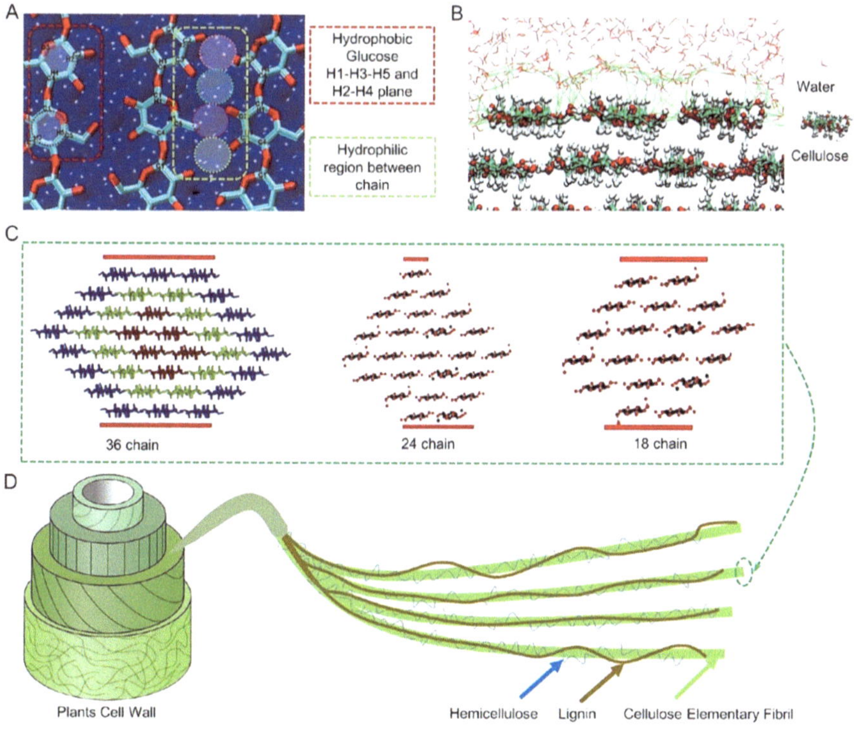

Fig. 3.13 Hydrophobic properties of cellulose chains and their effect on basic protofiber structure and plant cell walls (Jing et al. 2023). **a** Illustrating the molecular structure of a cellulose chain, with a depiction of its hydrophobic plane and hydrophilic domain (Miyamoto et al. 2014). **b** A molecular dynamics simulation snapshot captures the behavior of water molecules in proximity to the hydrophobic surface of cellulose I_α (Huang et al. 2020). **c** Three types of elementary fibers are densely arranged via the process involving hydrophobic cellulose accumulation (Cosgrove 2014). **d** The cell wall of the plant is made up of elementary cellulose fiber, lignin, and hemicelluloses: reproduced with permission (Rytioja et al. 2014)

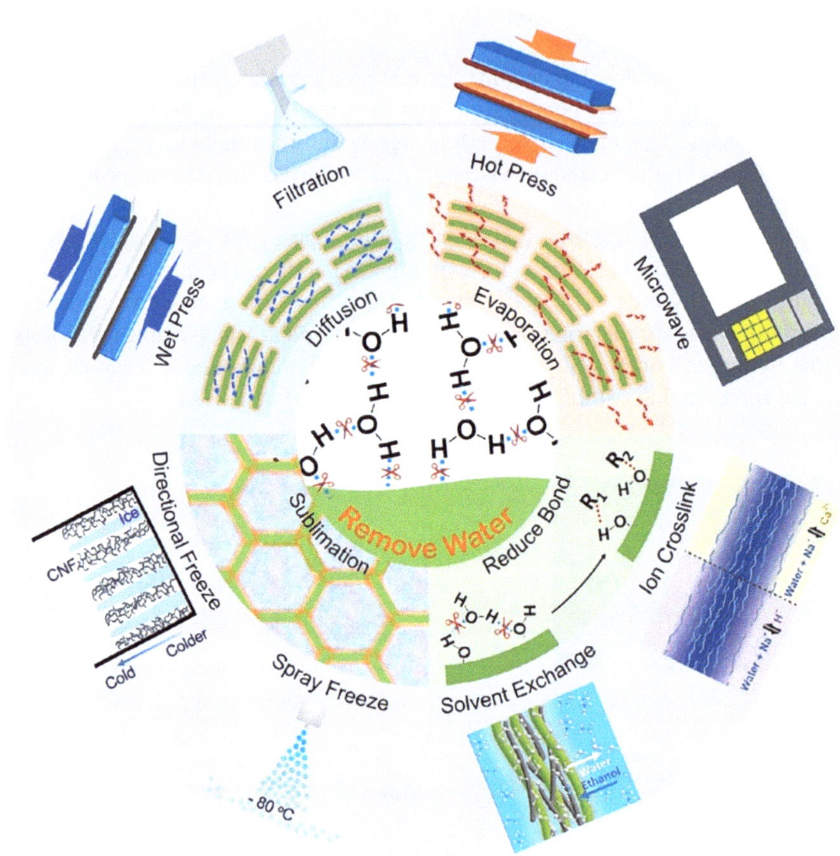

Fig. 3.14 Removal of water from cellulose nanomaterials by diffusion, evaporation, water sublimation, and weakening of the hydrogen bonds between cellulose and water: reproduced with permission (Jing et al. 2023)

Van der Waals Forces

Van der Waals forces are the short-range attraction between NC molecules. Owing to the nanoscale nature of NC and its high specific surface area, van der Waals forces serve as a critical factor in the aggregation behavior of NC. When other interactions (e.g., electrostatic repulsion) are suppressed or eliminated, van der Waals forces cause fibers to come together and aggregate. At higher concentrations and under external conditions, van der Waals forces can promote NC fibers to come closer to each other and form a denser network. This aggregation behavior is important for the structure formation of NC in solid materials and composites, which can increase the rigidity and density of materials.

Ion-Induced Aggregation and Charge Neutralization

Polyvalent ions (e.g., calcium and aluminum) can act as "bridges" to connect the negative charges of different NC chains, thereby triggering flocculation of the NC. This bridging action helps to form larger NC aggregates, which are suitable for water treatment and pollutant adsorption applications. At higher concentrations of applied cations, these ions can neutralize the negative charges on the surface of NC, weakening the electrostatic repulsion and causing the fibers to move closer to each other, flocculation, and sedimentation. This charge neutralization mechanism can exert a considerable influence on the stability of NC in solution and is used to control the flocculation process of NC in suspension systems.

Charge Reversal and Heterogeneous Flocculation

When sufficient multivalent cations are added to the NC solution, the NC surface may undergo a charge reversal, i.e., the surface changes from negatively charged to positively charged. The charge reversal results in heterogeneous flocculation of the NC with surrounding negatively charged colloidal particles, resulting in the formation of mixed aggregates. This charge reversal and heterogeneous flocculation mechanism play an important role in water treatment applications. NC can form aggregates with negatively charged contaminants or colloidal particles in water to accelerate settling or filtration, making it an effective water treatment method.

The flocculation effect is influenced by the chain length and surface charge of the flocculant, as indicated in various reports. The influence of flocculant's surface charge and the length of its chain is shown in Fig. 3.15. Flocculants with high charge density and extended chain length (Fig. 3.15a) are able to produce flocs with multiple suspended particles and colloids because the surface charge of the particles is partially neutralized (coagulated) and the repulsive forces within the particles are weakened. In this case, the mechanism is coagulation-flocculation, i.e., neutralization and bridging. In contrast, for flocculants that are long-chained but have a low charge (Fig. 3.15b), due to insufficient or non-existent localized surface charge neutralization, the number of particles and/or colloids that can bind together decreases, resulting in a lower density of the floc. This reduction in density leads to the formation of larger but structurally weaker flocs, which are prone to disintegration when subjected to mechanical stress, such as agitation. Additionally, the formation mechanism of short-chain flocs, as depicted in Fig. 3.15c, incorporates smaller particles and fewer colloids compared to the process illustrated in Fig. 3.15a. This structural difference impacts the stability and functionality of the flocs in various applications (Table 3.2).

Effect of Surface Chemical Modification on Aggregation

By introducing carboxyl or sulfonic acid groups to the NC surface, the surface charge density can be increased, and the electrostatic repulsion between fibers can

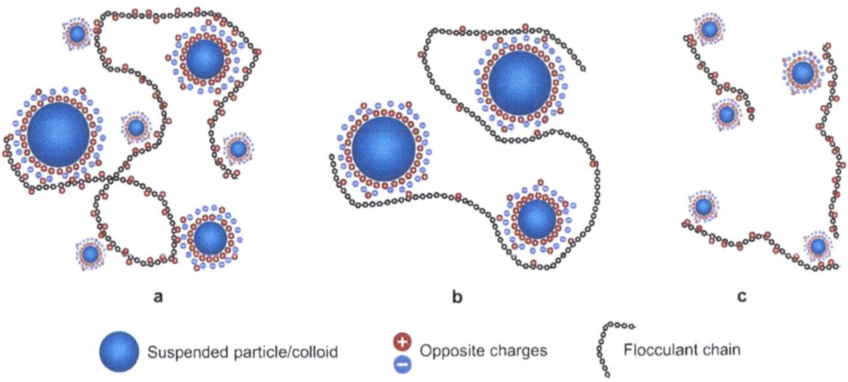

Fig. 3.15 Illustrative diagram depicting three types of flocculants: **a** those with a long chain and high charge, **b** those with a long chain but low charge, and **c** those featuring a short chain

be enhanced, thus reducing the tendency to agglomerate and making NC easier to disperse in a stable manner. Carboxylated NC is more stable in aqueous solutions and is suitable for low concentration suspension systems. Through hydrophobic modification, the hydrophobicity of the NC surface is increased, leading to phase separation in aqueous solutions. The ease with which hydrophobic NC aggregates into large clusters in polar solvents provides a unique advantage in oil–water separation, hydrophobic composites, and adsorbent applications.

Gel and Glass Transition at High Concentrations

At high concentrations, the transition of NC from the suspension state to the gel state is induced by the combined effects of hydrogen bonding, electrostatic interactions, and van der Waals forces among NC fibers. NC in the gel state exhibits a three-dimensional network structure with viscoelastic properties and is widely used as a thickener, soft material, and medical hydrogel. When the volume fraction is further increased, the NC structure becomes denser and more rigid, forming a glassy state. At this point, the molecular motion of NC is restricted, and it behaves as a solid. The glassy materials with a high mechanical strength and wear resistance and can be used for highly loaded composites and protective coatings.

Aggregation Regulation in Applications

NC is utilized as a reinforcing agent in composite materials, and its degree of dispersion and aggregation affects the material's mechanical characteristics. Appropriate surface modification can improve the dispersion of NC within the polymer matrix, create a uniform reinforcing phase, and improve the strength and elastic modulus

Table 3.2 Summary of main characteristics and applications of relevant cellulose-based flocculants documented over the past ten years (2013–2024)

Application	Encapsulation method	Compound	Microcapsule	Additive	Mean diameter (μm)	Encapsulation yield or efficiency (%)	References
Food industry	Spray drying	Calcium citrate and calcium lactate	CMC HPMC	–	13–19 7–489	84–91	Oneda and Ré (2003)
	Dripping + freeze-drying	Proanthocyanidins	Alg	$CaCl_2$ + NaCl	1296	86.5	Sheng et al. (2021)
	Dripping + freeze-drying	L. Plantarum	Alg–MC	+ Tween 80	1321	88.8	Fareez et al. (2018)
			Alg–HPMC		1338	92.7	
			Alg–MC		1322	97.1	
			Alg–CMC		1312	96.2	
			Alg–HPMC		1338	96.5	
	Filtration + freeze-drying	Quercetin	TEMPO–CNF	–	Nanopaper	88.8	Li et al. (2019)
	Dripping + filtration	Whey protein isolate	Alg–CNF	$CaCl_2$	488–559	–	Rungraung et al. (2022)
	Dripping + freeze-drying	L. Rhamnosus	Alg–CNC CNC	–	449–461 0.16–4.37	82.8–89.5	Maleki et al. (2020)
Agrochemicals	Spray drying	Gallic acid	HPMC phthalate Ethyl cellulose	Span 20	12.7 15.2	68.0 34.9	Cortesi et al. (2017)
	Dripping + filtration	Essential oil and pesticide	Polydopamine–CNC	Cinnamoyl chloride	0.08–0.11	40–93	Tang et al. (2019)
	Ionic gelation	Citronella essential oil	Chitosan–CNF	Tween 80	0.43	90.8	Ibrahim et al. (2022)

(continued)

Table 3.2 (continued)

Application	Encapsulation method	Compound	Microcapsule	Additive	Mean diameter (μm)	Encapsulation yield or efficiency (%)	References
Energy	Drying	Paraffin	TEMPO–CNF	–	Nanopaper	–	Li et al. (2017)
	Impregnation	PEG-ZPP	TEMPO–CNF	–	Aerogel	–	Yang et al. (2023)
Biomedical	Injection into hydrogel	Cell aggregates	TEMPO–CNF	–	Hydrogel	–	Das and Fernandez (2020)
	Dripping + freeze-drying	Ocimum basilicum L	Chitosan–TEMPO–CNF	–	Aerogel	–	Mishra et al. (2021)
	Dripping + centrifugation	Curcumin	Enzymatic–CNF	DMSO	12	99.14	Kumari et al. (2021)
	Dripping + filtration	L. Plantarum	Alg–TEMPO–CNF	$CaCl_2$	1250	> 95	Zhang et al. (2018)
	Dripping + freeze-drying		$ZnFe_2O_4$ CMC, PVA, AA	$FeCl_3 \cdot 6H_2O$ $ZnCl_2$	–	–	Zhao et al. (2024)

of composites. In water treatment applications, adjusting the surface charge of NC allows for the regulation of its interaction and aggregation with contaminants present in the water. For example, charge-neutralization-induced aggregation helps to remove dye molecules and heavy metal ions, improving the efficiency of water treatment.

The aggregation mechanisms and interactions of NC are fundamental to its functionality in different applications. The dispersion state, network architecture, and gel-forming behavior of NC can be manipulated by adjusting electrostatic forces, hydrogen bonding, van der Waals attractions, and aggregation driven by ions. These aggregation mechanisms not only affect the stability of NC in water, but also determine its performance in composites, adsorbents, and water treatment (Table 3.3).

3.3.5 Enhancement of NC Adsorption Properties

Nanocellulose (NC) can undergo various chemical modifications to enhance its adsorption efficiency. These modifications leverage different pathways, some of which are highlighted below: Firstly, a widely used method for producing carboxylated NC is TEMPO-mediated oxidation, which selectively oxidizes the primary hydroxyl group at the C6 position. While effective, this method is not without limitations. The process involves toxic reagents, extended reaction times, and a lack of versatility, as only the C6 hydroxyl groups are oxidized. Another approach involves oxidizing NC with periodate and chlorite. This dual-step process converts dialdehyde nanocellulose (DAC) into dicarboxylated NC (Suopajärvi et al. 2013). Despite its efficacy, this method cleaves glycosidic bonds within the cellulose chain, potentially reducing polymerization levels and altering the structural rigidity of the resulting NC (Sun et al. 2015). To address these drawbacks, citric acid/hydrochloric acid hydrolysis can be used to produce carboxylated CNC without significantly affecting their degree of polymerization (DP). The H_3O^+ ions generated either through the dissociation of carboxylic acids (RCOOH) or from hydrochloric acid play dual roles in the modification process. These ions facilitate the hydrolysis of the amorphous regions within microcrystalline cellulose and also act as catalysts for esterification reactions. Specifically, they enable the bonding of carboxylic acid groups from citric acid with the exposed hydroxyl groups on cellulose chains. At similar adsorbent dosage (18 mg/L) and dye concentration (200 mg/L), CNC-COOH (4 h) showed better removal of MB than TEMPO-oxidized NCs. The removal efficiency was significantly higher than that of CNCs functionalized with sulfuric acid, formic acid, and hydrochloric acid (Figs. 3.16 and 3.17).

In the second method, periodate oxidation has been extensively employed to synthesize dialdehyde nanocellulose (DAC) (Yao et al. 2016). This oxidative treatment cleaves the pyranose ring, resulting in DAC with aldehyde (–CHO) groups that are less hydrophilic compared to the primary hydroxyl (–OH) groups of native NC. Consequently, this derivative exhibits a lower water affinity and can effectively adsorb organic solvents. However, disrupting the pyranose structure reduces the crystallinity of NC, thereby increasing its amorphous regions, which enhances its capacity

Table 3.3 Advantages and disadvantages of various NC-based adsorbent materials as well as their preparation methods: reproduced with permission (Das et al. 2022)

Adsorbent	Processing/modification	Advantage	Disadvantage	Target Pollutants	References*
CNC-polypyrrole	Composite prepared by stirring	Polypyrrole is stable, low cost and eco-friendly, stirring is simple and easy to apply	Cooling and lyophilization are necessary, separability is difficult	Cr(VI) Congo Red	Das et al. (2022)
Sludge microfibers + CNF_{SL} (support layer)/ active layer (CNC_{BE} + gelatin)	Layered membranes fabricated by vacuum filtration	Layers of specific materials are feasible to apply, easy to operate, lab availability and low-cost method, quickly draws a filtrate, continuous operation, easy to control operating parameters, high surface charge in TEMPO oxidized NC, available—COOH group for adsorption	High power consumption by the vacuum pump, time consuming and support material is necessary	Cu(II) Fe(II) Fe(III)	Suopajärvi et al. (2013)
Chitosan and phosphorylated NC	Electrospun mat fabricated by electrospinning	Ease surface functionalization, good water permeability due to high porosities, interconnected network structures, and minimal pressure drops, tailor able to adsorb specific contaminants, good affinity of Cd (II) interaction with amine and phosphate groups	Expensive electrospinning setups, additional polymers requirement, difficult to achieve nanofiber with sub-nm diameter, necessary for parameters optimization etc	Cd(II)	Sun et al. (2015)

(continued)

Table 3.3 (continued)

Adsorbent	Processing/modification	Advantage	Disadvantage	Target Pollutants	References*
Sodium alginate and polyacrylamide—NC(BC/CNC/TEMPO oxidized NC)	Hydrogels fabricated by stirring	Crosslinked 3D network, transparent, high water-holding capacity, light-weight, excellent flexibility, superior homogeneity, easy access for dye adsorption	Water swollen, ions requirement for crosslinking (e.g. $CaCl_2$), lack of mechanical strength, covalently crosslinking is necessary	Methylene blue	Yao et al. (2016)
CNC-alginate-hydrogel	Beads prepared by mixing and gelation	Feasible for fixed bed columns and large scale implementation, effective for cyclic sorption/desorption, fast gelation method	Cross-linker needed ($CaCl_2$), acute toxicity of Alginic acid	Methylene blue	Ram and Chauhan (2018)
NC-graphene oxide	Aerogels fabricated by stirring and ultrasonication	Porous ultralight material, crosslinked 3D network, increased strength and stiffness, lightweight, excellent flexibility, easy access for solutes due to high porosity (>90%)	Necessary for freezing/subcritical drying	Methylene Blue Congo Red waste oil	Jin et al. (2015), Alsbaiee et al. (2016)
NC-aerogel-oleic acid-Fe_3O_4	Magnetic aerogel prepared by blending > mixing > stirring	Hydrophobic adsorbents, ease of separability after adsorption, preparation methods are easy to implement	Freeze-drying is needed but energy and time consuming process	Cyclohexane Ethyl acetate Vacuum Pump oil	Cova et al. (2018)

* The cited references are only for adsorbent, processing/modification and target pollutants columns, excluding advantage and disadvantage columns

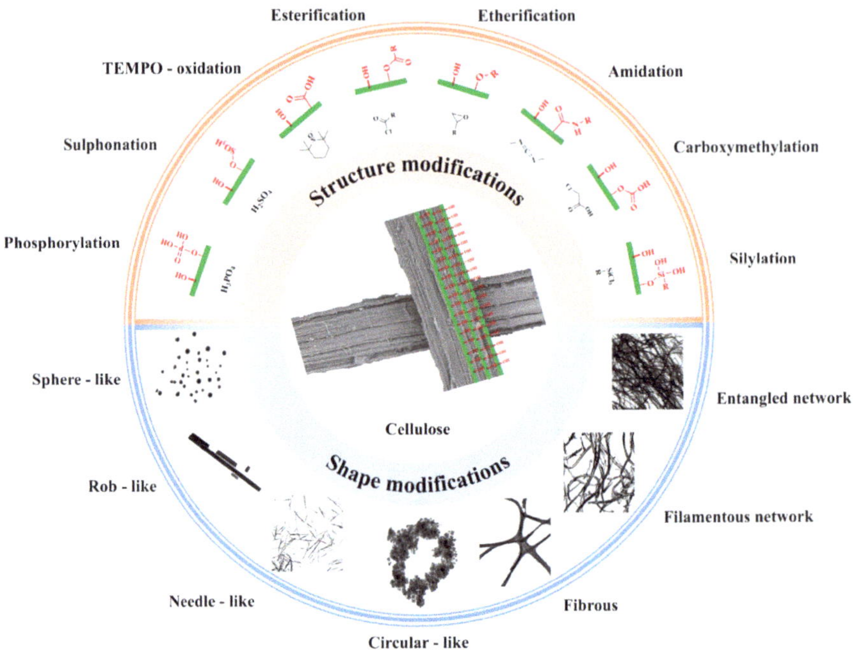

Fig. 3.16 Shape and structural modification of cellulose nanoparticles: reproduced with permission (Zhang et al. 2023)

to adsorb various pollutants (Yao et al. 2016). The aldehyde groups also provide redox functionality, enabling the reduction of absorbed metal ions (e.g., Cd^{2+}, Ag^{+}, Hg^{+}, and Ni^{2+}) into zero-valent nanoparticles. Moreover, these groups facilitate the attachment of amino acids or crosslinking with polymers like chitosan, opening pathways for functionalized DAC variants. For example, through reductive amination, dialdehyde cellulose (DAC) can be functionalized with L-cysteine amino acids. This modification enhances the adsorption performance of DAC for mercury ions (Hg^{2+}), achieving a maximum capacity of 923 mg/g at pH 5.71 (as estimated by the Langmuir model, with a theoretical value of 1034 mg/g). The removal efficiency was 93%, and even in the presence of competing metal ions such as Cu^{2+}, Pb^{2+}, Cd^{2+}, and Zn^{2+}, it remained at 87.4%. Another method involves functionalizing spherical nanocellulose (NC) with thiol (–SH) groups using 3-mercaptopropionic acid. This approach demonstrated a maximum Hg^{2+} adsorption capacity of approximately 190 mg/g. Moreover, selectivity for Hg^{2+} was notably high in various water types, achieving removal rates of 99%, 78%, and 88% in distilled water, ion-containing water, and tap water, respectively (Ram and Chauhan 2018). These thiol-functionalized adsorbents offer significant advantages over untreated cellulose, which typically has low specific surface area, minimal metal-binding groups, and limited Hg^{2+} selectivity (Jin et al. 2015). For anionic dyes, the maximum adsorption capacity of the NC system reached 555.6 mg/g for acid red GR, with removal rates of 100% and 98% for Congo

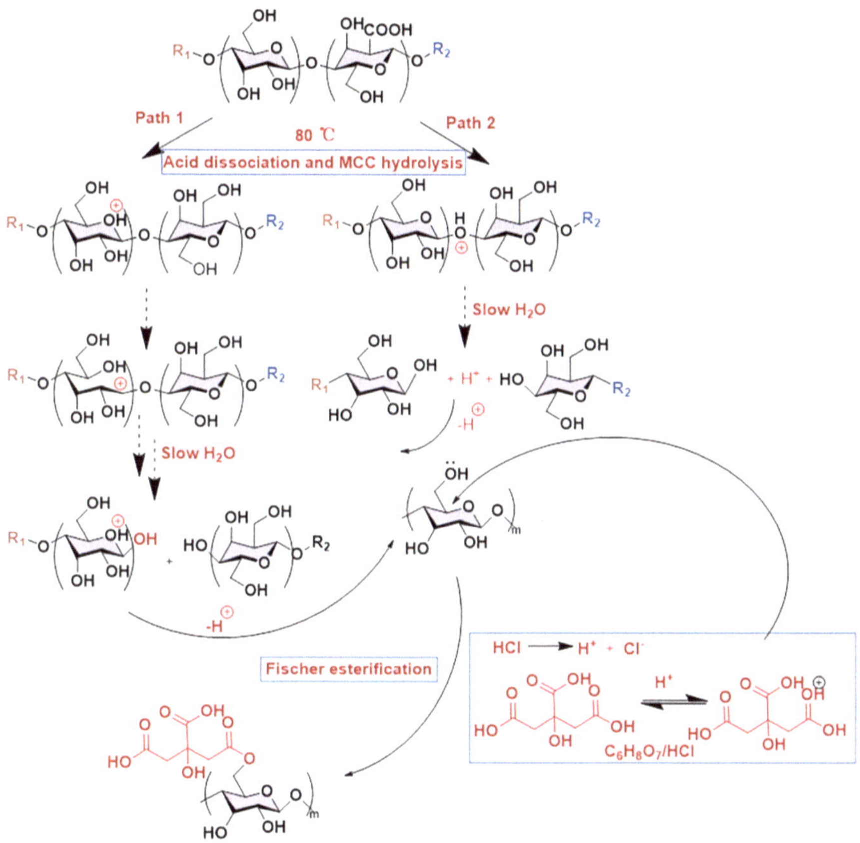

Fig. 3.17 Two possible routes (acidolysis and hydrolysis) for the manufacture of carboxylated CNC from microcrystalline cellulose

red 4BS and activated light yellow K-4G, respectively. The adsorption mechanism is primarily driven by electrostatic interactions and hydrogen bonding between the amino (–NH_2) and hydroxyl (–OH) groups on the grafted NC and the dye molecules (Huang et al. 2020). In another study, DAC with varying oxygenation levels (DAC-0.5, DAC-1, and DAC-2) was cationized using Girard's reagent T to enhance its dye removal efficiency. Among the variants, *c*-DAC-2 demonstrated the highest charge density and a maximum adsorption capacity of 80.5 mg/g, as predicted by the Langmuir model. The adsorption of hexavalent chromium by c-DAC primarily relied on electrostatic attraction between the quaternary ammonium groups and the Cr(VI) ions. Additionally, c-DAC exhibited consistently high adsorption capacities across a broad pH range (2–10).

The third approach involves grafting β-cyclodextrin (CD) onto nanocellulose (NC) or amine-functionalized NC. This combination produces an affordable and highly efficient biosorbent that can remove a wide variety of micropollutants (Alsbaiee et al. 2016). While CD-functionalized materials demonstrate improved adsorption capacities for diverse wastewater contaminants, the functionalization of NC with CD has not been extensively studied, despite the widespread recognition of cellulose's utility in pollutant removal (Cova et al. 2018). Both positively and negatively charged CDs are commercially available, suggesting that NC modified with positively charged CDs could effectively adsorb anionic and cationic pollutants.

The fourth pathway creates multifunctional cationic NC by incorporating quaternary ammonium-containing molecules (QAMs). These QAMs consist of a pentavalent nitrogen atom connected to four alkyl or heterocyclic groups and a small anion, such as iodide, chloride, or bromide. The cationic nitrogen interacts electrostatically with anionic pollutants, while the QAM structure also enables ion exchange with target contaminants. The length of the alkyl chain in the QAM can be adjusted to modulate lipophilicity, enhancing the adsorption of oils or organic solvents (Tavakolian et al. 2020). However, the full potential of QAM-modified NC as a multifunctional adsorbent has yet to be fully explored.

In the fifth method, peptides with strong metal-binding properties and excellent biocompatibility can be attached to nanocellulose (NC). The N-terminal amino group, C-terminal carboxylic acid group, and side chains of the peptide molecules provide ideal reactive sites for grafting onto NC (Brar et al. 2020). This functionalization primarily finds applications in the biomedical field, though there is limited research on its ability to selectively adsorb heavy metal ions or organic contaminants (Lin and Dufresne 2014). Expanding the understanding of its pollutant adsorption capabilities could unlock broader environmental applications.

Börjesson et al. (2018) demonstrated that sulfuric acid-pretreated NC could react with azetidinium ions. This reaction notably enhanced the thermal stability of NC by approximately 100 °C compared to untreated NC. Additionally, the modified NC displayed increased hydrophobicity, enabling it to better adsorb organic pollutants such as polycyclic aromatic hydrocarbons (PAHs). Given the versatility of azetidine derivatives, this modification pathway shows promise for creating NC materials with novel and diverse functionalities, warranting further exploration (Figs. 3.18 and 3.19).

3.3.6 Summary

This document provides a comprehensive review of cellulose nanofibers (NC), focusing on their structure–property relationships, functionalization strategies, and potential applications. The key points are:

(i) Structural characteristics

NC is found in three primary forms: cellulose nanocrystal (CNC), cellulose nanofibril (CNF), and bacterial nanocellulose (BNC). Each form shows unique morphologies, crystallinity, and properties tailored to specific applications. CNC, obtained via acid hydrolysis, features high crystallinity and strength, suitable for composites. CNF, derived through mechanical or enzymatic treatment, exhibits high aspect ratios and

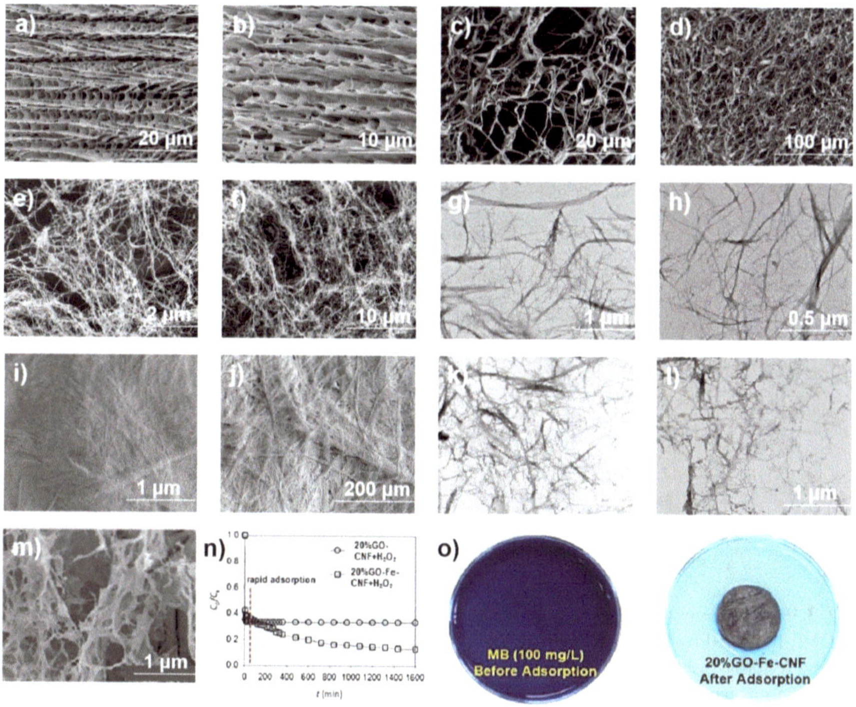

Fig. 3.18 SEM plots of TEMPO-CNF aqueous dispersion with different concentrations, after freeze-drying: **a** and **b** 1 wt%; **c** and **d** 0.05 wt%; **e** and **f** 0.005 wt%. **g** and **h** TEM plots of 0.001 wt% TEMPO-CNF. **i** and **j** 1 wt% SEM plots of mechanical CNF. **k** and **l** TEM plots of 0.001 wt% mechanical CNF. **m** SEM image of 20% GO-Fe-CNF; **n** Fenton oxidation of methyl bromide on GO-Fe-CNF nanocomposites. **o** Photographs of 20%GO-Fe-CNF before and after 30 min of adsorption of methyl bromide solution (100 mg/L): reproduced with permission (Das et al. 2022)

excellent network formation, ideal for thickeners and biomedical scaffolds. BNC, produced by bacterial fermentation, is pure, biocompatible, and structurally robust.

(ii) Surface characteristics

NC's surface hydroxyl groups enable diverse chemical modifications to enhance dispersibility, compatibility, and functionality. Common modifications include carboxylation and sulfonation for improved water dispersion and adsorption, hydrophobic modifications for oil–water separation, and multifunctionalization for specific applications like catalysis or biosensing.

(iii) Structure–Property Relationships:

Aspect ratio and volume fraction are critical in determining NC's phase behavior (suspension, gel, or glassy state) and rheological properties. High aspect ratios improve mechanical strength, network formation, and self-assembly into liquid crystals. Volume fraction influences stability, gelation, and optical properties. Increasing volume fraction leads to enhanced mechanical properties but reduced transparency.

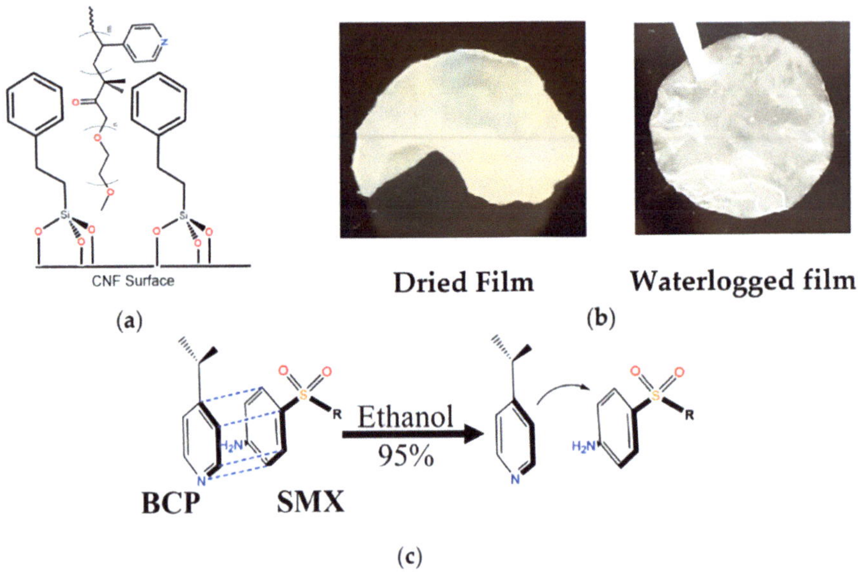

Fig. 3.19 Illustrates the following processes and mechanisms: **a** the covalent attachment of TMPES to CNF, followed by further modification using the block copolymer BCP (P4VP-PEO); **b** photographic comparison of CNF films treated with TMPES and BCP in their dried state and after immersion in water, highlighting differences in structural properties; and **c** the adsorption mechanism of SMX, driven by electron donor–acceptor interactions. These keywords highlight the role of surface functionalization and interaction mechanisms in the material's behavior and functionality: reproduced with permission (Das et al. 2022)

(iv) Aggregation mechanisms

The behavior of NC in solutions and composites is controlled by interactions like hydrogen bonding, van der Waals attractions, and electrostatic forces of repulsion. Chemical modifications can regulate these interactions to optimize dispersion and aggregation.

(v) Applications

Nanocellulose (NC) is a versatile material extensively applied across various fields, including water treatment, composite reinforcement, biomedical applications, and as adsorbents and thickeners. Notable applications include the adsorption of heavy metals, fabrication of optical films, and development of flexible electronics. The renewable and biodegradable nature of NC underscores its importance in green technologies. Its applications in water purification, lightweight composites, and sustainable packaging align closely with global sustainability objectives. Furthermore, NC's biocompatibility and tunable properties make it a prime candidate for advanced biomedical uses such as tissue engineering scaffolds, drug delivery systems, and biosensors. Efforts should focus on integrating therapeutic and diagnostic functionalities into NC-based materials.

In addition, NC exhibits self-assembly into liquid crystal phases, offering significant potential in optoelectronics, sensors, and energy storage devices. Expanding research into its electrical and thermal conductivity can unlock further applications in flexible electronics and thermal management solutions. To maximize NC's efficacy, continued research on optimizing key parameters such as volume fraction, aspect ratio, and surface chemistry is essential. These refinements will enhance control over its mechanical and rheological properties, making NC more effective in structural reinforcements and fluidic systems. However, challenges like production scalability, cost competitiveness, and performance under high-salt or extreme environmental conditions remain critical. Collaborative efforts across material science, engineering, and environmental disciplines will be pivotal in advancing NC's commercialization and unlocking its full potential.

3.4 Enabled Water Purification Technologies

3.4.1 Adsorption of Heavy Metal Ions

Due to the quick development of industrialization and the quick rise in energy demand, water pollution problems have arisen. The most prominent water pollution problems are the pollution of heavy metal ions, smelting, mining metals, fine chemicals, paints, pigments, lead batteries, pesticides, cosmetics, and alloy manufacturing, and other industrial activities will have much wastewater containing heavy metal ions, as well as pesticide spraying, the misuse of fertilizers, and other agricultural activities generate heavy metal ions that are discharged into the aquatic environment. Unlike total nitrogen, total phosphorus, and organic matter in wastewater, heavy metal ions such as Pb^{2+}, Cr^{6+}, Fe^{3+}, Ni^{2+}, Cu^{2+}, Hg^{2+}, and Cd^{2+} are very difficult to degrade and have the characteristics of continuous accumulation, easy migration, potential toxicity, and can be stable for a long period in the aquatic environment, which has already seriously threatened people's health and the ecosystems of the natural world, and impeded the sustainable economic and social development. Therefore, removing these toxic metal ions from wastewater is extremely urgent, and heavy metal pollution is a global environmental problem (Fig. 3.20).

So far, various techniques, including chemical precipitation, adsorption, reverse osmosis, solvent extraction, and electrochemical treatment, have been employed to eliminate radioactive and heavy metals from polluted water. Compared to additional techniques for wastewater treatment of heavy metal ions, nanosorption technology demonstrates the advantages of high efficiency, environmental protection, economy, and versatility in treating heavy metals, providing a new remedy for the heavy metal contamination issue. Among many nanomaterials, nanocellulose (NC), with a longer aspect ratio, higher mechanical strength, and excellent chemical accessibility, has received extensive research on the adsorption of heavy metal ions such as Cu^{2+}, Pb^{2+}, Cd^{2+}, Cr^{6+}, and Ni^{2+} in wastewater treatment. Yin et al. (2023a) developed

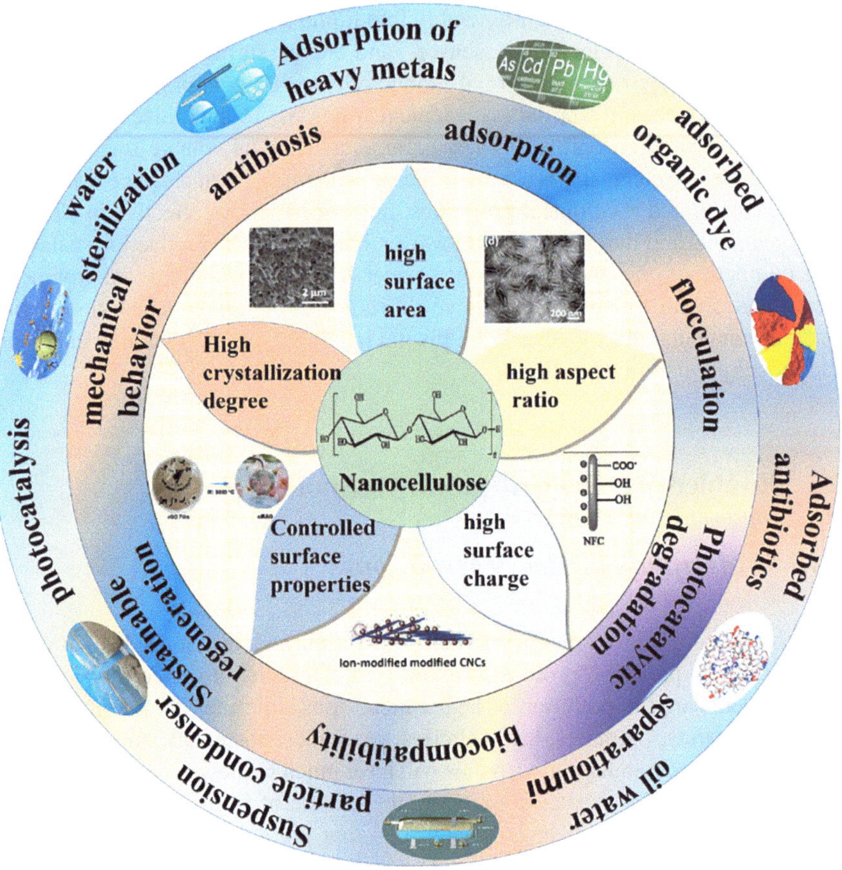

Fig. 3.20 Structure–property–application diagram of nanocellulose in boiling water treatment

a guanidine-functionalized silk gum/cellulose nanocrystal aerogel (GSNA) with a bionic skeletal structure by assembling silk gum protein (SS) into the 3D architecture of cellulose nanocrystal (CNC) aerogel and coupling polyhexamethylene biguanide (PHMB) to CNC aerogel, and the obtained aerogel exhibited excellent heavy metal adsorption capacity (Cu of 571.5 mg/g). Kaur and Pal (2023) synthesized novel NC (cellulose carbohydrate polymer) embedded in wheat straw. In the adsorption study, the ideal pH was 8, and the dose of nanosorbent was 0.6 g/50 mL. The removal efficiencies of Cu(II), Pb (II), and Zn(II) were 88.75%, 81.18%, and 86.84%, respectively. Wang et al. (2023) prepared fluorescent nanocellulose aerogel (FNA) by mixing carbon dots (CDs) and titanate nanofibers (TNs) with 2,2,6,6-tetramethylpiperidinyl-1-oxide (TEMPO)-oxidized cellulose nanocellulose (TOCNC) via Schiff base reaction. The highest adsorption potential of optimized FNA for Cr^{6+} was 543.38 mg/g.

Heavy metal ion adsorption on NC is mainly carried out through ion exchange and chemical complexation. In the ion-exchange mechanism, harmful metal ions (e.g., M^{n+}) replace other ions (e.g., K^+, Na^+, and H^+) already bound to the surface of NC to achieve adsorption, while in the chemical complexation mechanism, the carboxyl ($-COO^-$) and hydroxyl (–OH) groups of NC form stable complexes with specific metal ions to achieve the purpose of adsorption. NC itself has a weak adsorption capacity and cannot be recycled after adsorption. Also, due to the complexity of the constituents in industrial wastewater, a significant issue with the real-world implementation of adsorption is that the competitive kinetics between the target pollutants can alter the adsorption performance of the adsorbent and is not captured by a single analyte. Therefore, functionalized modification of NC has become an important means to achieve controllable adsorbents. The maximum absorption capacity of NC is influenced by the nature of the functional groups and their specific surface area. Usually, the functionalization of NC mainly includes three ways: surface functionalization, polymer grafting, and polymeric monomer complexation.

Surface Functionalization

NC has a lot of hydroxyl groups on its surface, and surface chemical modification can greatly increase its adsorption ability for heavy metal ions. The surface functionalization of nanocellulose (NC) typically involves the incorporation of functional groups into its cellulose framework. This process enhances the material's properties, enabling it to perform more effectively in various applications, including oxidation, esterification etherification, etc. These modifications can increase the binding sites and complexation ability with heavy metal ions, thus improving the adsorption performance.

TEMPO as the most commonly used reagent for the oxidation of NC not only enhances the hydrophilicity and ion-exchange capacity of cellulose molecules by selectively introducing carboxylic acid groups on their surfaces but also maintains the original cellulose crystalline structure and fibrous morphology. The advantages of TEMPO-catalyzed oxidations are their green and simple process, cost efficiency, and selectivity to primary alcohols, ensuring the preservation of the crystallinity in the final product, along with its morphological stability and excellent dispersibility in water, is crucial for maintaining the material's performance in various applications. The reaction mechanism is shown in Fig. 3.21. Zhan et al. (2020) oxidized rice husk using TEMPO oxidation to form carboxylated cellulose nanofiber (CNF-C), focusing on the comparison of three forms of NC adsorbents: suspension adsorbent, lyophilized adsorbent, and composites containing magnetic nanoparticles in the adsorption of Pb(II) and La(III). It was found that rice husk-based NC exhibited the highest maximum adsorption capacity in the form of suspension, which was 193.2 mg/g for Pb(II) and 100.7 mg/g for La(III), respectively. NC adsorbent in the suspension could be easily separated from treated water by gravity-driven filtration through the formation of gel flocs. The primary mechanism behind this adsorption process is the electrostatic attraction between negatively charged carboxylate

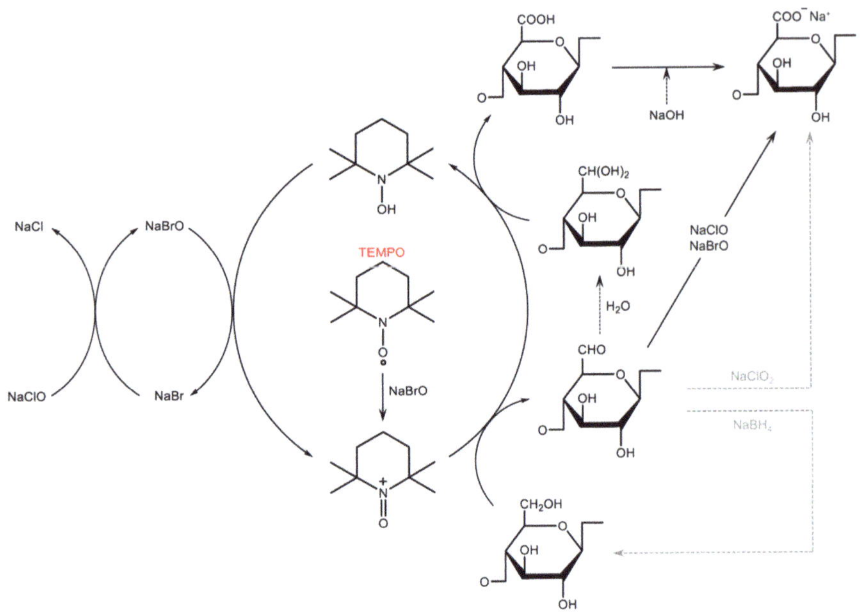

Fig. 3.21 Reaction mechanism of TEMPO-catalyzed oxidation of nanofibers: reproduced with permission (Isogai 2022)

groups and multivalent metal ions, which facilitates the binding and removal of these ions from the solution. Lanthanum oxychloride with 90% lanthanum content can be obtained by incinerating CNF/$LaCl_3$ gels. The research presents a viable and environmentally sustainable approach to converting agricultural waste into nanomaterials for the remediation and recovery of harmful heavy metal ions from polluted water.

Esterification modification refers to the action of cellulose by various acids or anhydrides, causing the change of hydroxyl groups in the structure to ester bonds. Citric acid is a commonly used esterification reagent, and Silva and Rosa (2022) prepared a foam material of citric acid-modified cellulose nanofiber (CNF) by freeze-drying method and investigated the ability of the substance to adsorb chromium in an aqueous solution. Modifying the citric acid concentration and subjecting the material to heat treatment resulted in a noticeable increase in pore size, alongside a reduction in the activation energy required for the thermal degradation of amorphous cellulose. In the subsequent adsorption experiments, the cellulose nanofiber (CNF) foam exhibited a two-phase kinetic behavior. Initially, during the first 2000 min, chromium adsorption adhered to a pseudo-second-order model. However, in the second phase, spanning from 2000 to 8000 min, the process became more complex, driven by a combination of endoparticle diffusion and the continued influence of pseudo-second-order kinetics. The highest adsorption capacity of the CNF foam was about 5 mg/g, and the maximum adsorption capacity was about 1 mg/g. The highest adsorption capacity of CNF foam was about 2 mg/g. The highest adsorption

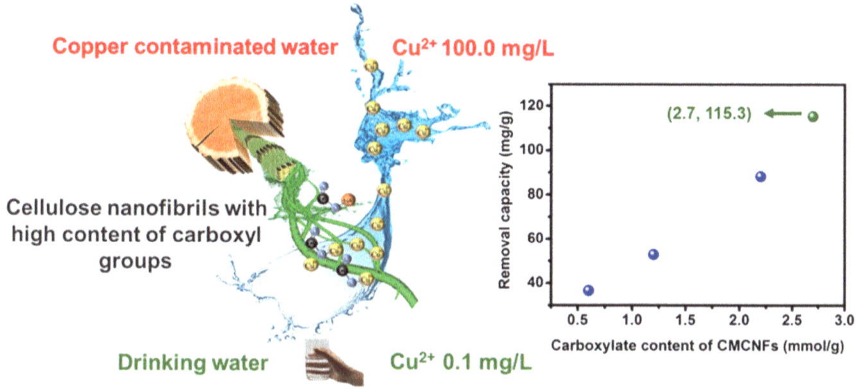

Fig. 3.22 Adsorption mechanism diagram of CMCNFs with Cu(II) adsorption capacity of CMCNFs with different carboxylate content: reproduced with permission (Matthysse et al. 2004)

of Cr(VI) by the CNF foam was about 5 mg/g, indicating that it was effective in removing chromium from water by reversible physical adsorption under neutral pH conditions.

Etherification modification refers to substituting hydrogen atoms in the hydroxyl group by using alkylating reagents (i.e., etherifying agents) to produce cellulose ethers under alkaline conditions. Among the many methods of etherifying cellulose, carboxymethylation is a common technique, where the most commonly used carboxymethylation reagent is monochloroacetic acid. Qin et al. (2019) also prepared carboxymethylated cellulose nanofibers (CMCNFs) modified with monochloroacetic acid (MCA) and explored their adsorption efficiency for Cu(II) in water. The adsorption mechanism is shown in Fig. 3.22, which demonstrates the ability of CMCNFs to purify copper ions from wastewater and reduce the copper ion content. The adsorption performance of CMCNFs with different carboxylate contents was comparatively investigated, and the results, as shown in the scatter plot on the right side of Fig. 3.22, indicated that CMCNFs with up to 2.7 mmol/g carboxylate content exhibited the best adsorption performance at pH 5.0, and its equilibrium adsorption amount reached the highest level of 115.3 mg/g. The electrostatic adsorption mechanism of carboxylate groups with Cu(II) was revealed by the fitting results of the Langmuir isothermal model and the pseudo-second-order kinetic model, combined with a series of characterization tools. After two treatments, the residual concentration of Cu(II) decreased dramatically from 100 to 3.1 mg/L, which was far below the national and World Health Organization standards for clean drinking water.

Polymer Grafting

Grafting of NC polymers refers to the attachment of various specific groups to the backbone of NC, e.g., through free radical polymerization, thus allowing NC-based adsorbents to adsorb pollutants. This can be done either by linking the reactive groups of the polymer to the cellulose or by polymerizing the monomers with cellulose as the starting point. In the former case, the polymer is directly grafted onto the cellulose, and in the latter case, the monomer forms a polymer on the cellulose. For the direct grafting of polymers onto cellulose, Park et al. (2020) successfully synthesized cellulose nanocrystal (CNC-PAH) particles grafted with polyacryloyl chloride (PAH) for the adsorption of hexavalent chromium (Cr(VI)) by atom transfer radical polymerization (ATRP).CNC-PAH particles showed the best adsorption capacity of Cr(VI) in pH 3 solution when the electrostatic attraction between amine groups on CNC-PAH and Cr(VI) was maximized. The electrostatic attraction was maximized. The experimental data revealed that the adsorption kinetics of Cr (VI) on CNC-PAH were predominantly influenced by the internal diffusion resistance arising from the PAH shell layer surrounding the CNC particles. Thermodynamic analysis further indicated that this adsorption process is spontaneous and endothermic, meaning it absorbs heat during the interaction. In particular, when higher molecular weight (approximately 50 kg/mol) PAH was grafted on CNC-PAH, it exhibited a more rapid adsorption rate and higher adsorption capacity for Cr(VI), which was as high as approximately 457.6 mg/g at 298.15 K. In addition, CNC-PAH showed excellent reusability after several adsorption–desorption cycles. Its high adsorption capacity was maintained during the cycles. These results confirm the potential of CNC-PAH as a high-performance, renewable adsorbent for the removal of hazardous pollutants from water.

For the polymerization of monomers starting from cellulose, Chen et al. (2022) developed a hydrogel composed of a cellulose nanofiber (CNF) and chitosan (CS) interpenetrating polymer network (IPN), incorporating nitrogen-doped carbon dots (NCDs) as crosslinking agents, was successfully developed for the dual-purpose detection and adsorption of copper (Cu(II)) and chromate (Cr(VI)) ions from aqueous solutions. The adsorption of Cu(II) primarily occurred via complexation with deprotonated amino and carboxyl groups present on the surface of both the NCDs and CS components, facilitating the removal of copper ions through these interactions, and part of Cu(II) formed copper hydroxide; while Cr(VI) was adsorbed through electrostatic attraction and partially reduced to Cr(III), which formed complexes with nitrogen- and oxygen-containing groups in the NCDs-CNF/CS gels, and at the same time, the addition of heavy metal ions resulted in the bursting of fluorescent characteristics of the NCDs. A sensitive fluorescence-responsive method was provided for the detection of these ions. The hydrogel was prepared by low-temperature hydrothermal method and solvent substitution method and has excellent fluorescence properties and heavy metal ion adsorption ability. The adsorption experiments showed that the maximum adsorption capacity of NCDs-CNF/CS gels for Cu(II) and Cr(VI) was 148.30 and 294.46 mg/g, respectively, and the adsorption process adhered to the pseudo-second-order kinetic model, as well as the Langmuir isotherm, indicating a well-defined relationship between adsorbate concentration and adsorption rate. In addition, the hydrogel was able to maintain the adsorption capacity of about 85 and 80% for Cu(II) and Cr(VI) after five adsorption–desorption cycles, showing good reusability. Meanwhile, NCDs-CNF/CS gels showed good detection performance for Cu(II) and Cr(VI). The findings suggest that NCDs-CNF/CS gels demonstrate significant potential for use in both the detection and removal of heavy metal ions from aqueous environments.

Monomer Compounding

NC composite adsorbent materials comprise cellulose/organic polymer nanocomposites, cellulose/metal nanocomposites, and cellulose/inorganic nanocomposites. Among them, NC metal composites can be synthesized by combining NC with metal-organic frameworks (MOFs), metal oxide nanoparticles, etc. Ma et al. (2019) work demonstrates that MOF aerogels with practical macroscopic moldability and graded porosity by in-situ growth of flexible MOF nanoparticles, namely zeolite imidazole skeleton-8 (ZIF-8), on bacterial cellulose (BC). In the process, cultured bacterial nanofibrillar cellulose was freeze-dried and immersed in $Zn(NO_3)_2{\cdot}6H_2O$/ ethanol solution for wetting and swelling, and then, 2-methylimidazole/ethanol solution was added under a stirred state before sonication and then washed with ethanol and freeze-dried to obtain BC@ZIF-8. Zhu et al. (2011) demonstrated that Fe_3O_4 nanoparticles could be prepared by the dissolution of $FeCl_3{\cdot}6H_2O$ and $FeCl_2{\cdot}4H_2O$ in water, with the addition of $NH_3{\cdot}H_2O$ (8 mol/L) and stirring, and by ultrasonication under conditions of deoxygenation. N_2 and BC spheres were obtained by stirring and fermentation of *Gluconacetobacter xylinum* by bioprocessing. Finally, the BC spheres were added to sterilised Fe_3O_4 nanoparticles to obtain the composite material of Fe_3O_4/BC microspheres by fermentation. For a visual representation of the preparation process, refer to Figure 3.23.

NC inorganic nonmetallic composites can be synthesized by combining NC with bentonite and graphene oxide, etc. Putro et al. (2017) utilized sulfuric acid hydrolysis to prepare CNC from deinked waste paper pulp, which was freeze-dried and then mixed with bentonite dispersed in water by stirring and ultrasonication, and

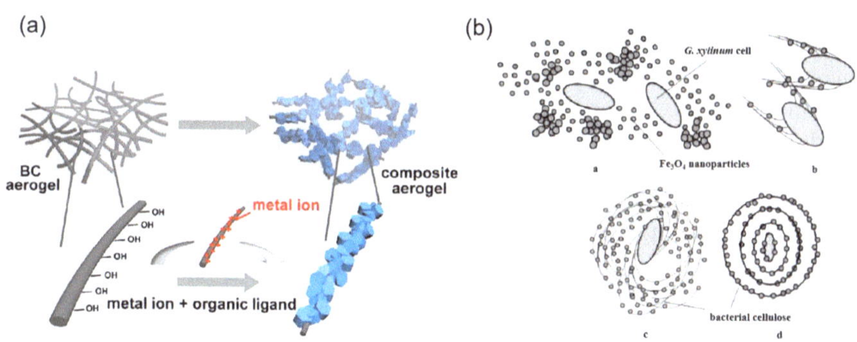

Fig. 3.23 **a** Flowchart to prepare BC@ZIF-8: reproduced with permission (Ma et al. 2019). **b** Flowchart for the preparation of Fe_3O_4/BC microsphere composites: reproduced with permission (Zhu et al. 2011)

finally, the obtained CNC-bentonite nanocomposites were poured into polystyrene Petri dishes and dried to obtain the composites. Yu et al. (2020) work founds that graphene oxide (GO) stock solution was stretched into GO fibers by rotary coagulation bath, which was mixed with carboxymethylated cellulose nanofibers (CMCNFs) and concentrated by centrifugation, and finally GO/CMCNFs-Ca^{2+} and GO/CMCNFs-Fe^{3+} composite fibers (CF) were prepared by a bath of coagulation using 5 wt% $CaCl_2$ and 0.01 M $FeCl_3$ solutions as coagulants, respectively. They were used for Pb^{2+} adsorption experiments.

NC organic polymer composites can be synthesized by combining NC with alginate and polypyrrole, among others. For example, Li et al. (2021) TEMPO-oxidized NC was mixed with sodium alginate (SA) and carboxymethyl chitosan (CMC) followed by crosslinking with Ca^{2+} for 24 h to obtain hydrogel beads. Subsequently, the material was subjected to a washing process using deionized water and ethanol to remove any impurities, followed by freeze-drying to preserve its structure and facilitate further analysis, and finally, a nanocellulose-SA-CMC(NSC) aerogel ball was obtained. Figure 3.24 shows the preparation and adsorption of NSC and its microscopic morphology. Alsaiari et al. (2021) obtained CNC through the hydrolysis of cellulose using sulfuric acid and then polymerized pyrrole by dissolving in $(NH_4)S_2O_8$ to form polypyrrole (PYY), and finally, PYY was mixed with CNC by sonication, solvent exchanged with n-butanol, and then freeze-dried to obtain PYY/CNC nanocomposite, which was used in the adsorption study of Cr^{6+}.

Other Materials

In addition to the functionalization of nanocellulose, there are many ways to further improve its adsorption properties, such as preparing the resulting NC into aerogel materials, hydrogel materials, membrane materials, and so on. Aerogels based on nanocellulose have a high ability to adsorb heavy metal ions because of their numerous active sites and special porous structure. Stephen et al. (2011) successfully synthesized a novel nanocellulose-based aerogel, incorporating both thiol and carboxyl functional groups, through a simple freeze-drying process. This method involved the use of TEMPO-oxidized nitrocellulose suspensions, with thiopropylsiloxanes acting as crosslinking agents, which exhibited excellent adsorption and good selectivity for mercury(II) ions.

Cellulose-based nanohydrogels are usually prepared by physical crosslinking or chemical crosslinking methods, which endow the hydrogels with good three-dimensional network structure and stability. Mautner et al. (2017) tackled this issue by designing innovative core-shell hydrogel beads composed of carboxymethylated cellulose nanofibers (CMCNFs) and sodium alginate (SA). The beads were engineered by incorporating CMCNFs (acting as the outer shell) onto the surface of SA hydrogel beads (serving as the core) through electrostatic interactions and hydrogen bonding, resulting in a mechanically stable configuration, and the Cu(II) removal capacity was as high as 221 mg/g, which exceeded that of the CMCNFs and most of the other nanocellulose-structured adsorbents.

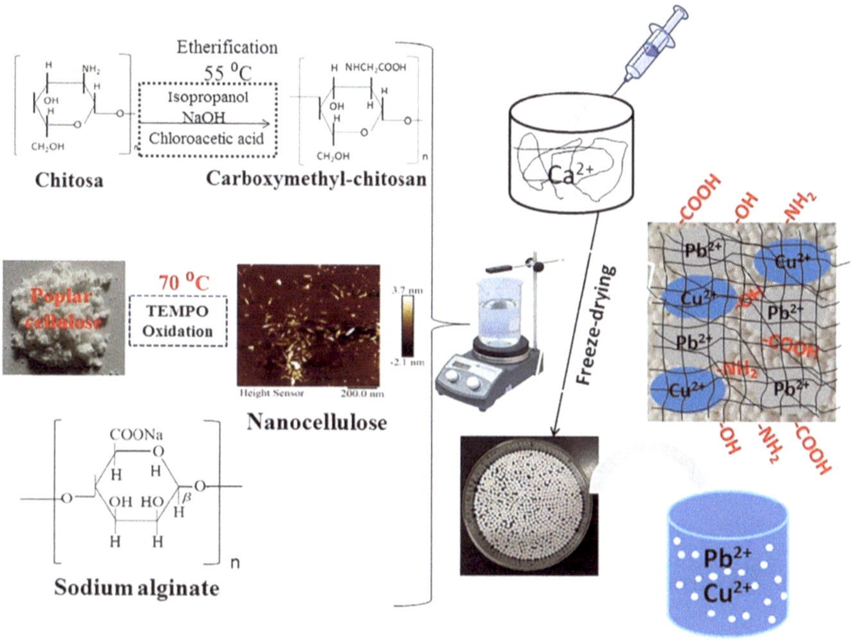

Fig. 3.24 Flowchart of NSC preparation and adsorption of heavy metals: reproduced with permission (Li et al. 2021)

NC has a very high porosity and particular surface area, which makes it provide sufficient adsorption space in a limited volume. Cellulose is difficult to prepare into NC fibers due to the presence of hydrogen bonding, and often NC fiber membranes are obtained by electrostatically spinning cellulose derivatives and post-processing, which not only increases the specific surface area but also introduces functional groups on the surface of the fiber membranes, which further improves the adsorption capacity of heavy metal ions. Stephen et al. (2011) The cellulose acetate nanofiber membrane was deacetylated using 0.08~1 mol/L NaOH solution to prepare a cellulose nanofiber membrane, compared with ordinary cellulose fibers, the specific surface area increased by 4.2 times, and through the post-succinic anhydride modified cellulose nanofiber membrane on the saturated adsorption of Cd^{2+}, Pb^{2+} in aqueous solution reached 0.59 mmol/g, 1.21 mmol/g.

Bacterial nanocellulose (BNC), a bacterial-derived extracellular cellulosic substance, free from lignin and hemicellulose, is produced by certain microorganisms through a biosynthetic process and is widely studied for the removal of heavy metal ions from water because of its low cost and environmentally friendly production. Gram-negative bacteria and microbial fermentation of carbohydrates are the two primary processes for producing BNC, such as *Acetobacter xylosoxidans* and *Gluconobacter xylosoxidans*. The presence of non-living cells and organic detritus within the bacterial nanocellulose (BNC) network offers additional metal-binding sites, owing to the rich array of functional groups, such as hydroxyl, carboxyl, amine, phosphate, and phosphoric acid, that are available for chelation. NC of bacterial origin effectively adsorbs Pb(II), Cu(II), and Cr(VI) and anionic organic dyes.

Because of its strong capacity for adsorption, selectivity, and efficient removal rate, NC can be further enhanced in wastewater treatment through its functionalization and compounding into other materials, showing enormous possibilities for commercial application in heavy metal ion removal. In the wastewater treatment of Riyadh Industrial Zone in Saudi Arabia, NC was utilized as an adsorbent to efficiently remove heavy metals, with removal efficiencies ranging from 97.52 to 100%, which far exceeded the heavy metal concentration levels required for crop production. However, NC still faces some challenges in practical applications, such as adsorption selectivity, interference of coexisting ions, and material applicability (mechanical properties and regeneration). Future research directions include the development of multifunctional, environmentally friendly, and efficient NC adsorbent materials and exploring their possible use in the treatment of industrial wastewater.

3.4.2 Treatment of Organic Dye Wastewater

Dyes are extensively utilized in various industries, including pharmaceuticals, textiles, paints, pulp and paper, and plastics. However, due to their persistent and non-biodegradable characteristics, dye discharge into aquatic systems represents a major environmental concern. Recent studies have highlighted the potential of both raw and functionalized cellulose nanomaterials for effectively adsorbing and removing dyes from contaminated wastewater.

NC has abundant surface hydroxyl groups and good water dispersibility, which enables it to effectively adsorb a wide range of organic dyes. In general, natural cellulose microfibers exhibit a significantly lower affinity for organic pollutants—ranging from 100 to 500 times less—compared to conventional nanomaterials like zeolite or activated carbon. This reduced adsorption efficiency is attributed to the limited number of active sites available for interaction with organic compounds. Furthermore, the intrinsic hydrophilic nature of native cellulose hampers its ability to effectively adsorb organic molecules. To enhance its adsorption and filtration performance, surface modification and the development of nanocomposite materials, such as porous membranes or aerogels with tunable porosity, are required. Surface-modified NC has been utilized as an effective carrier for adsorbing a wide range of organic pollutants. This is attributed to its large specific surface area, which facilitates the creation of active interaction sites upon functionalization. Additionally, the favorable mechanical properties and the fine pore size of its filters and membranes enhance its performance in adsorption applications. Mautner et al. (2017) prepared cationic cellulose nano-primary fibers (CCNFs) containing a large number of ammonium groups from ammoniated paper mill fibrous sludge and showed that nitrate ions can be adsorbed by CCNF nanopaper up to 2300 mg/m^2 filtration area, which equates

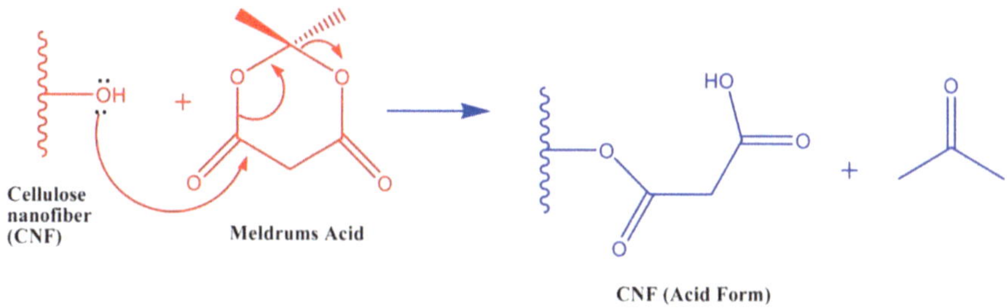

Fig. 3.25 Preparation of Meldrum acid-modified cellulose nanofibers: reproduced with permission (Gopakumar et al. 2017)

to an adsorption capacity of more than 290 mg of nitrate per gram of activated adsorbent. Bleached cellulose pulp extracted from sugarcane bagasse can be oxidized in two steps, TEMPO and chlorite, to form NC with tricarboxylic groups with a carboxyl group content of 3.5 mmol/g and a maximum adsorption capacity of 502 mg/g for methylene blue (Abou-Zeid et al. 2020). Gopakumar et al. (2017) employed a new surface treatment of NC as shown in Fig. 3.25; i.e., the adsorbent material prepared by an approach that uses Meldrum acid as an esterifying agent without the use of solvents was effective in removing 99% of crystalline violet dye from water. The enhanced adsorption capacity of the modified CNF can be attributed to the strong electrostatic interaction between the negatively charged carboxyl groups on the CNF surface and the positively charged crystalline violet dye. This interaction is two to three times greater than that observed with non-functionalized CNF. Yang et al. (2016) effectively created a bifunctional hairy nanomicrofilament with carboxymethyl chitosan and evaluated the ability of this adsorbent material to remove methylene blue dye at room temperature. The maximum adsorption capacity for methylene blue was found to be 785 mg/g. Desorption studies of the adsorbent material by hydrochloric acid and sodium hydroxide revealed that the removal of methylene blue by the regenerated adsorbent material was still maintained at 188 mg/g after five cycles, which indicates its good regeneration ability.

NC can effectively adsorb cationic and anionic dyes through electrostatic interactions. Camparotto et al. (2023) synthesized chitosan aerogel beads (CS@NC-mTA) containing crystalline NC and modified tannic acid, and in a binary system, CS@CNC-mTA microbeads demonstrated effective removal of RR250 and RO122 dyes, achieving maximum adsorption capacities of 298 mg/g and 311 mg/g, respectively. Synergistic effects were also observed in a multicomponent system containing cationic pollutants, citric acid sildenafil (SC) and cetylpyridinium chloride (CPC) increased the absorption by 209.6% and 40.9%, respectively. Gong et al. (2024) prepared carboxylated cellulose aerogels (CCA) with micro/nanostructures and oriented channels from cotton fabrics. The 2.0-CCA material demonstrated a high adsorption capacity for cationic dyes in aqueous solutions, particularly for MB. This adsorption process primarily occurred through electrostatic chemisorption involving the carboxyl groups present on the 2.0-CCA surface. The static maximum adsorption capacity (q_m) was measured at 751 mg/g. The prepared 2.0-CCA exhibited excellent adsorption capacity for cationic dyes in water. The adsorption of MB by 2.0-CCA was mainly based on electrostatic chemisorption with carboxyl groups, and its static maximum adsorption capacity (q_m) was 751 mg/g. Methylene blue (MB) has been shown to possess a dynamic breakthrough capacity of more than 700 mg/g, which is more effective than most of the reported cellulose-based adsorbents.

Nanocellulose-based materials such as aerogels and composites have also demonstrated excellent adsorption properties for organic dyes. Vijayan et al. (2023) synthesized NC, modified nanocellulose (pM-NC), and MOF composites (ZIF-8/pM-NC) from moringa seed pods and used them for dye adsorption, which resulted in better removal of rhodamine B dye by NC, pNC, and pNC-ZIF-8 composites as compared to earlier reported literature. The synthesized MOF(ZIF-8/NC) composites showed higher efficiency than other nanofibrillated cellulose in a short period. Tang et al. (2019) NC extracted from plants is readily transformed into aerogels with controllable densities as shown in Fig. 3.26. The crosslinked three-dimensional (3D) network with high porosity demonstrated significant effectiveness in adsorbing both methyl orange (MO) dye and copper (II) ions. The findings indicate that cellulose nanofiber (CNF)-based aerogels possess considerable potential for the removal of these contaminants. Specifically, the maximum adsorption capacities for MO and Cu^{2+} were 265.9 mg/g and 103.5 mg/g, respectively.

Magnetically functionalized NC also possesses the capacity to remove the dye. Amiralian et al. (2020) created an adsorbent substance with magnetic NC particles. Because of its strong magnetic behavioral properties (38.5 emu/g), it could activate sodium persulfate, which resulted in the removal of 94.9% of the organic dye Rhodamine B (RB) within 300 min at room temperature. In addition to this, NC can be compounded with other materials to enhance the dye removal ability. For example, Sajab et al. (2016) prepared a porous NC aerogel-loaded graphene oxide and Fe(III) composite nanomaterials and used them for the treatment of aqueous

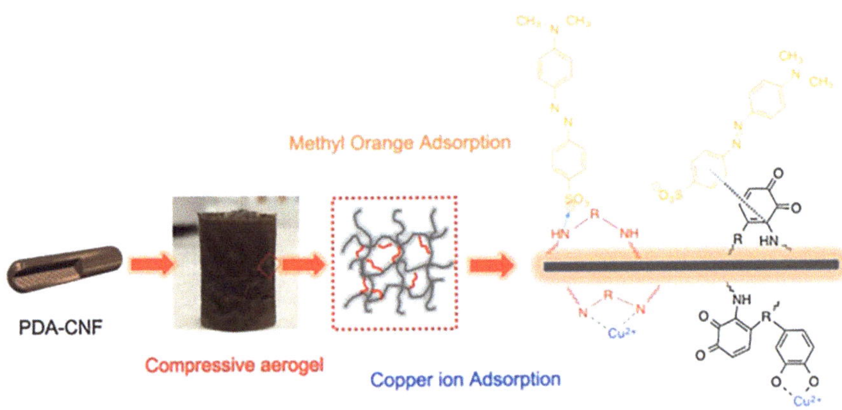

Fig. 3.26 Methyl orange (MO)/copper ions (Cu^{2+}) and synthesized compressed aerogel: reproduced with permission (Tang et al. 2019)

methylene blue solution. In this, the NC acted as an adsorbent for the dye, while the graphene oxide-Fe nanocomposite decomposed the dye through the reaction of Fenton oxidation. The material showed fast adsorption performance (< 10 min) with the biggest adsorption capacity of 142.3 mg/g. NC can be modified through chemical grafting, oxidation, or other processes to improve its adsorption capacity. The introduction of anionic groups, such as carboxyl groups, on cellulose-based adsorbents enhances their ability to adsorb cationic dyes, including MB. Jiang et al. (2017) carboxylated cellulose aerogels (CCAs) with a carboxyl group content of 1.27 mmol/g, synthesized via TEMPO oxidation, show an equilibrium adsorption capacity of 212.7 mg/g for malachite green.

In addition to organic dyes, NC has been investigated as a potential adsorbent for organic pollutants, including fertilizers, pesticides, and pharmaceuticals, due to its abundance, biocompatibility, and effective adsorption properties. Carboxymethyl cellulose (CMC) hydrogels prepared by filling with varying levels of NC, montmorillonite, or vermiculite and chemically crosslinked with citric acid have the potential to reduce fertilizer losses during application and improve water use in agriculture. Similarly, cationic cellulose nanopaper membranes, through ion exchange, can effectively adsorb and remove nitrate ions from fertilizers in agricultural soils and irrigation water, offering a sustainable method for mitigating nitrate pollution. In addition, several studies have demonstrated that NC can be used as a potential adsorbent for pesticide removal. Ma et al. (2020) developed a cellulose nanocrystalline/organic montmorillonite nanocomposite adsorbent (CNC/CTM), which showed that under optimal adsorption conditions, the adsorption of 5.86 mg/L of diuron by 0.07 g /L CNC/CTM achieved 82.32% efficiency in less than 318.68 min. Nikou et al. (2021) synthesized ZIF-8/GO hybrid nanocomposites at ambient temperature for the removal of diazinon and chlorpyrifos from aqueous solutions, demonstrating enhanced adsorption efficiency. NC has been explored as an adsorbent for removing pharmaceuticals from industrial wastewater, with surface modifications enhancing its adsorption capacity for effective contaminant removal. Cellulose extracted from grass NC was reduced to NC, which was effective in removing the antibiotic ciprofloxacin and the anti-inflammatory agent Diclofenac. Production of nanocomposites based on bacterial cellulose and garnet clay (BC/PLG) can be used for loading metronidazole (MTZ). Poly(acrylic-*co*-2-hydroxyoxyethyl methacrylate) grafted CNC composites exhibit potential for drug delivery and dye adsorption in wastewater treatment due to enhanced surface properties and multifunctional performance.

3.4.3 Oil-Water Separation

With the advancement of industrialization and urbanization, the use and discharge of oil and its derivatives have led to serious environmental pollution problems. Oil–water separation technology can not only effectively deal with these pollutions, reduce the damage to the ecosystem, and improve the recycling rate of resources, but also improve the operational efficiency and economic benefits of enterprises. Recently,

NC and its analogs have received extensive attention in filter membranes, aerogels, and porous sponges due to their large specific surface area, large porosity, and small density, which provide efficient adsorption abilities and sustainability in oil–water separation technology.

Filter Membranes

From the structural point of view of conventional filter membrane design (Fig. 3.27), it appears that porosity has to be sacrificed to obtain smaller pore diameters according to typical fabrication methods. However, this restriction can be somewhat circumvented by employing nanofiber technology, Sharma et al. (2020) created a novel class of water filtration nanofiber membranes with nonwoven membranes that, due to their high (up to 80%) porosity, can greatly increase the magnetic flux capacity, thus allowing for lower operating pressures but maintaining resistance to contamination. Incorporation of nanofibrillated cellulose (CNF and CNC) into membranes used for oil/water separation while (partially) altering the initial hydrophilic characteristics of the surface to balanced hydrophilicity or increased hydrophobicity provides a new way of thinking about oil/water separation technology.

Yin et al. (2023b) by using the chemical immersion approach to deposit $Co(OH)_2$ on nanocellulose-based membranes (NBMs), $Co(OH)_2$ @stearic acid nanocellulose-based membranes were created. This allowed for the effective separation and degradation of oil/water mixtures through photocatalysis in challenging conditions. Lou et al. (2024) prepared smart gated membranes with dual responsiveness using pH and temperature-responsive polymers glued onto bacterial cellulose (BC), and the separation efficiency of emulsified oil was > 99% because of the nanoscale pore structure and underwater superoleophobicity. Electrostatic spinning technology has also been

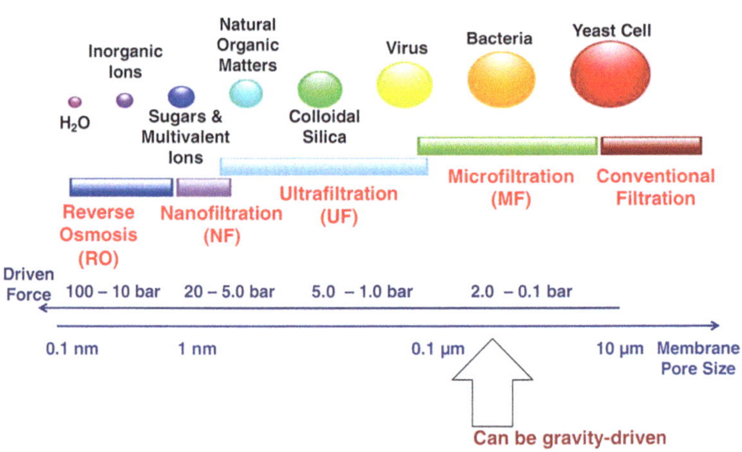

Fig. 3.27 Pressure-driven membrane filtration classification and the connections between pressure and pore size: reproduced with permission (Sharma et al. 2020)

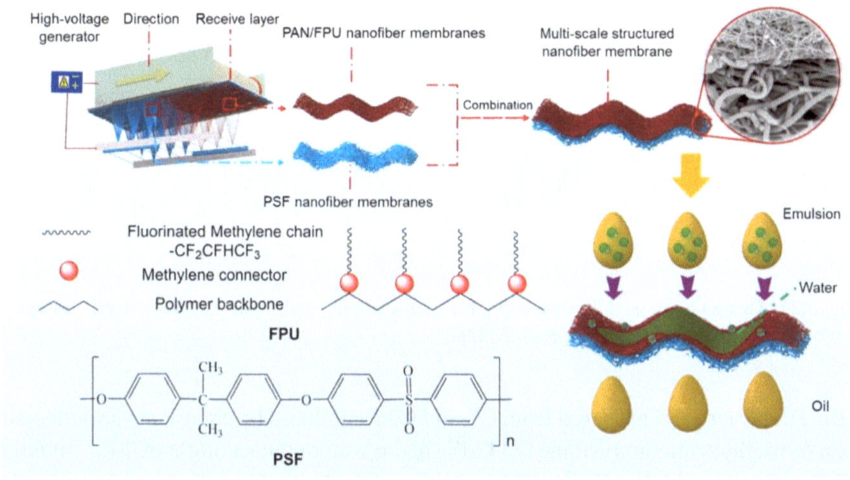

Fig. 3.28 Multiscale nanofiber membrane preparation process: reproduced with permission (Han et al. 2024)

widely used for the preparation of nanofiber membranes, which exhibit superhydrophobic and superoleophilic properties and are capable of effectively separating oil–water mixtures. Han et al. (2024) developed a dual-scale nanofiber membrane for oil/water separation, combining PAN-FPU and PSF layers with distinct fiber diameters (150 and 1200 nm) and pore sizes (3–4 μm and 8–9 μm). This membrane exhibited high separation efficiency and flux, fabricated in one step using electrostatic spinning equipment (Fig. 3.28).

Aerogel

As cellulose aerogels have a lot of hydroxyl groups on their surface, they exhibit amphiphilicity (both hydrophilic and lipophilic), which greatly limits the application of cellulose aerogels in the oil–water separation field. This greatly limits the application of cellulose aerogel in the field of oil–water separation. Cellulose has abundant hydroxyl groups and a large surface area, based on which the hydrophobic modification of cellulose aerogel can be realized. The modification methods of cellulose NC aerogels include chemical vapor deposition (CVD), an impregnation coating method, and an acid-hydrolyzed silane modification (AHSM) method.

Chemical vapor deposition is a method based on the chemical reaction of gaseous substances on a substrate and the synthesis of solid deposits. Li et al. (2024) developed ultralight, biodegradable, and highly porous methyltrimethoxysilane (MTMS)-modified polyvinyl alcohol (PVA)/cellulose aerogels (PCA) with excellent hydrophobicity and oil–water separation performance. These aerogels offer promising potential for applications in recyclable organic liquid capture and oil recovery. Sun et al.

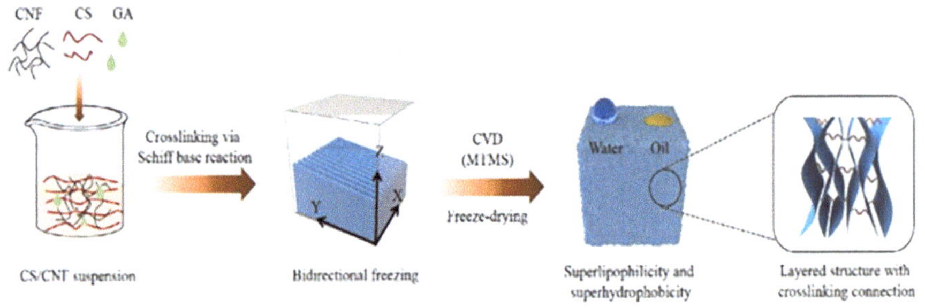

Fig. 3.29 Design approach for hydrophobic, anisotropic, and lightweight CS/CNF aerogels: reproduced with permission (Sun et al. 2022)

(2022) The aerogels prepared from CS and CNF modified by chemical vapor deposition of methyltrimethoxysilane (MTMS) had a water contact angle of 146°, making them extremely hydrophobic, and exhibited excellent oil–water separation capability. Chemical vapor deposition is among the most popular techniques, but it has the disadvantages of complicated operation, high energy consumption, and uneven hydrophobic modification (Fig. 3.29).

The dip-coating method offers a cost-effective and efficient means of applying hydrophobic coatings to substrates, utilizing intermolecular interactions. This technique ensures uniform coating deposition, making it suitable for applications in anti-corrosion, self-cleaning, and oil/water separation technologies. Hydrophobic cellulose aerogels are obtained by immersing the cellulose aerogel in an impregnation solution for several minutes to a few hours and then removing it into a constant temperature drying oven for drying. Chhajed et al. (2019) NFC/PVA aerogels for oil–water separation were prepared by physical crosslinking of NFC and PVA. To make the composite aerogel superhydrophobic, the aerogel was impregnated with stearic acid chloride (SAC) solution to obtain SAC-coupled aerogel with superhydrophobic and lipophilic properties, which had an angle of contact of 159° with water and almost 0° with oil. However, the dip-coating process may lead to pore collapse and reduction of particular surface region while drying, and most of the prepared cellulose aerogels have weak mechanical properties.

Acid hydrolysis silane modification refers to the hydrolysis of organosilanes to generate more reactive silanol groups under acidic conditions and slow down the condensation rate so that they can interact with cellulose's hydroxyl groups, which enhances the aerogel's porosity and mechanical characteristics in addition to helping to modify it hydrophobically. Superabsorbent, low-cost, and reusable cellulose NC aerogels can be prepared by hydrophobic modification of waste paper fibers with acid hydrolysis of two silylating reagents as raw materials. High-performance cellulose aerogels with excellent oil–water separation ability can also be made by freeze-casting polyvinyl alcohol and NC aqueous solutions. However, since the modification process of acid-hydrolyzed silane modification needs to be completed by acid hydrolysis, a few issues still need to be resolved for large-scale applications.

Porous Sponges

Cellulose-derived 3D porous materials, including aerogels, composite membranes, and hydrogels, are often limited by their insufficient mechanical strength, which restricts their practical applications. In recent years, sponge-like materials have gained increasing attention due to their advantageous properties, such as high surface area and tunable surface chemistry. These characteristics make them ideal candidates for a wide range of functional nanomaterials. Among these, cellulose-based sponges have emerged as a promising class of materials. However, a significant challenge remains: cellulose itself is hydrophilic, which hinders its oil selectivity. To overcome this limitation, cellulose sponges must undergo hydrophobic modification to enhance their ability to selectively separate oil from water, thereby improving their utility in oil/water separation applications and other oil-related processes. To address this issue, Ahuja et al. (2021) prepared sponges from waste jute bags containing 60–65% cellulose by the sol-gel method as shown in Fig. 3.30. Surface modification of cellulose sponges was achieved using tetraethyl orthosilicate (TEOS)/hexadecyltrimethoxysilane (HTMS) to impart superhydrophobicity. Ethyl cellulose (EC), a hydrophobic cellulose derivative, was further modified by altering its residual -OH groups to enhance its water-repellent and oil-selective properties, improving its suitability for oil/water separation applications. Lu et al. (2017) prepared a superhydrophobic magnetic EC sponge that could be reused up to 50 times from EC. Fe_3O_4 nanoparticles were combined with EC to provide the sponge more surface roughness and magnetic qualities.

Spongy materials have great potential for application in several fields due to their unique physical and chemical properties and have become a new trend in functional nano-3D materials, but they still have drawbacks such as insufficient physical strength. NC sponges show good compression energy absorption in compression stress–strain curves, but brittle materials are prone to fracture during compression. In addition, the physical strength of cellulose sponges was not appreciably increased by the addition of CNF.

3.4.4 *Flocculants for Suspended Particles*

Flocculants are essential agents in water treatment processes, used to aggregate colloidal substances and particulate matter suspended in liquids, forming larger flocs that can easily settle out of suspension. Among various types of flocculants, biopolymer-based versions are particularly notable. These flocculants are composed of long-chain molecules produced either by living organisms or from bio-derived monomers. Compared to traditional inorganic and organic coagulants, biopolymer-based flocculants offer several advantages. These include abundant natural availability, ease of modification, and environmental benefits, such as low toxicity or non-toxicity and biodegradability. Additionally, they typically require lower dosages to achieve effective flocculation and are adaptable to a wide range of pH conditions.

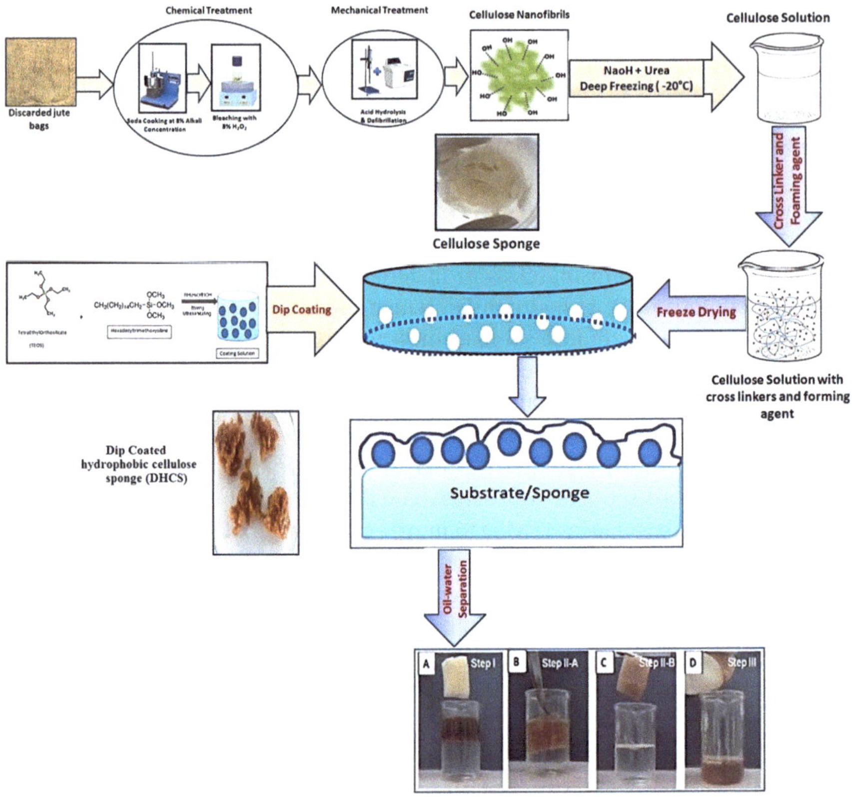

Fig. 3.30 Overall approach to developing 3D superhydrophobic sponges from discarded jute sacks: reproduced with permission (Ahuja et al. 2021)

Their high flocculation efficiency makes them suitable for various industrial and environmental applications, particularly those seeking sustainable and eco-friendly solutions in water treatment. Table 3.4 outlines the primary benefits and drawbacks of the most widely used flocculants.

Because of its high specific surface area, high tensile strength, high reactivity, and low density, NC has caught the interest of numerous researchers. NC can effectively promote the flocculation of suspended particles by neutralizing the charge on the surface of the suspended particles or forming bridges between the particles. Yu et al. (2016) developed a simple method to isolate carboxylated cellulose nanocrystal (CNC) from microcrystalline cellulose using carboxyl functional groups. And looked into how CNC's morphology, microstructure, and thermal stability were affected by response time for separation. The rod-shaped CNC was roughly 15–20 nm in width and 200–250 nm in length, which were obtained by one-step hydrolysis of MCC by citric acid/hydrochloric acid ($C_6H_8O_7$/HCl). Extraction of CNC for 4 h exhibited the highest carboxyl functional group-containing fraction, resulting in an absolute potential value as high as 46.63 mV. In addition, the CNC obtained by this isolation

Table 3.4 Overview of the primary benefits and drawbacks of the most widely used flocculants: reproduced with permission (Blockx et al. 2021)

Flocculant type	Main benefits	Main drawbacks
Metal salts	Low cost Ideal for media with a high ionic strength	High dosage needed Biomass contamination
Synthetic polymers	Low dosage needed	Biomass contamination Not ideal for media with a high ionic strength
Bio-based polymers	Low dosage needed No biomass contamination	Not ideal for media with a high ionic strength
Existing CNC-based flocculants	Low dosage needed Ideal for media with a high ionic strength Regulation of the parameters of the response flocculant	Synthetic grafts

method can be served as an adsorbent for cationic dyes (e.g., methylene blue) and a highly efficient flocculating agent, with a 99.5% removal of turbidity from kaolin suspension.

In recent years, the cationic modification of cellulose nanocrystals as flocculants in wastewater treatment has also gained extensive research. Clay suspensions can be flocculated by cationic hairy cellulose nanocrystals throughout a broad dosage range. Morantes et al. (2019) used cellulose nanocrystals and epoxypropyltrimethylammonium chloride to react. They created a cationic flocculant based on cellulose using this technique to treat 0.25 weight percent SiO_2 suspensions. The findings indicated that turbidity in the suspension was reduced by 99.7% when the concentration of flocculant was only 2 ppm. Showed advantages over commercially available homologous flocculants and offered a sustainable and naturally occurring substitute. Blockx et al. (2021) created the first glycine betaine cellulose nanocrystal flocculant with a grafted cationic point charge (i.e., betaine) grafted CNC, which effectively flocculated kaolin (a model system for wastewater treatment), freshwater microalgae Chlorella vulgaris, and marine microalgae nannochlorophyllum. Glycine betaine was successfully grafted onto CNC using a one-pot reaction, using p-methyl chloride to activate the esterification reaction, having substitution levels between 0.078 ± 0.003 and 0.152 ± 0.002. The degree of substitution of the biopolymer-based flocculant was dependent on reaction conditions. At a dosage of 2 mg/L, it effectively flocculated kaolin (0.5 g/L), comparable to commercial polyacrylamide-based alternatives. Additionally, it demonstrated improved flocculation efficiency for freshwater and marine microalgae (300 mg/L dry biomass), highlighting its versatility in treating both inorganic and organic contaminants.

In addition to nanofibrillated cellulose, Teixeira et al. (2024) used cationic nanofibrillated cellulose (cNFC) as a novel bio-based flocculant, as shown in Fig. 3.31. Different water characteristics were used as coagulants and flocculants to evaluate two cNFC samples with varying charge densities. The use of cNFC as a flocculant

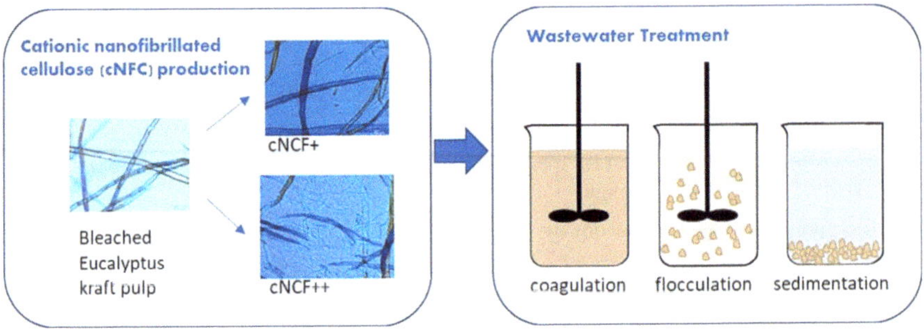

Fig. 3.31 Production of cationic protofibrillated cellulose nanoparticles and treatment of wastewater with different characteristics: reproduced with permission (Teixeira et al. 2024)

effectively reduced turbidity in medium to high dissolved organic carbon (DOC) waters, with removal efficiencies ranging from 66.0 to 85.7%, without increasing DOC levels. Additionally, cNFC demonstrated DOC removal capabilities (22.1–65.5%) in domestic wastewater, highlighting its potential for sustainable water treatment solutions, and high-density charged cNFC showed greater removal capacity than the commercial coagulant $FeCl_3$ at lower doses.

Jiang et al. (2024) developed an injection-driven filtration system, shown in Fig. 3.32, which combines a syringe to remove ultrafine suspended particulates with a totally bio-based, biodegradable nanofiber hydrogel membrane for portable, environmentally friendly water filtration. During operation, the injection-driven filtration system achieved a flux of 90.6 $g/cm^2{\cdot}h$, which, while operating under the same conditions, is 7.2 times more than that of a commercial polycarbonate ultrafiltration membrane. This work offers a sustainable water purifying technology that is easy to use and reasonably priced.

NC has been investigated for use as a flocculant due to its unique physicochemical properties that effectively promote flocculation of suspended particles, but its high hydrophilicity, ease of agglomeration, surface hydrogen bonding network, and potential environmental concerns during modification have limited its application. In addition, its commercial application faces challenges such as modification and functionalization, cost and production efficiency, and toxicity and safety. Future studies ought to concentrate on the development of green and environmentally friendly novel preparation methods and clean production modes to promote NC from the laboratory to industrial or large-scale production.

3.4.5 Photocatalytic Materials for Hazardous Pollutants

Photocatalytic materials mainly generate electron–hole pairs through semiconductors under light, thus triggering a series of redox reactions to realize the degradation

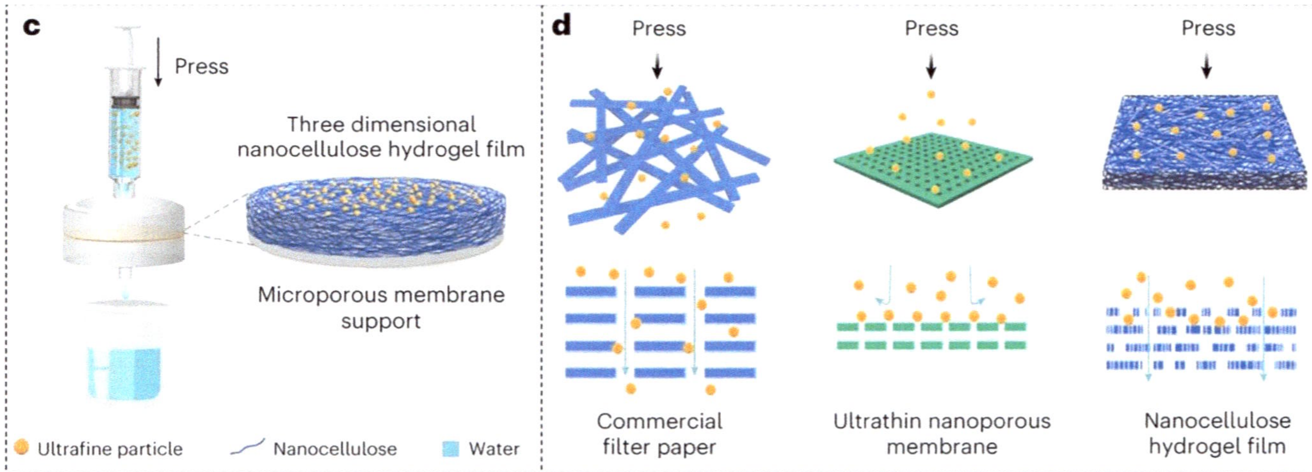

Fig. 3.32 **c** Diagrammatic illustration of SS-contaminated water filtration and purification using BNHF; **d** Schematic demonstrating the advantages of purifying SS-contaminated water using hydrogel membranes in BNHF compared to commercial filter paper and nanopore membranes: reproduced with permission (Jiang et al. 2024)

of pollutants and energy conversion. Common photocatalytic materials include titanium dioxide (TiO_2), zinc oxide (ZnO), cadmium sulfide (CdS), and so on. These materials have found widespread application in air purification, water treatment, and other fields due to their good photocatalytic performance and stability. However, the aggregation of these materials is difficult to avoid and not easy to recover. NC has significant potential for application in photocatalytic materials, especially in the degradation of harmful pollutants. Although NC itself has restricted photocatalytic activity in the UV and visible ranges, its photocatalytic performance can be significantly enhanced by the addition of semiconducting materials such as metal oxides, e.g., TiO_2, and ZnO.

NC is often utilized to support metal oxide nanoparticles due to its high specific surface area, high mechanical strength, and superhydrophilicity. Voisin et al. (2021) reported that CNC-TiO_2 hybrids exhibited improved photocatalytic properties in suspension over pure anatase TiO_2, attributed to their enhanced dispersibility and colloidal stability, demonstrating the potential of CNC-based hybrids in photocatalysis. Additionally, CNC's interfacial characteristics were maintained, allowing organic dyes in Pickering emulsions to degrade. It is interesting to note that these emulsions can be freeze-dried into multipurpose aerogels for effective organic and aqueous dye degradation. Liu et al. (2021) synthesized anatase TiO_2/NC composites, which exhibited excellent morphological properties and maintained anatase crystal structure. The composites achieved 99.72% methyl orange degradation within 30 min, with no significant loss of activity after five cycles. The degradation efficiency was directly correlated with the amount of CNF used, highlighting the potential of these composites as effective photocatalysts for organic dye removal in wastewater. Otherwise, Nahi et al. (2020) described the environmentally friendly production of visible-light-driven photocatalysts composed of zinc oxide/nanocellulose composites (ZnO/NC) for the photodegradation of enrofloxacin (EF), an antibiotic frequently used in veterinary medicine to treat animals with bacterial infections. The ZnO/NC composite exhibited enhanced photocatalytic degradation of EF under visible light due to an increased band gap and efficient electron–hole interactions. The optimal conditions were a pH of 5 and a ZnO/NC dosage of 0.20 g/L, achieving a 97% degradation efficiency within 120 min.

In addition, as the aerogel has a highly developed pore structure, which allows the incident light to be reflected and scattered several times within the pores, the utilization of light by the catalyst material can be enhanced. Su et al. (2017) prepared Cu_2O-loaded cellulose-based aerogels (Cu_2O/CBA) by depositing nanoparticles of Cu_2O onto cellulose via in-situ synthesis, which has a three-dimensional macroporous structure containing abundant catalytically active sites (Fig. 3.33). In the reaction, the aerogel with a multistage pore structure could act as a microreactor and effectively prevent the Cu_2O nanoparticles from polymerizing. Subsequently, the performance of the composite aerogel catalyst was evaluated using methylene blue (MB) as a model. It was found that the photodegradation rate of MB was catalyzed by the Cu_2O/CBA composite with an efficiency of 95.79%, which was much higher than the catalytic degradation degree of pure Cu_2O (73.59%). Saeed et al. (2019) prepared composite photocatalytic aerogel CuS/CBA by loading visible light responsive CuS

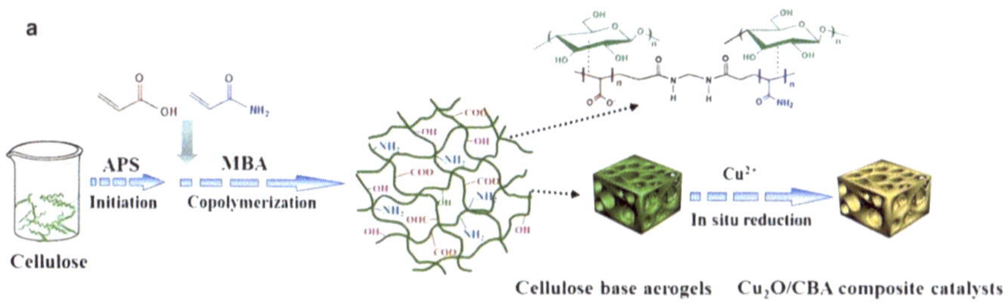

Fig. 3.33 Cu_2O/CBA preparation process: reproduced with permission (Su et al. 2017)

photocatalysts in cellulose-based aerogel containing three-dimensional macroporous structure through an in-situ deposition strategy. The degradation rate of CuS/CBA for methylene blue (MB) reached 94.1% under visible light, while for CuS, the degradation degree was only 67.4%. The CuS/CBA composite catalyst's degradation rate in MB could still be kept above 80% after five cycles, indicating its excellent catalyst stability. Although NC can significantly enhance the photocatalytic performance as a support material for metal oxide nanoparticles. However, NC/TiO_2 and NC/ZnO photocatalysts have poor absorption of visible light due to their wide energy gaps, and it is also difficult to separate them from the reaction mixtures due to their nonmagnetic properties. Therefore, researchers are exploring new ways to improve the performance of these composites, such as covering the cellulose fibers with inert UV-absorbing materials to protect them from UV fading.

3.4.6 *Water Disinfection for Pathogenic Microorganisms*

Water disinfection technology plays a crucial role in modern water treatment. Traditional water disinfection technologies such as chlorination, ozone disinfection, ultraviolet disinfection, and emerging water disinfection technologies such as photocatalytic disinfection, electrochemical disinfection, ultrasonic and magnetically enhanced disinfection mostly suffer from the problems of by-product generation, technological limitations, economics, and environmental impacts. NC is considered an emerging renewable, sustainable, and cost-effective material for microbial removal, and with unique physicochemical properties, such as high specific surface area, hydrophilic surface chemistry, and good biocompatibility, it exhibits significant potential in water disinfection applications for pathogenic microorganisms.

NC by itself has a limited inhibitory effect on microorganisms, so it is usually necessary to enhance its antimicrobial properties by chemical modification or surface modification. Kokol et al. (2023a) showed that NC with quaternized containing hydrophobic and amino groups increased hydrophobic and electrostatic interactions with nonpolar bacterial cells, resulting in excellent bacteriostatic and bactericidal activities against Gram-positive (G^+) Staphylococcus aureus and Gram-negative (G^-) *E. coli* (up to 100% inhibition at $\geq$ 8 mg/L), and the nanofibers were further prepared impregnated with quaternized (qCNF) and/or aminohydrophobic (aCNF) cellulose nanofibers. Water permeability and the bacterial suspension flow rate were decreased by the aCNF level, and the bacterial reduction was higher (>5.6) logarithmic even after the second to the fourth filtration cycle at comparatively high and consistent filtration rates (>50 L/h·bar·m^2), approaching the effectiveness of industrial polymer filters in removing germs. Kokol et al. (2023b) further prepared aldehydes (CNF/CNC-ox) by periodate oxidation of CNF/CNC and further functionalized hexamethylenediamine/HMDA by the Schiff base reaction, changing their morphological and physicochemical characteristics to prepare miniature CNF/CNC-ox-HMDA as shown in Fig. 3.34. The results showed that CNC-ox-HMDA increased the interaction with the bacterial cell wall and accessibility to its cell membrane and exhibited the highest antimicrobial efficiency but was potentially toxic, whereas the high concentration-dependent antimicrobial activity of CNF/CNC-ox had low toxicity and could be considered safe. Since CNF/CNC-ox-HMDA possesses several targeted ways of operation, it may be used for the microbiological treatment of household and commercial wastewater and for reusing it.

In addition, NC can be endowed with antimicrobial properties by introducing surface modifications such as antimicrobial polymers, metal nanoparticles (e.g., silver and gold), and metal oxide nanoparticles (copper oxide, zinc oxide, titanium dioxide, magnesium oxide nanoparticles, and copper/titanium dioxide nanocomposites). Kupnik et al. (2024) incorporation of 14 mg/g of avocado seed extract uniformly

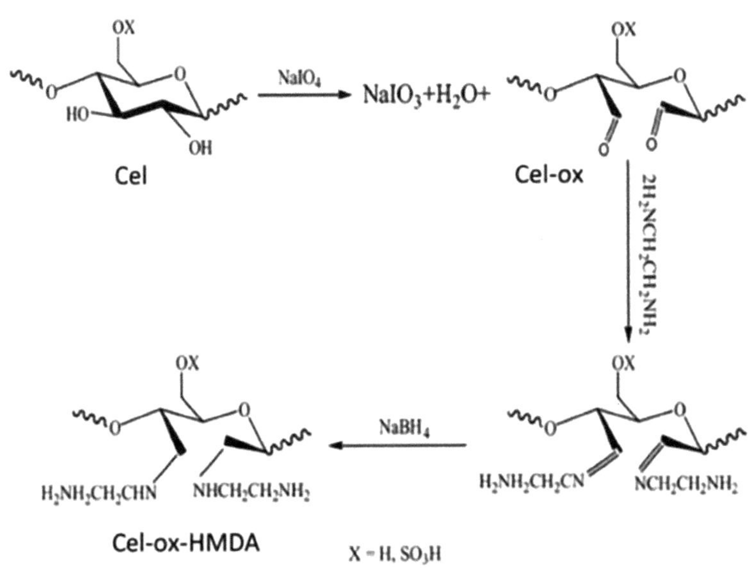

Fig. 3.34 CNF/CNC modification steps for natural (CNF/CNC), oxidized (CNF/CNC-ox), and HMDA functionalized (CNF/CNC-OX-HMDA): reproduced with permission (2023b)

into bacterial NC film reduced G$^-$ *E. coli* and G$^+$ *Staphylococcus aureus* by 100%, with a logarithmic reduction of up to 2.56, and 100% bactericidal efficacy after 24 h of exposure. Syukri et al. (2024) BC acetate-based nanocomposite membranes, enhanced with ferrous oxide nanoparticles derived from pineapple peel bio-waste, demonstrated effective removal of Gram-negative bacteria and anionic dyes from water. The addition of ferrous-iron oxide improved the membrane's performance, highlighting their potential for environmental water treatment applications. Ge et al. (2024) A dual network hydrogel synthesized by polymerization of the presence of quaternized cellulose and acrylic acid (AA) nanocellulose protofibrils (QCNF) and Fe^{3+} showed that the AA/QCNF/Fe^{3+} (PQFe) hydrogel showed strong antibacterial action against *Staphylococcus aureus* and *E. coli*. These modifications not only improved the antimicrobial activity of NC but also enhanced its application in water treatment.

Nanofibrillar cellulose-based membrane filters utilize a size exclusion mechanism to effectively capture most aquatic bacteria with diameters greater than 0.2 μm. The development of multilayer nanofiber microfiltration systems by impregnating ultrafine NC into electrospun polyacrylonitrile nanofiber scaffolds showed complete retention of several types of bacteria. In addition, the development of cationic CNF compound foams by low-temperature templating technique also showed excellent anti-*E. coli* activity. Furthermore, the application of NC in combination with other materials (e.g., chitosan, MOFs, etc.) further improves their antimicrobial and filtration properties. For example, cellulose acetate (CA) and polysulfone (PSF) composite nanofibers combined with ZnO exhibited high antimicrobial activity against *E. coli*. NC has significant advantages in water disinfection applications for pathogenic microorganisms. Its antimicrobial properties and filtration efficiency can be significantly improved by surface modification and compounding with other materials, providing a new solution for the provision of safe drinking water.

3.4.7 Antibiotic Adsorption

Antibiotics can be defined as chemotherapeutic agents that inhibit or suppress the growth of microorganisms. They can be divided into 15 sections based on different mechanisms of action: beta-lactams, carbapenems, aminoglycosides, and others. From the discovery of penicillin to the present day, antibiotics have played a crucial part in combating a variety of bacterial infections while helping to protect human health. However, bacterial resistance caused by the misuse of antibiotics has been ranked by the World Health Organization (WHO) as one of the greatest health threats facing mankind. Treatment methods for antibiotics include traditional methods for treating water, biological treatment processes, advanced oxidation, and adsorption. Conventional water treatment processes and biological therapy processes are ineffective in removing most antibiotics.

The application of NC for antibiotic adsorption is mainly characterized by its efficient adsorption properties and versatility. NC is an efficient adsorbent material because of its excellent pore structure and large specific surface area. Lu et al. (2021) prepared bamboo fiber nanocellulose composites (Fe(III)@CNFs) loaded with trivalent iron. The adsorbent could realize the simultaneous adsorption of three tetracycline antibiotics: tetracycline (TC), chlortetracycline (CTC), and oxytetracycline (OTC) over a wide pH range, with maximum adsorption amounts of 294.12, 232.56, and 500 mg/g, respectively. The potential mechanism of adsorption of tetracycline (TC) by Fe(III)@CNFs is shown in Fig. 3.35. After five cycles of adsorption experiments, the adsorption of tetracycline antibiotics remained above 92%, which has a good prospect for industrial application. Tetracycline can be adsorbed via four different adsorption mechanisms: surface complexation, hydrogen bonds, electrostatic reactions, and van der Waals forces. Tao et al. (2020) prepared cellulose nanocrystalline-graphene oxide composites (CNCs-GO) for levofloxacin hydrochloride (Levo-HCl) adsorption in water using a three-dimensional structure. The composites possessed a substantial specific surface area. Furthermore, strong electrostatic attraction was observed, and when the initial concentration of the pollutant was 10.0 mg/g, pH = 4, and the contact time was 4 h, the adsorption rate was up to 80.12%, and the adsorption isotherm was consistent with the three-parameter Sips model. Kim (2023) prepared a doped iron oxide (FeO) incorporated carbide nanofiber (IOC) composite material doped with iron oxide, which has the two roles of adsorption and photocatalytic degradation. The Act-IOC process, involving water vapor activation of carbide nanofibers, created surface carbon structures and active sites that efficiently adsorbed tetracycline (TC), ciprofloxacin (CIP), and lead (Pb). Fe_3O_4 nanoparticles in the material also promoted photocatalytic oxidation of TC and CIP, with the 4-h synthesized Act-IOC complex demonstrating the highest photodegradation activity. This easy-to-control synthesis method positions Act-IOC as a promising material for removing micropollutants from water.

NC aerogels have shown promising applications in water treatment. For example, they have been used to remove antibiotic contaminants such as ciprofloxacin and other emerging pollutants from water bodies. Wei et al. (2021) prepared bacterial cellulose (BC) aerogels using freeze-drying technology. After that, it was carbonized in a tube furnace to produce the bacterial carbon cellulose aerogel (BCCA). At 298 K, BCCA's highest adsorption of sulfamethoxazole (SMX), norfloxacin (NOR), and chloramphenicol (CAP) was 1264 mg/g, 1926 mg/g, and 525 mg/g, respectively. BCCA was able to adsorb a maximum of 525 mg/g of chloramphenicol (CAP), 1926 mg/g of norfloxacin (NOR), and 1264 mg/g of sulfamethoxazole (SMX). The application of NC for antibiotic adsorption has significant advantages, including efficient adsorption properties, good regeneration properties, a wide range of promising environmental applications, and the ability to enhance antimicrobial properties when combined with other materials. These properties make NC a promising material for addressing antibiotic contamination and providing new solutions.

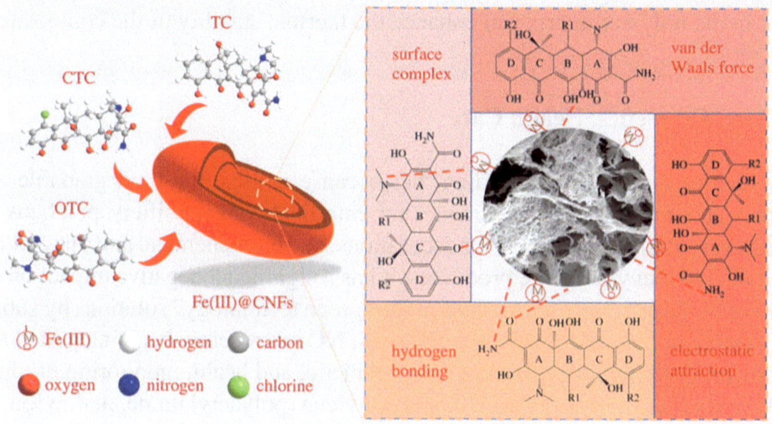

Fig. 3.35 Potential mechanism of adsorption of tetracycline (TC) by Fe(III)@CNFs: reproduced with permission (Lu et al. 2021)

3.4.8 Other Aspects

Paper Production

NC, with its high specific surface area, excellent mechanical strength, and tunable surface chemistry, brings unprecedented opportunities for innovation in traditional industries, especially in the paper industry, demonstrating a high degree of application potential. Due to the presence of a large number of hydroxyl groups on NC's active surface, which strengthens the fibers' hydrogen bond, NC can lead to an increase in the tensile strength, splitting index, and internal cohesion of paper. TEMPO-oxidized cellulose nanofibers can significantly improve the tensile and tear strength of paper for their high aspect ratio and the introduction of carboxyl groups on the surface.

Despite the promising applications of NC in papermaking, its commercialization on an industrial scale still faces several challenges. The main problems include high production costs, insufficient dispersion and compatibility, and high energy consumption. Its production cost can be reduced in several ways, such as optimizing the preparation process, enzymatic production, acid recovery, and equipment optimization. In addition, by modifying NC, its binding ability can be further improved to solve the problem of insufficient dispersibility and compatibility. The integration of chemically functionalized NC into diverse matrices of polymers, including thermosets, thermoplastics, and biopolymers, can enhance the thermal stability of the resulting polymer nanocomposites. Through chemical grafting reactions, the dispersion of NC in polymer matrices can be improved and can lead to an increase in the composites' thermal stability. In addition, numerous theoretical investigations have shown that by slowing down the rate of thermal deterioration, increasing the filler content in the polymer matrix can enhance the thermal stability of the composites.

Skin Care, Cosmetics, Health Care

The cosmetics industry is expanding, but because it uses non-biodegradable and non-renewable resources—like silicones for emollients or emulsifiers, parabens for preservatives, plastic microbeads for decorations, and even chemical and mineral UV filters added to many cosmetic products—it has a significant negative impact on the environment. Thus, it is necessary to provide "green technology" solutions by substituting natural elements for manufactured ones. NC has recently been integrated as a green alternative biopolymer in skin care, cosmetic, and health monitoring products to replace synthetic polymers such as polyethylene, polyacrylamide, and nylon. To date, NC has been utilized as an anti-wrinkle agent, compatibilizer, moisturizer, and rheology agent or thickener. They have been used in cleansing formulas to eliminate debris, lower sebum, and external contaminants, and manage microbial communities and skin odor.

NC hydrogels have a lot of potential for use in many cosmetic and medical applications. When used for face masks and basic cosmetics, they give off a viscous but non-sticky texture. The elasticity and mechanical strength (in the wet state) of bacterial NC make it especially impervious to handling, easy to apply, and able to adhere perfectly to the skin, enhancing the penetration of actives into the skin. Biocellulose (BNC) sheet masks, known for their moisture retention and skin compatibility, have gained popularity in the skincare market. Notable examples include products from BOWIL-Biotech Ltd (CELMAT), Elizabeth Arden (Intense Brightening Mask), DHC (Bio-Cellulose Mask), Lancôme (Blank Expert Whitening Mask), and CORYMER (Second Skin Mask with Aloe Vera). These masks showcase the diverse applications of BNC in cosmetic treatments aimed at enhancing skin health.

NC hydrogels maintain a high water content to keep wounds warm and wet, which is best for recovery. New wearable biosensors that track biomarker levels for illness diagnosis and health monitoring are also being developed using them. Due to their biocompatibility, solubility, and responsiveness, hydrogels have revolutionized drug delivery systems, allowing for sustained, controlled, targeted drug delivery by minimizing dosage. NC is also commonly utilized for a range of other medical applications, including precision drug design and drug delivery targeting cells, early diagnosis of diseases, and cardiovascular diseases. By modifying NC with carboxymethylation and silylation, the modified NC drug carriers can be applied to the selective distribution of anticancer drugs, which can improve the targeting and efficacy of drugs. Composite hydrogel membrane was prepared by polyacrylamide-bacterial NC, which can improve the imaging of superficial blood vessels and can be used for the early diagnosis of superficial vascular diseases. Nanofibrillar cellulose can also be used to prepare antithrombotic drug carriers with unique fiber morphology and biodegradability, which can effectively improve the solubility of the drug in the blood and ensure the effectiveness and stability of the drug.

Textile Industry

Textiles are inseparable from people's lives. To meet the demand for textiles in terms of quality and style, it is necessary to continuously increase the research and development of textile materials. NC has a large specific surface area and good mechanical properties, and its application in textile materials can give the textile materials more functionality, such as better moisture absorption and breathability, wear resistance, and antibacterial function. As cellulose has the characteristics of degradable, the application of NC in textile materials can reduce the pollution of textile materials in the production process, to provide support for the green development of the textile industry. NC is widely utilized to improve the mechanical characteristics of textiles due to its excellent mechanical strength and high specific surface area. For example, NC can improve the tensile strength, abrasion resistance, and tear strength of textiles. In addition, NC can be paired with other biodegradable materials such as polylactic acid (PLA) and polyhydroxybutyrate (PHB) to enhance the thermal and physical properties of the materials. NC performs well in the dyeing process, especially in dye fixation and prevention of dye loss. Permanent grafting of dye-labeled polysaccharides can be achieved through the use of polycarboxylic acid (PCA) as a post-treatment strategy, significantly improving dye fixation rates and reducing dye loss. NC can also stabilize natural dyes, simplify the dyeing process, and enhance the value of natural dyes. NC has also shown environmental benefits in textile surface treatment. For example, EcoaTEX has developed an environmentally friendly coating technology based on NC that saves water and electricity and eliminates harmful chemicals in the dyeing and finishing process.

NC has a lot of potential for use in the smart textiles industry. By adding NC to masks, illness prevention and environmental issues are addressed while biodegradability, renewability, and functional adaptability are provided. In addition, NC can improve the UV protection of cotton fabrics.

Foodstuffs

There are numerous uses for NC in the food industry. Because of its superior mechanical strength, gas barrier qualities, and biodegradability, NC is frequently utilized in food packaging materials. NC can be blended with other biodegradable polymers such as polylactic acid (PLA) and polybutylene succinate (PBS) to replace traditional petroleum-based plastics. Food packaging must be resistant to oxygen permeation since bacteria start to taint food and reduce its nutritional value when oxygen is present. Food products must be kept in an environment with low oxygen levels to extend their shelf life. To make a dense network that prevents gas permeation, NC can form hydrogen bonds with other biopolymers to form a dense network in the film used for packaging to provide a barrier to gases. Nanoclay and NC can be used as reinforcing materials in a polylactic acid (PLA) matrix to produce nanocomposite films for food packaging. The composite film's oxygen and water barrier is further improved by the addition of NC and nanoclay.

NC has good emulsifying and dispersing properties and is therefore often used as a food stabilizer and thickener. NC can also be used to stabilize Pickering emulsions, thus maintaining the stability of the emulsion in the absence of surfactants. This knowledge may be valuable in designing food products with novel rheological or stabilizing properties. For example, it is possible to create low-calorie or slow-digesting pastes, coatings, dressings, or sauces with new textures. In addition, NC has applications in a variety of areas such as nutraceuticals and functional food ingredients.

Energy, Electronics

NC is being utilized more and more in the production of electrochemical energy storage devices and high-performance energy storage devices such as solar cells, lithium-ion batteries, and supercapacitors. In addition, NC is used in electronic devices like actuators, sensors, displays, and sensing, mainly stemming from the evolution of conductive nanofibrillated cellulose sheets, which offer properties features such as excellent printability, foldability, heat stability, flexibility, and transparency. NC needs to be surface conductive to be suitable for most energy applications. For this reason, they are usually surface-modified using other conductive polymers, such as nanocarbon. In the last few years, there has been a boom in research aimed at developing methods for storing energy with high performance. Cellulose-based battery electrode materials are inexpensive, flexible, and perform well. NC, as a component, has also shown great promise in the construction of ultrathin solar cells. It is anticipated that these ongoing developments will result in the creation of next-generation energy storage technologies. Optoelectronics is a new field of NC applications. For example, NC can be used to prepare moisture-responsive nanopaper, electrochemical gas sensors, flexible organic light-emitting diodes, and environmentally friendly piezoelectric composite films.

Packaging

Conventional plastic packaging is largely non-biodegradable and has limited recyclability. Nanocellulose (NC) enhances polymer composites, reducing oxygen and water vapor permeability, and improving mechanical and barrier properties for packaging. Surface-modified NC increases hydrophobicity, while functionalized NC disperses bioactive agents, offering antimicrobial and antioxidant benefits. These attributes make NC-based hybrid composites promising for sustainable food packaging.

3.4.9 Summary and Outlook

In both academic and industrial domains, the creation of effective and eco-friendly water treatment materials has gained significant attention. Faced with one of the major challenges facing mankind in the twenty-first century—access to clean water—numerous innovative approaches have been proposed to cope with it. Among many nanomaterials, NC stands out due to its outstanding performance and efficiency and is considered a sustainable material for wastewater treatment. Compared to inorganic nanomaterials, NC exhibits greater mechanical strength, chemical stability, diverse surface chemistry, lower production cost, and operational safety, showing its great potential as a sustainable wastewater treatment material. This essay is centered on the employment of NC in various fields such as heavy metal ion adsorption, organic dye wastewater treatment, oil–water separation, flocculants, photocatalytic materials, pathogenic microbial water disinfection, and antibiotic adsorption. Numerous researchers have explored the preparation of various forms of cellulose-based nanomaterials (e.g., membranes, sponges, hydrogels, aerogels, and nanoparticles) and their application to wastewater treatment, and bacterial cellulose, cellulose fibers, and cellulose nanocrystals, which are of different morphologies, have been effectively used in the water treatment industry, especially in membrane filtration and adsorbent development as well as adsorbent to remove heavy metal ions, contaminants such as dyes, drugs, pesticides, and fertilizers. Functionalized nanocellulose (NC), modified with inorganic materials, polymers, or carbon-based compounds, enables the large-scale production of low-cost, renewable cellulose composites. NC's unique properties make it suitable for diverse water treatment applications, such as adsorption, photocatalysis, disinfection, and filtration.

Although NC shows great potential in wastewater treatment, it still faces some challenges such as cost, regeneration conditions, and environmental friendliness. Future research needs to further optimize its preparation process and functionalized modification to improve its efficiency and application scope in wastewater treatment. The research direction should focus on functionalized modification, composite with other materials, application of membrane technology, development of novel materials, and optimization of structural design.

3.5 Perspectives Summary and Outlook

3.5.1 Summary of Nanocellulose for Sustainable Water Purification

Nanocellulose (NC) has become a new material with great potential for development in the field of water treatment due to its wide range of sources, green environmental protection, functionalization, and excellent water treatment performance. Although there are still challenges in large-scale production, stability of complex water systems, and material recycling, with the continuous advancement of science and technology and the deepening of interdisciplinary research, NC-based materials are expected to achieve large-scale applications in the future, promoting the development of global water treatment technology toward high efficiency, low cost, and sustainability, and providing green and environmentally friendly new solutions to solve the global water crisis. Although NC has advantages such as: (i) wide availability and green renewability, (ii) excellent physical and chemical properties, (iii) multifunctionality and composite material design, and (iv) environmental friendliness and degradability. Although nanocellulose exhibits excellent performance in the field of water treatment, its practical application still faces some challenges, mainly including the following aspects: (i) Large-scale preparation and cost control. At present, the preparation processes of NC (such as mechanical, chemical, and enzymatic) have not fully achieved low-cost and high-efficiency industrial production, especially in large-scale applications. How to reduce costs remains a key challenge. In addition, the uniformity and stability of the material also need to be further optimized. (ii) Performance stability in complex water systems. In practical water treatment applications, the composition of the water system is complex, containing various ions, particles, and organic pollutants, which may affect the adsorption performance and durability of NC. Therefore, how to improve the stability and selectivity of NC-based materials in complex environments is an urgent problem to be solved. (iii) Material recycling and reuse. Although NC-based materials have good adsorption capacity, achieving efficient recovery and regeneration of materials remains a challenge. Especially for water systems with low pollutant concentrations, how to improve material reuse efficiency and reduce operating costs is an important research direction. (iv) Standardization of composite material design and application. Currently, there are various types of composite material designs for NC, but there is a lack of unified standards and evaluation systems in practical applications. There are differences in the evaluation indicators and methods of material properties in different studies, which affected the promotion and application of technology.

3.5.2 Outlook of Nanocellulose for Sustainable Water Purification

Based on the unique advantages and existing research progress of NC in the field of water treatment, future development directions can focus on the following aspects.

(i) Design of high-performance NC-based composite materials

Future research should further develop functionalized NC composite materials, with a focus on: (a) Intelligent response materials: Combined with temperature pH, developing water treatment materials with adaptive and intelligent response functions to external stimuli such as electromagnetic fields. (b) Multifunctional composite materials: By combining with nano metal oxides, carbon nanotubes, graphene, and other materials, they achieve synergistic removal of multiple pollutants, photocatalytic degradation, and efficient filtration functions.

(ii) Improve the practical application performance of materials

To further improve the stability, selectivity, and anti-pollution performance of NC-based materials for complex water systems, ensuring the long-term effectiveness of the materials in practical working conditions. In addition, optimizing the regeneration and recycling processes of materials, reducing operating costs, and achieving sustainable applications.

(iii) Promote large-scale production and application demonstration

In the future, we should strengthen the research and development of large-scale preparation technology for NC, reduce production costs, and promote the commercial application of materials. At the same time, carry out application demonstration projects to verify the actual effects of NC-based materials in drinking water treatment, industrial wastewater treatment, and environmental remediation.

(iv) Establish a standardized evaluation system

Develop performance evaluation standards and safety assessment systems for NC-based water treatment materials to ensure their reliability, controllability, and ecological safety, laying the foundation for industrial applications.

(v) Interdisciplinary Collaboration and Technological Innovation

Future research requires interdisciplinary collaboration, combining the advantages of materials science, environmental engineering, chemical engineering, and other fields to promote innovative development of NC-based water treatment technology. In addition, by combining technologies such as big data and artificial intelligence, intelligent monitoring and optimization control of water treatment processes can be achieved.

References

Abe K, Iwamoto S, Yano H (2007) Obtaining cellulose nanofibers with a uniform width of 15 nm from wood. Biomacromol 8(10):3276–3278

Abol-Fotouh D, Hassan MA, Shokry H, Roig A, Azab MS, Kashyout AEHB (2020) Bacterial nanocellulose from agro-industrial wastes: low-cost and enhanced production by Komagataeibacter saccharivorans MD1. Sci Rep 10(1):3491

Abou-Zeid RE, Salama A, Al-Ahmed ZA, Awwad NS, Youssef MA (2020) Carboxylated cellulose nanofibers as a novel efficient adsorbent for water purification. Cellul Chem Technol 54(3–4):237–245

Ahuja D, Dhiman S, Rattan G, Monga S, Singhal S, Kaushik A (2021) Superhydrophobic modification of cellulose sponge fabricated from discarded jute bags for oil water separation. J Environ Chem Eng 9(2):105063

Almeida IF, Pereira T, Silva NHCS, Gomes FP, Silvestre AJD, Freire CSR, Sousa Lobo JM, Costa PC (2014) Bacterial cellulose membranes as drug delivery systems: an in vivo skin compatibility study. Eur J Pharm Biopharm 86(3):332–336

Alsaiari NS, Katubi KM, Alzahrani FM, Amari A, Osman H, Ben Rebah F, Tahoon MA (2021) Synthesis, characterization and application of polypyrrole functionalized nanocellulose for the removal of Cr(VI) from aqueous solution. Polymers 13(21):3691

Alsbaiee A, Smith BJ, Xiao L, Ling Y, Helbling DE, Dichtel WR (2016) Rapid removal of organic micropollutants from water by a porous β-cyclodextrin polymer. Nature 529(7585):190–194

Amiralian N, Mustapic M, Hossain MSA, Wang C, Konarova M, Tang J, Na J, Khan A, Rowan A (2020) Magnetic nanocellulose: a potential material for removal of dye from water. J Hazard Mater 394:122571

Banavath HN, Bhardwaj NK, Ray AK (2011) A comparative study of the effect of refining on charge of various pulps. Biores Technol 102(6):4544–4551

Belton PS, Tanner SF, Cartier N, Chanzy H (1989) High-resolution solid-state 13C nuclear magnetic resonance spectroscopy of Tunicin, an animal cellulose. Macromolecules 22(4):1615–1617

Blockx J, Verfaillie A, Deschaume O, Bartic C, Muylaert K, Thielemans W (2021) Glycine betaine grafted nanocellulose as an effective and bio-based cationic nanocellulose flocculant for wastewater treatment and microalgal harvesting. Nanoscale Adv 3(14):4133–4144

Börjesson M, Sahlin K, Bernin D, Westman G (2018) Increased thermal stability of nanocellulose composites by functionalization of the sulfate groups on cellulose nanocrystals with azetidinium ions. J Appl Polym Sci 135(10):45963

Brar SK, Wangoo N, Sharma RK (2020) Enhanced and selective adsorption of cationic dyes using novel biocompatible self-assembled peptide fibrils. J Environ Manage 255:109804

Cacicedo ML, Castro MC, Servetas I, Bosnea L, Boura K, Tsafrakidou P, Dima A, Terpou A, Koutinas A, Castro GR (2016) Progress in bacterial cellulose matrices for biotechnological applications. Biores Technol 213:172–180

Camparotto NG, Neves TD, Pereira GMP, Mastelaro VR, Prediger P (2023) Synergistic effect in simultaneous adsorption of cationic and anionic emerging contaminants by chitosan aerogels containing nanocellulose-modified tannic acid. J Water Process Eng 53:103749

Cao Y, Chen X, Li Y, Wang Y, Yu H, Li Z, Zhou Y (2023) Regulating and controlling the microstructure of nanocellulose aerogels by varying the intensity of hydrogen bonds. ACS Sustain Chem Eng 11(4):1581–1590

Castro C, Zuluaga R, Álvarez C, Putaux J-L, Caro G, Rojas OJ, Mondragon I, Gañán P (2012) Bacterial cellulose produced by a new acid-resistant strain of Gluconacetobacter genus. Carbohyd Polym 89(4):1033–1037

Chanthathamrongsiri N, Petchsomrit A, Leelakanok N, Siranonthana N, Sirirak T (2021) The comparison of the properties of nanocellulose isolated from colonial and solitary marine tunicates. Heliyon 7(8):1–7

Chávez-Guerrero L, Vazquez-Rodriguez S, Salinas-Montelongo JA, Roman-Quirino LE, García-Gómez NA (2019) Preparation of all-cellulose composites with optical transparency using the banana pseudostem as a raw material. Cellulose 26(6):3777–3786

Chávez-Guerrero L, Silva-Mendoza J, Toxqui-Terán A, Vega-Becerra OE, Salinas-Montelongo JA, Pérez-Camacho O (2021) Direct observation of endoglucanase fibrillation and rapid thickness identification of cellulose nanoplatelets using constructive interference. Carbohyd Polym 254:117463

Chen L, Wang Q, Hirth K, Baez C, Agarwal UP, Zhu JY (2015) Tailoring the yield and characteristics of wood cellulose nanocrystals (CNC) using concentrated acid hydrolysis. Cellulose 22(3):1753–1762

Chen YW, Hasanulbasori MA, Chiat PF, Lee HV (2019) Pyrus pyrifolia fruit peel as sustainable source for spherical and porous network based nanocellulose synthesis via one-pot hydrolysis system. Int J Biol Macromol 123:1305–1319

Chen XQ, Song ZH, Yuan BN, Li XJ, Li SP, Nguyen TT, Guo MH, Guo ZH (2022) Fluorescent carbon dots crosslinked cellulose Nanofibril/Chitosan interpenetrating hydrogel system for sensitive detection and efficient adsorption of Cu (II) and Cr (VI). Chem Eng J 430:133514

Chhajed M, Yadav C, Agrawal AK, Maji PK (2019) Esterified superhydrophobic nanofibrillated cellulose based aerogel for oil spill treatment. Carbohydr Polym 226:115286

Chu Y, Sun Y, Wu W, Xiao H (2020) Dispersion properties of nanocellulose: a review. Carbohyd Polym 250:116892
Cortesi R, Quattrucci A, Esposito E, Mazzaglia A, Balestra GM (2017) Natural antimicrobials in spray-dried microparticles based on cellulose derivatives as potential eco-compatible agrochemicals. J Plant Dis Prot 124(3):269–278. https://doi.org/10.1007/s41348-016-0055-7
Cosgrove DJ (2014) Re-constructing our models of cellulose and primary cell wall assembly. Curr Opin Plant Biol 22:122–131
Cova TF, Murtinho D, Pais AACC, Valente AJM (2018) Combining cellulose and cyclodextrins: fascinating designs for materials and pharmaceutics. Front Chem 6:271
Da Silva DJ, Rosa DS (2022) Chromium removal capability, water resistance and mechanical behavior of foams based on cellulose nanofibrils with citric acid. Polymer 253:125023
Das R, Lindström T, Sharma PR, Chi K, Hsiao BS (2022) Nanocellulose for sustainable water purification. Chem Rev 122(9):8936–9031
Ding Q, Xu Z, Yang H (2024) Recent research progress on hairy cellulose nanocrystals: preparation and applications. Cellulose 31(17):10087–10113
Fareez IM, Lim SM, Zulkefli NAA, Mishra RK, Ramasamy K (2018) Cellulose derivatives enhanced stability of alginate-based beads loaded with Lactobacillus plantarum LAB12 against low pH, high temperature and prolonged storage. Probiotics Antimicrob Proteins 10(3):543–557. https://doi.org/10.1007/s12602-0179284-8
Filson PB, Dawson-Andoh BE, Schwegler-Berry D (2009) Enzymatic-mediated production of cellulose nanocrystals from recycled pulp. Green Chem 11(11):1808–1814
Ge WJ, Cao S, Yu H, Wang XH (2024) Tough polyacrylic acid hydrogels with stable swelling and active functionalities enabled by quaternized cellulose nanofibrils and iron ions for absorbent pad interlayers. Carbohydr Polym 345:122491
George J, Ramana KV, Bawa AS (2011) Bacterial cellulose nanocrystals exhibiting high thermal stability and their polymer nanocomposites. Int J Biol Macromol 48(1):50–57
Gibson LJ (2012) The hierarchical structure and mechanics of plant materials. J R Soc Interface 9(76):2749–2766
Gilmore SP, Henske JK, O'Malley MA (2015) Driving biomass breakdown through engineered cellulosomes. Bioengineered 6(4):204–208
Gómez HC, Serpa A, Velásquez-Cock J, Gañán P, Castro C, Vélez L, Zuluaga R (2016) Vegetable nanocellulose in food science: a review. Food Hydrocoll 57:178–186
Gong SJ, Zhang XY, Chen SW, Ma YM, Mao ZP, Ji BL (2024) Micro/nanostructure and directional channel of recyclable carboxylated aerogels from waste cotton for wastewater treatment. Cellulose 31(10):6403–6419
Gopakumar DA, Pasquini D, Henrique MA, de Morais LC, Grohens Y, Thomas S (2017) Meldrum's acid modified cellulose nanofiber-based polyvinylidene fluoride microfiltration membrane for dye water treatment and nanoparticle removal. ACS Sustain Chem Eng 5(2):2026–2033
Han PJ, Yu B, Chen YF, Nie GW, Wang K, Sun XY, Li X, Shao WL, Liu F, He JX (2024) Electrospun polyacrylonitrile-fluorinated polyurethane/polysulfone nanofiber membranes for oil-water separation. ACS Applied Nano Materials 7(4):4336–4348
Helbert W, Nishiyama Y, Okano T, Sugiyama J (1998) Molecular imaging of Halocynthia papillosa cellulose. J Struct Biol 124(1):42–50
Hoeng F, Denneulin A, Neuman C, Bras J (2015) Charge density modification of carboxylated cellulose nanocrystals for stable silver nanoparticles suspension preparation. J Nanopart Res 17(6):244
Hu W, Chen S, Yang J, Li Z, Wang H (2014) Functionalized bacterial cellulose derivatives and nanocomposites. Carbohyd Polym 101:1043–1060
Huang X, Dognani G, Hadi P, Yang M, Job AE, Hsiao BS (2020) Cationic dialdehyde nanocellulose from sugarcane bagasse for efficient chromium(VI) removal. ACS Sustain Chem Eng 8(12):4734–4744
Ibrahim SS, Abou-Elseoud WS, Elbehery HH, Hassan ML (2022) Chitosan-cellulose nanoencapsulation systems for enhancing the insecticidal activity of citronella essential oil against the cotton

leafworm Spodoptera littoralis. Ind Crops Prod 184:115089. https://doi.org/10.1016/j.indcrop.2022.115089
Isogai A (2021) Emerging nanocellulose technologies: recent developments. Adv Mater 33(28):2000630
Isogai A (2022) TEMPO-catalyzed oxidation of polysaccharides. Polym J 54(4):387–402
Isogai A, Saito T, Fukuzumi H (2011) TEMPO-oxidized cellulose nanofibers. Nanoscale 3(1):71–85
Iwamoto S, Kai W, Isogai A, Iwata T (2009) Elastic modulus of single cellulose microfibrils from tunicate measured by atomic force microscopy. Biomacromol 10(9):2571–2576
Jiang F, Dinh DM, Hsieh YL (2017) Adsorption and desorption of cationic malachite green dye on cellulose nanofibril aerogels. Carbohyd Polym 173:286–294
Jiang MH, Jing CY, Lei CX, Han X, Wu YQ, Ling SJ, Zhang YY, Li Q, Yu HP, Liu SX, Li J, Chen WS, Yu GH (2024) A bio-based nanofibre hydrogel filter for sustainable water purification. Nat Sustain 7(2):168–178
Jin L, Li W, Xu Q, Sun Q (2015) Amino-functionalized nanocrystalline cellulose as an adsorbent for anionic dyes. Cellulose 22(4):2443–2456
Jing S, Wu L, Siciliano AP, Chen C, Li T, Hu L (2023) The critical roles of water in the processing, structure, and properties of nanocellulose. ACS Nano 17(22):22196–22226
Jozala AF, de Lencastre-Novaes LC, Lopes AM, de Carvalho Santos-Ebinuma V, Mazzola PG, Pessoa A Jr, Grotto D, Gerenutti M, Chaud MV (2016) Bacterial nanocellulose production and application: a 10-year overview. Appl Microbiol Biotechnol 100(5):2063–2072
Kaur M, Pal J (2023) Evaluation of efficiency of Wheat straw nanocellulose as nanoadsorbent for the removal of divalent copper, lead and zinc from aqueous solution. Carbohydr Polym Technol Appl 6:100350
Kim CY, Kim SH, An HR, Park JI, Jang Y, Seo J, Kim H, Son B, Jeong Y, Jeong B, Lee HU (2023) Iron oxide incorporated carbide nanofiber composites for removal of organic pollutants and heavy metals from water. Ceram Int 49(11):17984–17992
Kokol V, Novak S, Kononenko V, Kos M, Vivod V, Gunde-Cimerman N, Drobne D (2023a) Antibacterial and degradation properties of dialdehyded and aminohexamethylated nanocelluloses. Carbohyd Polym 311:120603
Kokol V, Kos M, Vivod V, Gunde-Cimerman N (2023b) Cationised fibre-based cellulose multi-layer membranes for sterile and high-flow bacteria retention and inactivation. Membranes 13(3):284
Kumari P, Raza W, Meena A (2021) Lemongrass derived cellulose nanofibers for controlled release of curcumin and its mechanism of action. Ind Crops Prod 173:114099
Kupnik K, Primozic M, Kokol V, Knez Z, Leitgeb M (2024) Antibacterial *Komagataeibacter hansenii* nanocellulose membranes with avocado seed bioactive compounds. Cellulose 31(7):4305–4327
Li Y, Yu S, Chen P, Rojas R, Hajian A, Berglund L (2017) Cellulose nanofibers enable paraffin encapsulation and the formation of stable thermal regulation nanocomposites. Nano Energy 34:541–548. https://doi.org/10.1016/j.nanoen.2017.03.010
Li X, Liu Y, Yu Y, Chen W, Liu Y, Yu H (2019) Nanoformulations of quercetin and cellulose nanofibers as healthcare supplements with sustained antioxidant activity. Carbohydr Polym 207:160–168. https://doi.org/10.1016/j.carbpol.2018.11.084
Li WQ, Zhang LP, Hu D, Yang R, Zhang J, Guan Y, Lv FX, Gao H (2021) A mesoporous nanocellulose/sodium alginate/carboxymethyl-chitosan gel beads for efficient adsorption of Cu^{2+} and Pb^{2+}. Int J Biol Macromol 187:922–930
Li Z, Shen K, Liu M, Hu M, Yao SJ, Liu Q, Fan CG, Wu XX (2024) Ultralight poly(vinyl alcohol)/cattail fiber cellulose aerogels as green and recyclable organic liquid captors. ACS Appl Polym Mater 6(22):13690–13700
Lin N, Dufresne A (2014) Nanocellulose in biomedicine: current status and future prospect. Eur Polymer J 59:302–325
Liu GQ, Pan XJ, Li J, Li C, Ji CL (2021) Facile preparation and characterization of anatase TiO_2/nanocellulose composite for photocatalytic degradation of methyl orange. J Saudi Chem Soc 25(12)

Liu E, Meng H, Zhang H, Liang H, Wu Z (2024a) Nanocellulose, a renewable biobased functional/reinforcement additive for composite coating materials: a review. ACS Appl Polym Mater 6(1):28–48

Liu S, Ji W, Wu T, He Y, Huang Y, Yu Y, Yu W (2024b) Unraveling the intricate interaction: bonding mechanism between tannic acid and wood fibers. ACS Sustain Chem Eng 12(10):4224–4235

Lou YL, Xi JF, Jiang S, Chu YL, Deng W, Bian HY, Xu ZY, Xiao HN, Wu WB (2024) Nanocellulose-based membranes with pH- and temperature-responsive pore size for selective separation. Int J Biol Macromol 263:130176

Lu YQ, Wang Y, Liu LJ, Yuan WZ (2017) Environmental-friendly and magnetic/silanized ethyl cellulose sponges as effective and recyclable oil-absorption materials. Carbohyd Polym 173:422–430

Lu LX, Liu M, Chen Y, Luo Y (2021) Effective removal of tetracycline antibiotics from wastewater using practically applicable iron(III)-loaded cellulose nanofibres. R Soc Open Sci 8(8):210336

Ma XT, Lou Y, Chen XB, Shi Z, Xu Y (2019) Multifunctional flexible composite aerogels constructed through in-situ growth of metal-organic framework nanoparticles on bacterial cellulose. Chem Eng J 356:227–235

Ma CX, Yi LJ, Yang J, Tao JH, Li JF (2020) Nanocellulose-organic montmorillonite nanocomposite adsorbent for diuron removal from aqueous solution: optimization using response surface methodology. RSC Adv 10(51):30734–30745

Maleki O, Khaledabad MA, Amiri S, Asl AK, Makouie S (2020) Microencapsulation of Lactobacillus rhamnosus ATCC 7469 in whey protein isolate-crystalline nanocellulose-inulin composite enhanced gastrointestinal survivability. Lwt 126:109224. https://doi.org/10.1016/j.lwt.2020.109224

Matthysse AG, Deschet K, Williams M, Marry M, White AR, Smith WC (2004) A functional cellulose synthase from ascidian epidermis. Proc Natl Acad Sci USA 101(4):986–991

Mautner A, Kobkeatthawin T, Bismarck A (2017) Efficient continuous removal of nitrates from water with cationic cellulose nanopaper membranes. Resour Effic Technol 3(1):22–28

Meng X, Pu C, Cui P, Xiao Z (2020) Preparation, thermal and sensitivity properties of nano-sized spherical nitrocellulose composite crystal. Propellants Explos Pyrotech 45(8):1194–1203

Mishra D, Khare P, Singh DK, Yadav V, Luqman S, Kumar PA, Shanker K (2021) Synthesis of Ocimum extract encapsulated cellulose nanofiber/chitosan composite for improved antioxidant and antibacterial activities. Carbohydr Polym Technol Appl 2:100152. https://doi.org/10.1016/j.carpta.2021.100152

Miyamoto H, Schnupf U, Brady JW (2014) Water structuring over the hydrophobic surface of cellulose. J Agric Food Chem 62(46):11017–11023

Miyashiro D, Hamano R, Umemura K (2020) A review of applications using mixed materials of cellulose, nanocellulose and carbon nanotubes. Nanomaterials 10(2):186

Morantes D, Muñoz E, Kam D, Shoseyov O (2019) Highly charged cellulose nanocrystals applied as a water treatment flocculant. Nanomaterials 9(2):272

Naderi A (2017) Nanofibrillated cellulose: properties reinvestigated. Cellulose 24(5):1933–1945

Nagarajan KJ, Ramanujam NR, Sanjay MR, Siengchin S, Surya Rajan B, Sathick Basha K, Madhu P, Raghav GR (2021) A comprehensive review on cellulose nanocrystals and cellulose nanofibers: pretreatment, preparation, and characterization. Polym Compos 42(4):1588–1630

Nahi J, Radhakrishnan A, Beena B (2020) Green synthesis of zinc oxide incorporated nanocellulose with visible light photocatalytic activity and application for the removal of antibiotic enrofloxacin from aqueousmedia. In: International conference on management and recycling of metallurgical wastes (MetWaste-2020). Varanasi, India, pp 583–589

Nikou M, Samadi-Maybodi A, Yasrebi K, Sedighi-Pashaki E (2021) Simultaneous monitoring of the adsorption process of two organophosphorus pesticides by employing GO/ZIF-8 composite as an adsorbent. Environ Technol Innov 23:101590

Noremylia MB, Hassan MZ, Ismail Z (2022) Recent advancement in isolation, processing, characterization and applications of emerging nanocellulose: a review. Int J Biol Macromol 206:954–976

Oneda F, Ré MI (2003) The effect of formulation variables on the dissolution and physical properties of spray-dried microspheres containing organic salts
Pääkkö M, Ankerfors M, Kosonen H, Nykänen A, Ahola S, Österberg M, Ruokolainen J, Laine J, Larsson PT, Ikkala O, Lindström T (2007) Enzymatic hydrolysis combined with mechanical shearing and high-pressure homogenization for nanoscale cellulose fibrils and strong gels. Biomacromol 8(6):1934–1941
Park SH, Shin SS, Park CH, Jeon S, Gwon J, Lee SY, Kim SJ, Kim HJ, Lee JH (2020) Poly(acryloyl hydrazide)-grafted cellulose nanocrystal adsorbents with an excellent Cr(VI) adsorption capacity. J Hazard Mater 394:122512
Pelissari FM, Sobral PJDA, Menegalli FC (2014) Isolation and characterization of cellulose nanofibers from banana peels. Cellulose 21(1):417–432
Portela R, Leal CR, Almeida PL, Sobral RG (2019) Bacterial cellulose: a versatile biopolymer for wound dressing applications. Microb Biotechnol 12(4):586–610
Putro JN, Santoso SP, Ismadji S, Ju YH (2017) Investigation of heavy metal adsorption in binary system by nanocrystalline cellulose—bentonite nanocomposite: improvement on extended Langmuir isotherm model. Microporous Mesoporous Mater 246:166–177
Qin FM, Fang ZQ, Zhou J, Sun C, Chen KH, Ding ZX, Li GH, Qiu XQ (2019) Efficient removal of Cu^{2+} in water by carboxymethylated cellulose nanofibrils: performance and mechanism. Biomacromol 20(12):4466–4475
Rafieian F, Jonoobi M, Yu QL (2019) A novel nanocomposite membrane containing modified cellulose nanocrystals for copper ion removal and dye adsorption from water. Cellulose 26(5):3359–3373
Ram B, Chauhan GS (2018) New spherical nanocellulose and thiol-based adsorbent for rapid and selective removal of mercuric ions. Chem Eng J 331:587–596
Rungraung N, Jain S, Mitbumrung W, Khomein P, Suphantharika M, McClements DJ, Winuprasith T (2022) Controlling the in vitro gastrointestinal digestion of emulsified lipids by encapsulation within nanocellulose-fortified alginate beads. Food Struct 32:100266. https://doi.org/10.1016/j.foostr.2022.100266
Rytioja J, Hildén K, Yuzon J, Hatakka A, de Vries RP, Mäkelä MR (2014) Plant-polysaccharide-degrading enzymes from Basidiomycetes. Microbiol Mol Biol Rev 78(4):614–649
Saeed RMY, Bano Z, Sun JZ, Wang FY, Ullah N, Wang QQ (2019) CuS-functionalized cellulose based aerogel as biocatalyst for removal of organic dye. J Appl Polym Sci 136(15):47404
Saito T, Nishiyama Y, Putaux J-L, Vignon M, Isogai A (2006) Homogeneous suspensions of individualized microfibrils from TEMPO-catalyzed oxidation of native cellulose. Biomacromol 7(6):1687–1691
Saito T, Kimura S, Nishiyama Y, Isogai A (2007) Cellulose nanofibers prepared by TEMPO-mediated oxidation of native cellulose. Biomacromol 8(8):2485–2491
Saito T, Kuramae R, Wohlert J, Berglund LA, Isogai A (2013) An ultrastrong nanofibrillar biomaterial: the strength of single cellulose nanofibrils revealed via sonication-induced fragmentation. Biomacromol 14(1):248–253
Sajab MS, Chia CH, Chan CH, Zakaria S, Kaco H, Chook SW, Chin SX, Noor AM (2016) Bifunctional graphene oxide-cellulose nanofibril aerogel loaded with Fe(III) for the removal of cationic dye via simultaneous adsorption and Fenton oxidation. RSC Adv 6(24):19819–19825
Sarafraz-Yazdi A, Rokhian T, Amiri A, Ghaemi F (2015) Carbon nanofibers decorated with magnetic nanoparticles as a new sorbent for the magnetic solid phase extraction of selected polycyclic aromatic hydrocarbons from water samples. New J Chem 39(7):5621–5627
Sehaqui H, Mautner A, Perez de Larraya U, Pfenninger N, Tingaut P, Zimmermann T (2016) Cationic cellulose nanofibers from waste pulp residues and their nitrate, fluoride, sulphate and phosphate adsorption properties. Carbohydr Polym 135:334–340
Sharma C, Bhardwaj NK (2019) Bacterial nanocellulose: Present status, biomedical applications and future perspectives. Mater Sci Eng C Mater Biol Appl 104:109963
Sharma PR, Sharma SK, Lindström T, Hsiao BS (2020) Nanocellulose-enabled membranes for water purification: perspectives. Adv Sustain Syst 4(5):1900114

Sheng K, Zhang G, Kong X, Wang J, Mu W, Wang Y (2021) Encapsulation and characterisation of grape seed proanthocyanidin extract using sodium alginate and different cellulose derivatives. Int J Food Sci Technol 56(12):6420–6430. https://doi.org/10.1111/ijfs.15299

Stephen M, Catherine N, Brenda M, Andrew K, Leslie P, Corrine G (2011) Oxolane-2,5-dione modified electrospun cellulose nanofibers for heavy metals adsorption. J Hazard Mater 192(2):922–927

Su XP, Liao Q, Liu L, Meng RJ, Qian ZQ, Gao HY, Yao JM (2017) Cu_2O nanoparticle-functionalized cellulose-based aerogel as high-performance visible-light photocatalyst. Cellulose 24(2):1017–1029

Sun B, Hou Q, Liu Z, Ni Y (2015) Sodium periodate oxidation of cellulose nanocrystal and its application as a paper wet strength additive. Cellulose 22(2):1135–1146

Sun Y, Ye WJ, Xi JF, Chu YL, Xiao HN, Wu WB (2022) Ultralight and shape recovery bio-based aerogel for oil-water separation. J Environ Chem Eng 10(6):108822

Suopajärvi T, Liimatainen H, Hormi O, Niinimäki J (2013) Coagulation–flocculation treatment of municipal wastewater based on anionized nanocelluloses. Chem Eng J 231:59–67

Syukri D, Suryanto H, Kurniawan F, Hari PD, Fiana RM (2024) Rini, bacterial reduction in river water using nanocellulose membrane from pineapple biomass with ferrous-ferric oxide reinforcement. Glob J Environ Sci Manag (GJESM) 10(2):643–656

Tang JT, Song Y, Zhao FP, Spinney S, Bernardes JD, Tam KC (2019) Compressible cellulose nanofibril (CNF) based aerogels produced via a bio-inspired strategy for heavy metal ion and dye removal. Carbohyd Polym 208:404–412

Tao JH, Yang J, Ma CX, Li JF, Du KQ, Wei Z, Chen CZ, Wang ZY, Zhao C, Deng XY (2020) Cellulose nanocrystals/graphene oxide composite for the adsorption and removal of levofloxacin hydrochloride antibiotic from aqueous solution. R Soc Open Sci 7(10):200090

Tavakolian M, Jafari SM, van de Ven TGM (2020) A review on surface-functionalized cellulosic nanostructures as biocompatible antibacterial materials. Nano Micro Lett 12(1):73

Teixeira MR, Ismail A, Medronho B, Alves L, Pedrosa JFS, Ferreira PJT, Sousa VS, da Costa AMR (2024) Nanofibrillated cationic cellulose derivatives as flocculants for domestic wastewater treatment. J Water Process Eng 58:104817

Thomas P, Duolikun T, Rumjit NP, Moosavi S, Lai CW, Bin Johan MR, Fen LB (2020) Comprehensive review on nanocellulose: Recent developments, challenges and future prospects. J Mech Behav Biomed Mater 110:103884

Thomas B, Raj MC, B AK, H RM, Joy J, Moores A, Drisko GL, Sanchez C (2018) Nanocellulose, a versatile green platform: from biosources to materials and their applications. Chem Rev 118(24):11575–11625

Tonoli GHD, Teixeira EM, Corrêa AC, Marconcini JM, Caixeta LA, Pereira-da-Silva MA, Mattoso LHC (2012) Cellulose micro/nanofibres from Eucalyptus kraft pulp: preparation and properties. Carbohyd Polym 89(1):80–88

Varanasi S, Batchelor WJ (2013) Rapid preparation of cellulose nanofibre sheet. Cellulose 20(1):211–215

Vijayan JG, Prabhu TN, Asthana N (2023) Optimization of rhodamine-b dye adsorption using *Moringa olifera* seed pod based NC, phosphorylated NC and ZIF-8/pNC composites. J Inorg Organomet Polym Mater 33(12):3716–3731

Voisin H, Falourd X, Rivard C, Capron I (2021) Versatile nanocellulose-anatase TiO_2 hybrid nanoparticles in Pickering emulsions for the photocatalytic degradation of organic and aqueous dyes. JCIS Open 3:100014

Wang HY, Chen YH, Mo MQ, Dorsel PKP, Wu CJ (2023) Visualized adsorption and enhanced photocatalytic removal of Cr6+by carbon dots-incorporated fluorescent nanocellulose aerogels. Int J Biol Macromol 253:127206

Wei MD, Zheng HB, Zeng TN, Yang J, Fang XB, Zhang C (2021) Porous carbon aerogel derived from bacterial cellulose with prominent potential for efficient removal of antibiotics from the aquatic matrix. Water Sci Technol 84(8):1896–1907

Xu Y, Atrens A, Stokes JR (2020) A review of nanocrystalline cellulose suspensions: rheology, liquid crystal ordering and colloidal phase behaviour. Adv Coll Interface Sci 275:102076

Yang H, Tejado A, Alam N, Antal M, van de Ven TGM (2012) Films prepared from electrosterically stabilized nanocrystalline cellulose. Langmuir 28(20):7834–7842

Yang H, Sheikhi A, van de Ven TGM (2016) Reusable green aerogels from cross-linked hairy nanocrystalline cellulose and modified chitosan for dye removal. Langmuir 32(45):11771–11779

Yang Y, Liu S, Zhang G, Hu Z, Jin L, Quan B, Hao X, Wu H, Ji X, Lu X (2023) Cellulose nanofiber encapsulated polyethylene glycol phase change composites containing AIE-gen for monitoring leak process. Compos Part A: Appl Sci Manuf 167:107452. https://doi.org/10.1016/j.compositesa.2023.107452

Yao C, Wang F, Cai Z, Wang X (2016) Aldehyde-functionalized porous nanocellulose for effective removal of heavy metal ions from aqueous solutions. RSC Adv 6(95):92648–92654

Yin JA, Huang GH, Xiao HN, Chen N, An CJ, Chao TC, Feng RF, Read S (2023a) Bioinspired and dual-functional nanocellulose aerogels for water disinfection and heavy metal removal. Nano Today 51:101918

Yin ZZ, Li M, Li ZH, Deng YT, Xue MS, Chen YH, Ou JF, Lei S, Luo YD, Xie C (2023b) A harsh environment resistant robust $Co(OH)_2$@stearic acid nanocellulose-based membrane for oil-water separation and wastewater purification. J Environ Manag 342:118127

Yu HY, Zhang DZ, Lu FF, Yao JM (2016) New Approach for single-step extraction of carboxylated cellulose nanocrystals for their use as adsorbents and flocculants. ACS Sustain Chem Eng 4(5):2632–2643

Yu H, Hong HJ, Kim SM, Ko HC, Jeong HS (2020) Mechanically enhanced graphene oxide/carboxymethyl cellulose nanofibril composite fiber as a scalable adsorbent for heavy metal removal. Carbohyd Polym 240:116348

Zhan CB, Sharma PR, He HR, Sharma SK, McCauley-Pearl A, Wang RF, Hsiao BS (2020) Rice husk based nanocellulose scaffolds for highly efficient removal of heavy metal ions from contaminated water. Environ Sci Water Res Technol 6(11):3080–3090

Zhang F, Shen R, Li N, Yang X, Lin D (2023) Nanocellulose: An amazing nanomaterial with diverse applications in food science. Carbohyd Polym 304:120497

Zhang H, Yang C, Zhou W, Luan Q, Li W, Deng Q, Dong X, Tang H, Huang F (2018) A pH-responsive gel macrosphere based on sodium alginate and cellulose nanofiber for potential intestinal delivery of probiotics. ACS Sustain Chem Eng 6(11):13924–13931. https://doi.org/10.1021/acssuschemeng.8b02237

Zhao X, Zhang X, Zhang E, Shen L, Liu W, Ouyang Y (2024) J Jilin Univ (Sci Ed)/Jilin Daxue Xuebao (Lixue Ban), 62. https://doi.org/10.13413/j.cnki.jdxblxb.2024079

Zhu HX, Jia SR, Wan T, Jia YY, Yang HJ, Li J, Yan L, Zhong C (2011) Biosynthesis of spherical Fe_3O_4/bacterial cellulose nanocomposites as adsorbents for heavy metal ions. Carbohyd Polym 86(4):1558–1564

Chapter 4
Fate of Plastics in the Ocean Environment: The Plastic Odyssey

Jayjit Majumdar, Bipradeep Mondal, Surajit Ghosh, Sukhendu Dey, Shubhajit Saha, Ahamadul Hoque Mandal, Subhrajyoti Roy, Jayanta Kumar Biswas, and Debojyoti Moulick

Abstract Plastic pollution has emerged as a critical threat to marine ecosystems. Plastic debris enters the aquatic environment primarily through land-based sources, including rivers and coastal regions, but also from marine-based activities and atmospheric transport. Microplastics are subject to weathering processes, including photodegradation, oxidation, and mechanical fragmentation, resulting in further size reduction and changes in physical properties. These transformations impact microplastics' ability to adsorb pollutants and their buoyancy, influencing their distribution within the water column. Additionally, the colonization of microplastics by microorganisms forms a biofilm, altering their physical and chemical characteristics. Microplastic poses significant risks to marine life. These tiny particles are readily ingested by many organisms, leading to harmful effects such as digestive issues, malnutrition, and mortality. Furthermore, microplastics act as vectors for persistent organic pollutants, bioaccumulating in the food chain and potentially impacting human health. Ocean currents contribute to the accumulation of plastic debris in

---

J. Majumdar (✉) · B. Mondal · J. K. Biswas
Department of Ecological Studies, University of Kalyani, Kalyani, West Bengal, India
e-mail: jayjitmajumdar@gmail.com

B. Mondal
Department of Environmental Sciences, Central University of Jharkhand, Ranchi, Jharkhand, India

S. Ghosh · S. Saha · A. H. Mandal
Fisheries and Ecotoxicology Research Laboratory, Department of Zoology, The University of Burdwan, Bardhaman, West Bengal, India

S. Dey
Department of Environmental Science, The University of Burdwan, Bardhaman, West Bengal, India

S. Roy
Department of Zoology, University of Gour Banga, Malda, West Bengal, India

D. Moulick (✉)
Department of Environmental Science, University of Kalyani, Kalyani, West Bengal, India
e-mail: drubha31@gmail.com

S. K. Panda and V. P.M. (eds.), *Bionanocomposites from Marine Origin*, Advanced Structured Materials 247, https://doi.org/10.1007/978-3-032-10749-7_4

gyres, forming vast areas of marine pollution. Additionally, plastic particles can sink to the seabed, affecting deep-sea ecosystems. The impacts of plastic pollution on marine biodiversity and ecosystem function are profound, necessitating urgent global action to reduce plastic production and consumption, improve waste management, and develop effective clean-up strategies. This study investigates the complex pathways and consequences of plastics in the ocean. Understanding the transformation of microplastics is essential for assessing their environmental impact and developing effective mitigation strategies. This research contributes to the growing body of knowledge on microplastic pollution, providing insights into the complex processes governing their fate in the ocean and informing future studies on their ecological consequences. Addressing this complex issue requires interdisciplinary research, international cooperation, and public awareness to mitigate the long-term consequences for marine life and human well-being.

Keywords Plastic waste · Microplastics · Marine environment · Transport · Transformation · Degradation · Biofilm · Ecotoxicological impact

Abbreviations

MNP	Micro and nano plastic
MP	Microplastic
MMT	Million metric tonnes
PPE	Personal protective equipment
PC	Polycarbonate
PCB	Polychlorinated biphenyl
PAH	Polycyclic aromatic hydrocarbons
PE	Polyethylene
PP	Polypropylene
PPA	Polypthalamide
PS	Polystyrene
PVC	Polyvinyl chloride
SUP	Single-use plastics

4.1 Introduction

Plastic pollution has become a significant as well as critical environmental concern in the current Anthropocene epoch, characterized by humanity's substantial influence on Earth's ecosystems (Rangel-Buitrago et al. 2023). Plastic pollution in the marine environment manifested as a critical challenge that poses momentous threats to aquatic ecosystems, human health, and economic stability. The scale of the problem is staggering; it is estimated that annual production of plastics exceeds 300 million tons, with around 150 million tons currently residing in the oceans (Tyllianakis and Ferrini 2021). Plastic products' extensive production, consumption, and disposal have precipitated a global crisis. In the current era, known as the Anthropocene, the cumulative weight of plastic produced surpasses the cumulative mass of all living land and sea animals (Elhacham et al. 2020). As of 2017, the cumulative production of all plastics reached 8300 million metric tonnes (MMT), comprising 6800 MMT of plastic resins, 500 MMT of additives, and 1000 MMT of PPA (polyphthalamide) fibers. About 76% (6300 MMT) had already been discarded as waste. Of this waste, 12% was incinerated, and 9% was recycled, with only 10% of the recycled material being reprocessed more than once (Geyer et al. 2017). Rivers transport between 0.41 and 4.0 MMT/year (Lebreton et al. 2017; Schmidt et al. 2017) Of this was improperly handled waste into the global oceans. This accumulation of plastic waste, which has increased by a factor of 25 from 1960 to 2000, now constitutes 60–80% of all marine debris and can account for nearly 100% of floating debris (Tyllianakis & Ferrini 2021). The perseverance of plastics in the marine environment and their tendency to fragment into microplastics exacerbates the issue, leading to widespread ecological consequences (Schmidt et al. 2017). The United Nations designated it a "planetary crisis" in 2019 (MacLeod et al. 2021; Schmidt et al. 2024).

The sources of marine plastic pollution are multifaceted, primarily stemming from land-based activities. Research indicates that approximately 80% of marine plastic debris is of terrestrial sources, including improper waste management and industrial runoff (Yap et al. 2022). Rivers play a pivotal function in delivering plastic waste to the oceans, with studies revealing that over 1000 rivers account for 80% of global riverine plastic emissions (Meijer et al. 2021). This transboundary nature of plastic pollution complicates mitigation efforts, as it requires coordinated action across multiple jurisdictions and stakeholders (Zhou and Luo 2024). Moreover, the economic drivers behind plastic production and consumption, such as the agricultural and fishing sectors, contribute significantly to the problem, with practices like plastic mulching and abandoned fishing gear being key contributors (Abalansa et al. 2020). Plastic pollution in the marine environment is significantly influenced by domestic usage of plastic products, which contributes to the vast amounts of plastic waste entering the marine environments. A primary source of this pollution stems from single-use plastics (SUPs), which have gained widespread popularity due to their convenience and low cost. Items such as plastic bags, food packaging, and straws are often discarded improperly, leading to their eventual entry into waterways and, subsequently, the ocean (Canning-Clode et al. 2020). The COVID-19 pandemic has exacerbated this issue, as the increased use of personal protective equipment (PPE) and other SUP items has resulted in a surge of plastic waste, further complicating waste management efforts and increasing marine litter (Canning-Clode et al. 2020).

Suboptimal waste management systems play a crucial role in the transition of domestic plastic waste into marine environments. Many coastal cities, particularly in developing nations, struggle with inefficient waste disposal practices, contributing to significant amounts of plastic waste being washed into rivers and oceans during rainfall or flooding (Chukwuone et al. 2022). Studies have shown that land-based sources make up for approximately 80% of marine plastic pollution, with rivers acting as conduits for transporting this waste from inland areas to the sea (Zhang et al. 2020; Emmerik et al. 2019). Moreover, consumer behaviours and societal attitudes toward plastic usage significantly influence the plastic waste load. Research indicates that a lack of awareness regarding the environmental ramifications of plastic pollution leads to continued reliance on disposable plastic products (Chukwuone et al. 2022). Efforts to educate the public about sustainable practices and the importance of reducing plastic consumption are essential in fostering a culture of responsibility and environmental stewardship. This serves to illustrate the significance of addressing waste management practices at the domestic level to mitigate the flow of plastics into marine ecosystems. By promoting recycling and encouraging the use of substitute materials, it is possible to reduce the domestic sources of marine plastic pollution and protect marine biodiversity (Shi 2023).

Marine plastic pollution has profound ecological repercussions. Marine organisms, from the smallest zooplankton to large marine mammals, are increasingly exposed to plastic debris, which can lead to ingestion, entanglement, and increased susceptibility to diseases (Focardi et al. 2022). Microplastics present a distinct hazard due to their capacity to vector harmful pollutants and infiltrate the food chain, with subsequent implications for human health (Meaza et al. 2021). The leachates from plastics have been shown to impair the growth and oxygen production of crucial marine microorganisms, such as *Prochlorococcus marinus*, which are vital for oceanic primary production (Tetu et al. 2019). This disruption of marine microbiomes can have cascading effects on marine food webs and overall ecosystem health (Focardi et al. 2022).

The socioeconomic ramifications of marine plastic pollution are equally alarming. Annual economic losses to marine ecosystems due to plastic pollution are estimated to exceed $13 billion, with significant impacts on fisheries, tourism, and coastal communities (Tetu et al. 2019). The tourism industry suffers from the aesthetic and environmental degradation caused by plastic litter, leading to increased shoreline cleaning efforts that further strain local economies (Meijer et al. 2021). In regions like Southeast Asia, where countries such as Indonesia have been identified as major contributors to plastic waste, the challenge is compounded by inadequate waste management infrastructure and policies (Ma'ruf 2019). The neoliberal economic framework, which prioritizes short-term gains over environmental sustainability, has been identified as a core factor perpetuating this crisis (Saavedra 2024).

Addressing marine plastic pollution requires a comprehensive and multifaceted approach. Strategies should prioritize the curtailment of plastic production and consumption, the enhancement of waste management practices, and the advancement of recycling initiatives (Zhao 2023). The European Union's commitment to a sustainable blue circular economy exemplifies a proactive strategy aimed at tackling marine litter through enhanced recyclability and responsible plastic use (Pestana 2023). Furthermore, international cooperation is essential, as the global nature of plastic pollution necessitates collaborative governance frameworks that can effectively manage its transboundary impacts (Zhou and Luo 2024). This chapter aims to analyze the ecological impact of plastic waste on marine organisms, to ascertain significant research voids, and to provide guidance for future scientific inquiry, thereby improving our comprehension of the extensive ecological consequences of plastic pollution within marine environments.

4.2 The Scientific Basis of Plastic Pollution

4.2.1 Ocean-Based Sources of Plastic Pollution

The problems of ocean plastic originated as a prominent environmental issue in 1975 when a fishing fleet sank, releasing 135,400 tons of fishing equipment made up of plastic and 23,600 tons of packaging material, made with synthetic materials, into marine ecosystems. This incident earned widespread recognition for this highly noticeable form of ecological contamination (Cawthorn 1984). The rise in pollution has drawn attention from the public and researchers. Visiting coastal areas, where discarded plastics and debris are visible, shows the scale of this environmental problem. Research findings from multiple studies suggest that rivers annually transport approximately 1.15–2.41 MMT of plastic to the oceans, with major Asian rivers contributing approximately 67% of the total plastic input to oceanic environments (Lebreton et al. 2017). Most ocean-bound plastic waste originates from coastal regions within 50 km of the coastline (Brooks et al. 2020). Twenty percent of total plastic waste in the oceans originates from sea-based sources, with the main human activity contributing to this issue being commercial fishing operations (Van Dyck et al. 2016). Studies show a strong correlation between coastal plastic debris from marine sources and commercial fishing activity (Carson et al. 2013; Deshpande et al. 2020; Simeonova and Chuturkova 2020). Abandoned fishing equipment, such as single-strand fishing lines and nets made of nylon, is suspended at various levels in the ocean, leading to "ghost fishing" and potentially ensnaring marine life (Gilman 2015; Stelfox et al. 2016). Eriksen et al. (2014) reported an estimated 5 trillion pieces of floating plastic in the oceans, weighing over 250,000 tons. Half of global annual plastic production is single-use plastic in the oceans (Chen et al. 2021). Surveys using continuous plankton recorders in the North Atlantic and its neighboring seas showed a dramatic rise in plastic contamination between 1957 and 2016, with a minor reduction observed in the most recent decade (Bancone et al. 2020). Sedimentary layers on the ocean floor, where particles continuously accumulate, offer a chronological view of plastic pollution through the study of embedded microplastics (Dasgupta et al. 2022; Grundlehner et al. 2023; Cau et al. 2024). Stratigraphic analysis revealed a 40-year increase in microplastics in the South China Sea and Indian Ocean (Chen et al. 2020; Dasgupta et al. 2021). Sediments from mangrove ecosystems in the Arabian Gulf and the Red Sea exhibited an exponential increase in microplastic accumulation, correlating with the rise in plastic production (Martin et al. 2020). Likewise, coastal sediments in North West Europe and the Mediterranean Sea had rising microplastic

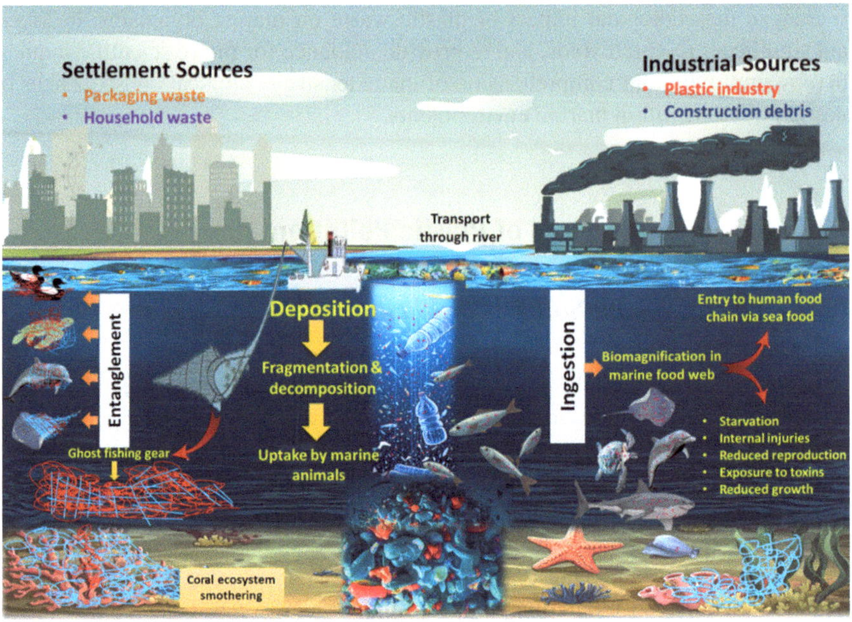

Fig. 4.1 Dynamics of plastic waste movement in the ocean and its interactions with marine organisms

deposition tied to global plastic production (Maes et al. 2018; Cesarini et al. 2022). Figure 4.1 shows the pathways of plastic deposition in marine ecosystems.

4.2.2 Chemical Composition of Marine Plastic

Marine plastic pollution is a pressing environmental issue characterized by the widespread chemical composition of marine plastics, which is complex, comprising various synthetic polymers and additives that can pose significant risks to marine life and human health. Plastics' physical and chemical characteristics contribute to their persistence in the environment. The primary constituents of marine plastic pollution are synthetic polymers, which are derived from petrochemical sources. The prevalent plastic polymers identified within marine ecosystems include polyethylene (PE), polypropylene (PP), polystyrene (PS), and polyvinyl chloride (PVC) (Rochman 2015). These materials are favoured for their durability, lightweight nature, and versatility, making them prevalent in consumer products, packaging, and industrial applications. Microplastics, defined as plastic particles less than 5 mm in size, dominate marine debris and are derived from the fragmentation of larger plastic items or primary microplastics used in products like cosmetics (Amaral-Zettler et al. 2015; Frère et al. 2018). The composition of these microplastics varies, with common types including fragments of PE, PP, and PS, which have different densities affecting their buoyancy and their dissemination throughout marine environments (Summers et al. 2018; Frère et al. 2018). PE and PP are considered the most abundant groups of microplastics and nanoplastics found in marine waters around the globe. It is not unusual for PE as it is the most widely produced and plastic material in the world (Expósito et al. 2021). From day-to-day items to personal care products to medical utensils, the abundance of PE is widely recognizable (Lei et al. 2017). Not only that, the low-density nature and therefore their tendency to float in water are also the reasons behind the abundance of both PE and PP in marine waters. Apart from type, the shape of the polymer is also important in predicting the source of micro-nano plastics in the oceanic environment. Studies revealed that fiber is the most abundant type of MP shape found in marine waters and comes mainly from the washing of synthetic clothes as well as discharges from wastewater treatment plants. Fibers may also meet marine waters from discarded fishing equipment like nylon ropes (Acarer 2023; Welden and Cowie 2017). Apart from fibers, fragment-shaped MPs are also abundant in marine waters and beach sediments (Urban-Malinga et al. 2020). Although, in contrast, some researchers have advocated for pellet-shaped MPs to be the most abundant MNP found in seawaters (Expósito et al. 2021). However, their persistence in the marine environment poses significant ecological challenges, as plastics can take hundreds of years to degrade (Cole et al. 2011). Proper detection techniques for MNPs in marine waters are essential to trace their origin. The most popular method of choice for MP detection is Fourier transform infrared (FTIR) spectroscopy. In addition to the base polymers, marine plastics often contain a variety of chemical additives that enhance their performance and longevity. These additives include plasticizers, stabilizers, flame retardants, and colorants, which can leach into the surrounding environment over time. For instance, phthalates, commonly used as plasticizers, have been detected in marine organisms and are known to disrupt endocrine functions (Rochman et al. 2013). Other hazardous chemicals associated with plastics include polycyclic aromatic hydrocarbons (PAHs) and polychlorinated biphenyls (PCBs), which can adhere to plastic surfaces and accumulate in marine food webs (Rochman et al. 2013). The presence of these toxic substances raises concerns about their potential bioaccumulation in marine biota and subsequent transfer to humans using the consumption of seafood (Landrigan et al. 2020). Different types of plastics in marine environments and their source are summarized in Table 4.1.

4.3 Bioaccumulation of Micro- and Nanoplastics (MNPs)

Upon release into the marine environments, plastic debris/wastes undergo fragmentation into MPs mainly due to water turbulence, fluctuations in water pressure, photochemical reactions, and/or microbial interactions. In marine organisms like crustaceans, mollusks, and especially fishes, MNPs enter the system through their mouth, skin, and gill as food, accumulate in their gastrointestinal (GI) tract, then distributed throughout the body via circulatory or lymphatic systems, and finally

Table 4.1 Forms of plastics and their sources in the marine environment

Type	Usage	Source	References
Sources from the land			
Polyethylene (PE) including both LDPE and HDPE	Packaging and single-use items (most abundant marine plastic)	Urban runoff; tourism	Li et al. (2015), Khan (2021), Lebreton et al., (2017), Schmidt et al. (2017)
Polypropylene (PP)	Food containers and automotive parts (,	Tourism; inadequate waste management systems	Li et al. (2015), Cole et al. (2016)
Polystyrene (PS)	Disposable cutlery and foam containers (contribute the maximum in marine debris)	Tourism; inadequate waste management systems	Li et al. (2015), Summers et al. (2018), Calcar and Emmerik (2019), Geraeds et al. (2019)
Polyethylene terephthalate (PET)	Commonly found in beverage bottles	Tourism; inadequate waste management systems	Li et al. (2015), Calcar and Emmerik (2019), Zhang et al. (2022), Biagi et al. (2021)
Polyvinyl chloride (PVC)	Used in construction and plumbing (known for its durability)	Urban runoff	Li et al. (2015), Lebreton et al. (2017)
Polycarbonate (PC): slow degradation, formation of bisphenol A (BPA)	Automotive, medical, and manufacturing of lenses. and the construction industries	Boat windows, porthole covers, marine equipment, improper disposal	Staniszewska et al. (2016), Banik et al. (2022), Biswal et al. (2020), Yuan et al. (2023)
Maritime source			
Fishing gear			
Fishing dredges (drags along the ocean floor) contain polycarbonate, acrylic (Polyethyl methacrylate, PMMA), and HDPE	Catch shellfish, sea urchin, and other bottom-dwelling species	Lost in operation; discarded in mid-ocean (ghost gear)	Khan (2021), García-Gómez et al. (2021)
Gillnets: consist of marine environment-resistant nylon monofilament, buoyants made up of HDPE	A type of net that catches fish by entangling them, usually by their gills	Lost in operation; discarded in mid-ocean (ghost gear)	Ghauoar et al., (2024), Do and Armstrong (2023)
Trawls or Trawl net: composed of Polyethylene Nylon; Braided twine	A cone-shaped net with multiple panels that is towed through the water to catch fish	Lost in operation; discarded in mid-ocean (ghost gear)	Ghauoar et al. (2024), Do and Armstrong (2023), Hardesty et al. (2021)

(continued)

Table 4.1 (continued)

Type	Usage	Source	References
Soft plastic bait consists of elastomers/thermoelastomers like silicone rubber and butyl rubber; neoprene, fluoroelastomers, nitrile rubber, polyurethane, styrene-butadiene, plastisol (PVC), vinyl plastic (PVC), polyolefins, etc.	Intended to look like the small creatures that fish typically feed on	Lost in operation; discarded in mid-ocean (ghost gear)	Dąbrowska et al. (2024), Watson et al. (2022), Bettler and Mitchell (2021)
Shipping activities: PET, HDPE, LDPE, PVC, PP, PS, and various forms of microplastics	During transport, accidental or intended disposal	Cargo spills; shredding of plastic from paint, coat, hull, sail, and other parts; waste disposal	Lebreton et al. (2017), Roager and Sonnenschein (2019)

eliminated through urine and feces. But sometimes, MNPs, after being absorbed in the GIT, can be stored in the adipose tissues. The bioaccumulation of MNPs is influenced by several factors, viz. the osmotic pressure of water, the particle size of MNPs, its dose and duration of exposure, species variation in fishes, and their rate and amount of food consumption (Ding et al. 2018). Among these factors, the most significant is the size of the MNPs. Typically, MPs less than 5 μm in size possess the greatest potential to pass through the gut epithelial cells and accumulate into various tissues via the bloodstream more readily than large-sized MNPs (Jovanovic et al. 2018). Sometimes, the 5 μm polystyrene (PS) is capable of penetrating and accumulating in the gut as well as the liver. Furthermore, the nanoscale PS (size: 50 nm) can penetrate the blood–brain barrier, leading to brain accumulation, central nervous system (CNS), and other organs of fish. In spotted sea bass (*Lateolabrax maculatus*), comparatively smaller-sized MPs were detected in the gills compared to intestines (Su et al. 2019). The bioaccumulation of MNPs is not only dependent on the size of the particles but also tissue specific. MNPs are transported and accumulated into a specific organ via ingestion and absorption, and gills in fishes, having vast surface area, can accumulate many MNPs. MNPs are usually accumulated specifically in the gills and gut than in the brain and liver (Lei et al. 2018). The gut plays important roles both in digestion and absorption, and direct accumulation of many MNPs occurs during feeding. The ingested and absorbed MNPs then reach the liver and kidney, are detoxified or partially detoxified, and excreted from the circulation (Kim et al. 2021). Sometimes, these excreted MNPs get settled at the bottom of the sea/ocean and may produce toxic effects on the benthic species (Yin et al. 2018). Microplastics can change the sinking rates of zooplankton fecal pellets, which may disrupt nutrient cycling in marine ecosystems (Cole et al. 2016; Summers et al. 2018).

MP accumulation is also influenced by different feeding strategies and the position of a particular fish species at the trophic level, their size/body weight, and their growth rate. The minute body size of zooplankton is a possible determinant of small number of MPs in zooplankton which is thought to prevent the consumption and subsequent bioaccumulation of MPs larger than the food they consume. Moreover, these larger-sized MPs may induce digestive tract obstruction, leading to a reduction in feeding activity (Cole et al. 2013; Suwaki et al. 2020; Welden and Cowie 2016). Clupeids such as sprats and sardines, which are dependent on phytoplankton and zooplankton, may ingest twice the number of MPs than other fish species. MNP accumulation is also found to be positively correlated with body size, as larger animal requires more food and therefore a higher amount of MP intake (Covernton et al. 2021). Moreover, the gradual growth of an animal during its life cycle also plays a crucial role in the MP accumulation rate (Proki´c et al. 2021). It has been hypothesized that species occupying higher trophic levels may exhibit increased microplastic accumulation through trophic transfer, specifically the ingestion of contaminated prey (Hartmann et al. 2019).

4.4 Ecological Consequences of Plastic Pollution in Marine Environments

4.4.1 Ingestion of Marine Plastics

Contemporary research confirms plastic particles in the oceans and their entry into sinking clusters (Pojar et al. 2021; Vega-Herrera et al. 2021). Various marine organisms, ranging from iconic species like whales, turtles, and seabirds to fish, crustaceans, mollusks, and worms, as well as plankton at the foundation of ocean food chains, consume these sinking particles either partially or completely (Duncan et al. 2017; Thiel et al. 2018; Jepsen and de Bruyn 2019; Costa et al. 2020; Prampramote et al. 2022; Daudt et al. 2023). Plastic bags and utensils present the highest risks of ingestion, despite the consumption of numerous other plastic objects (Provencher et al. 2017; Isangedighi et al. 2020). The pathway of plastic ingestion by different organisms is displayed in Fig. 4.1. Bordbar et al. (2018) conducted a study in the eastern Mediterranean and discovered that 5.93% of narwhal shrimps (*Plesionika narval*) had consumed plastic materials, specifically nylon. A total of 10.3% of females and 4.8% of males had empty stomachs with tangled balls of nylon fibers, suggesting blockage in their digestive systems. Analysis of the gastrointestinal tracts of all fish specimens collected within the Bahía Blanca Estuary, Argentinean Sea, revealed the presence of diverse microplastic types. Fibers were the most common microplastics (60.8%), followed by pellets (28.9%), fragments (8.6%), and laminas (1.4%). The size of these microplastics ranged from 0.98 to more than 5 mm (Arias et al. 2019). Within the digestive tracts of six commercially important fish species from Campeche Bay, Mexico, researchers discovered a total of 316 microplastic particles. These particles were identified across 240 individual fish examined during the study (Borges-Ramírez et al. 2020). *Haemulon plumierii*, a demersal species, had the most microplastics, with 115 items total—73 fibers and 42 fragments (Borges-Ramírez et al. 2020). A study by Garcés-Ordóñez et al. (2020) investigated the occurrence of microplastic consumption among fish species in the mangrove ecosystems of Cispata, located in the Colombian Caribbean. The researchers examined the gastrointestinal contents of 302 fish samples, encompassing 22 different species. Their findings revealed the presence of 69 microplastics, with 7% of the specimens analyzed showing evidence of microplastic ingestion. The composition of ingested microplastics: 55% filaments, 23% fragments, 19% films, and 3% foam. In a study conducted by Jonathan et al. (2021) at Magdalena Bay on the Pacific Coast of Mexico, researchers identified a total of 64 microplastics (MPs) in 87 specimens across six fish species. The study revealed that *Eucinostomus dowii* and *Paralabrax maculatofasciatus* contained the highest concentrations of MPs, with 30 and 27 particles, respectively. Bessa et al. (2018) conducted a study to determine the presence of microplastics in three commercially relevant fish species, namely *Dicentrarchus labrax*, *Diplodus vulgaris*, and *Platichthys flesus*, sampled from the Mondego estuary in Portugal. They found 157 particles in 38% of the fish, with 1.67 microplastics per fish on average. The uptake of microplastics is influenced by their dimensions. Caribbean coral demonstrated that they actively consumed large microplastics, but did not ingest particles smaller than 250 μm (Hankins et al. 2018). Lam et al. (2023) investigated the presence of microplastics in 249 clams across six species and sampled from two prominent recreational clamming sites in Hong Kong. The research revealed that microplastic contamination affected more than half (56.6%) of the examined clams, with an average of 1.04 particles per gram and 0.98 particles per individual clam. This suggests that Hong Kong residents may be ingesting an estimated 14,307 microplastic particles annually through their diet. A wide variety of marine life can consume microplastics, enhancing their potential to enter food chains and ascend through trophic levels via predation (Setälä et al. 2018; Kangas et al. 2023). Filter-feeding animals ranging from tiny mussels to enormous whale sharks have the potential to amass considerable amounts of microplastics (Germanov et al. 2019). The occurrence of plastic consumption among various marine species is documented in Table 4.2.

4.4.2 *Physiological Impacts of Plastic Ingestion*

The ingestion of large and medium-sized plastic items by marine mammals, seabirds, and turtles poses grave dangers. One of the most pressing and apparent consequences is blockage of the digestive system. When swallowed, plastic waste can amass within the gut, obstructing the movement of food and impeding nutrient uptake. This can result in malnutrition, stunted growth, and ultimately, starvation to death (Pierce et al. 2004; De Stephanis et al. 2013; da Silva Mendes et al. 2015; Alexiadou et al. 2019).

Table 4.2 Reported occurrences of plastic material ingestion within marine fauna

Type of organism	Species	Percentage with plastic (%)	Number of particles/ individuals	Location of study	Source
Seabird	*Alle alle*	14	1.11	White Bay, Newfoundland	Fife et al. (2015)
Seabird	*Larus michahellis*	10.48	0.80	Portugal	Basto et al. (2019)
Seabird	*Larus fuscus*	18.69	1.85	Portugal	Basto et al. (2019)
Turtle	*Chelonia mydas*	91	-	Brazilian coast	Wedemeyer-Strombel et al. (2015)
Turtle	*Lepidochelys olivacea*	82	-	North Pacific Ocean	Wedemeyer-Strombel et al. (2015)
Mussel	*Mytilus galloprovincialis*	41	0.51	East Mediterranean Sea	Digka et al. (2024)
Cetacean	*Pontoporia blainvillei*	100	9.6	South-eastern Buenos Aires, Argentina	Padula et al. (2025)
Elasmobranch	*Scyliorhinus canicula*	6.5	0.14	Southwest Coast of the United Kingdom	Morgan et al. (2021)

Furthermore, indigestible items can create a misleading feeling of fullness, potentially causing the animal to consume insufficient food. This inadequate nutrition may ultimately impact the animal's fitness, growth, and reproductive capabilities (Baak et al. 2020; Roman et al. 2021). Research shows plastic exposure disrupts hormones, impairs egg production, and reduces fertility in fish, invertebrates, and marine mammals. Studies indicate that seabirds consuming much plastic experience weight loss and reduced breeding success due to a lack of nutrients (Roman et al. 2019; Robuck et al. 2022). Contact with plastic particles can upset the sensitive equilibrium between antioxidants and reactive oxygen species (ROS) in organisms, triggering oxidative stress. This stress can inflict damage on cellular components, such as DNA, proteins, and lipids, and has been associated with metabolic changes and negative impacts on the nervous system (Hu and Palić 2020; Das 2023). Plastic ingestion is positively correlated with the hepatosomatic index, indicating probable stress in their health condition caused by plastic ingestion (Arias et al. 2019). Experiments revealed young glassfish (*Ambassis dussumieri*) fed diets laced with microplastics exhibited stunted growth compared to counterparts provided natural sustenance (Naidoo and Glassom 2019). Microplastics' effects depend on size. Rotifers exposed to nanoplastics measuring 0.07-µm in size experienced growth reduction, survival rates, and reproductive capabilities at elevated concentrations. In contrast, larger plastic particles did not show any significant effects on these organisms (Sun et al. 2019). Research on the spiny chromis (*Acanthochromis polyacanthus*), a type of damselfish, revealed that ingestion of smaller microplastic particles, but not larger ones, negatively impacted their growth and overall physical condition (Critchell and Hoogenboom 2018).

4.4.3 Entanglement by Marine Plastics

When a rope-like or looped plastic object wraps around a body, it results in entanglement. Such incidents are uncommon and do not require extensive examination. Reports on this rarity are lacking. In recent times, research focusing on the interaction between marine life and debris, particularly plastics, has grown significantly. Marine organisms often become entangled in discarded fishing gear, such as ghost nets, fishing lines, as well as plastic materials like sheets and rings, leading to increased incidents of entanglement (Katsanevakis 2008; Parga Martínez et al. 2020; Richardson et al. 2021). Fishing gear is the most commonly reported debris material, originating from fishing or shipping activities (Gilardi et al. 2010). The greatest threats of entanglement come from debris related to fishing activities, as well as balloons and plastic bags (Richardson et al. 2022). Figure 4.1 shows the pathway of plastic entanglement by different organisms. A review by Gall and Thompson (2015) found that entanglement with plastic (55%) was much more common than plastic ingestion (31%). Behavioral characteristics of a species play a role in determining its susceptibility to entanglement (Karamanlidis et al. 2008). Due to their slow movements and tendency to feed near the water's surface, sea turtles are especially susceptible to becoming entangled in abandoned fishing gear (Wilcox et al. 2015). Research shows that larger mesh sizes in fishing nets create major problems, as they greatly increase the risk of sea turtles becoming trapped (Wilcox et al. 2015; Ennen et al. 2021). Items like packing loops can be particularly appealing to young sea lions and seals, potentially capturing their attention and curiosity (Franco-Trecu et al. 2017; Allyn and Scordino 2020). Their large size and power can make them struggle harder, making the entanglement worse. Juvenile sea lions and seals easily get caught in plastic debris because of their small size. Their necks and limbs pass through the material easily. Their heads may get trapped in openings, making the neck the main point of entanglement. Once caught, they struggle to free themselves as their long fur prevents the material from sliding off (Lawson et al. 2015; Allyn and Scordino 2020; Curtis et al. 2021; Salazar-Casals et al. 2022). Another group particularly susceptible to entanglement is seabirds. They can become entangled in lines and nets, both at the surface and while diving for prey (Bond et al. 2012). Votier et al. (2011) found that 66 birds annually get tangled in plastic, like netting, in Grassholm, Wales, UK. The plunge diving feeding technique of seabirds often results in them confusing floating plastic with fish or other food sources, leading to a high number of entanglement incidents (Isangedighi et al. 2020). Rodríguez et al. (2013) observed 34 entangled birds, making up 0.93% of all counted gannets. Entanglement rates were exceptionally high in Mauritania in late spring. Most affected birds were immature, representing 52.4% of the total.

Air-breathing animals trapped underwater are unable to surface for oxygen, resulting in death by asphyxiation. Gill-breathing organisms are also impacted: prolonged entanglement leads to starvation or predation. The olfactory stimuli emanating from decomposing organic matter attract additional fauna, which may become entrapped, particularly in discarded fishing gear (Butterworth 2016). This continuous process, known as "ghost fishing," can persist for extended periods. Consequently, numerous entrapped animals succumb undetected, indicating that the figures likely represent only a fraction of the "real impact." The true extent of this phenomenon remains largely obscured from observation (Parton et al. 2019). Entanglement can be identified as the cause of death or injuries when a carcass exhibits severe recent wounds. A case study conducted in northern Patagonia documented 27 instances of kelp gulls (*Larus dominicanus*) exhibiting entanglement with fishing lines. Of these birds, 81% had recently perished, while some of the surviving gulls suffered from serious injuries (Yorio et al. 2020). Moreover, entanglement can result in malnutrition, as affected organisms are impeded in their foraging activities, and can disrupt migration patterns, consequently impacting breeding and feeding territories (Gall and Thompson 2015; Perez-Venegas et al. 2021; Clark et al. 2023). Entanglement incidents pose a significant threat to various marine life forms, including sea turtles, mammals, seabirds, and crustaceans. These events can lead to severe consequences such as drowning, suffocation, and laceration, potentially resulting in death for the affected organisms. The frequency of entanglement incidents for various marine species is presented in Table 4.3. Plastic entanglement adds materials to organisms' bodies, raising the energy needed to move and increasing food needs. According to Derraik (2002), Northern fur seals (*Callorhinus ursinus*) experienced a quadruple increase in their daily energy requirements to sustain regular activities when burdened with an additional 200 g of material. Entanglement materials like fishing nets or packing bands can cause wounds that become infected or develop necrosis. When the lymphatic system fails to manage these infections effectively, the respiratory organs may become compromised, potentially resulting in fatal consequences. Microbes entering the bloodstream can also cause secondary infections (Carretta et al. 2017). Entanglement can impact the populations of marine organisms. Perez-Venegas et al. (2021) predicted that entanglement at a rate of 1.2×10^{-3} will reduce the *Arctocephalus australis australis* population by 20.34% in 30 years. Entanglement by ghost gears can smother coral and damage seagrass ecosystems. (Gilman et al. 2023).

Table 4.3 Entanglement records for marine animals due to plastic pollution from the scientific literature

Organisms	Study period	Entangled material	Location of study	Source
Seabirds	1997–2017	Line and balloons	Central California	Donnelly-Greenan et al. (2019)
Marine mammals	2001–2005	Fishing material	United States	Moore et al. (2009)
Marine mammals	1995–2005	Trawl net and strapping tape	Kaikoura region, New Zealand	Boren et al. (2006)
Marine mammals	2018–2020	Fishing materials	Namibia	(Curtis et al. 2021)
Elasmobranch *(Prionace glauca)*	2016	Polypropylene strapping bands	Atlantic and Mediterranean	Colmenero et al. (2017)
Elasmobranch *(Carcharhinus falciformis)*	2012	Fishing materials	Western/central Pacific Ocean	Hutchinson et al. (2015)

4.5 Effects of MNPs on Fishes and Other Marine Organisms

4.5.1 Effects of MNPs on Fishes

Bioaccumulation of MNPs in the GI tract of fish causes ulcers, perforation, and distortion in the gut. Sometimes, these damages may be fatal, and death soon follows. Apart from this, MNPs can cause endocrine, metabolic, and reproductive disorders, oxidative stress-mediated neurological problems, immunotoxicity, and genotoxicity. Besides MNPs, plastic additives, plasticizers, other organic pollutants like metals and metalloids, polychlorinated biphenyls (PCBs), polycyclic aromatic hydrocarbons (PAHs), and pesticides, especially organochlorine use MPs as vehicle/carriers for transport and may also cause detrimental health effects on aquatic fauna (Kim et al. 2021). The potential toxicity caused by MNPs on fishes and other aquatic organisms is discussed in the following section.

Effects of MNPs on Hematological System

About 1–4% of ingested MNPs (size: < 5 μm) in fishes penetrate the circulatory system and cause localized inflammation. These MNPs may generate oxidative stress and cause lipid peroxidation in RBCs that may lead to hemolysis. Moreover, these MNPs may interfere with the oxygen-carrying capacity of hemoglobin, leading to impaired function in several tissues. They can also alter the physico-chemical properties of the RBCs, resulting in hemolysis (Kim et al. 2021). Additionally, MPs may alter transmembrane gradient regulation of enzymes and hormones required for lipid metabolism, thereby modifying cholesterol and triglyceride content in the liver and muscle (Banaee et al. 2018; Jovanovic 2017). MPs can interact and make complexes with albumin and globulin, affecting blood circulation (Yong et al. 2020). MPs (plasticizers) are also reported to alter the biosynthesis of glucose. Besides, several MP particles can also alter the levels of protein and cholesterol. For example, polymethylmethacrylate (acrylic glass), a nanoplastic, is demonstrated to significantly increase the blood cholesterol level in *Sparus aurata* (Brandts et al. 2021). Alanine aminotransferase (ALT), aspartate aminotransferase (AST), and alkaline phosphatase (ALP) are considered as the important biomarkers of liver injury. Polyethylene (PE) exposed fishes were shown to exhibit higher levels of ALT, AST, and ALP, indicating liver damage by this MP particle (Hamed et al. 2019). On the contrary, significantly reduced levels of ALT, AST, and ALP were estimated in European seabass (*Dicentrarchus labrax*) exposed to polymethylmethacrylate, indicative of hemotoxicity (Brandts et al. 2018).

Effects of MNPs on the Nervous System

While research on the neurotoxicological effects of MNPs has been conducted on freshwater fishes, only a few such studies were conducted on marine fishes. These studies have demonstrated the acetylcholinesterase (AChE) inhibitory activity of MPs in marine fishes. The inhibition of AChE has been demonstrated to hamper cholinergic neurotransmission, which may lead to impaired neuromuscular and neurological function (Wen et al. 2018). Researchers have reported the AchE inhibitory activity of MPs in Common Goby (*Pomatoschistus microps*) and *D. labrax*. In *P. microps*, polyethylene (PE), MP particles are presumed to be responsible for AChE suppression, resulting in motor dysfunction (Oliveira et al. 2013; Chen et al. 2017).

Effects of MNPs on the Immune System

Bioaccumulated MPs demonstrated to trigger the immune system of marine fishes. MPs, after accumulation in the GIT, can enter the cell, alter the signaling pathways, and thereby hamper cellular immunity (Li et al. 2024). MP exposure to *S. aurata* has been demonstrated to decrease the level of phagocytosis and increase immunoglobulin and peroxidase contents (Espinosa et al. 2017). Similarly, in *D. labrax*, the exposure of PE-MPs causes the synthesis of higher levels of immunoglobulins, which affects the innate immune system (Espinosa et al. 2019). Exposure of polystyrene (PS) (0.8–20 μm) MP particles to *Oncorhynchus mykiss* has been reported to elevate the number of phagocytic B cells (Zwollo et al. 2021).

Effects of MNPs on Oxidative Stress and Antioxidant Activity

MNPs when interacting with cellular components like DNA, protein, and lipid generate free radicals which can alter the redox homeostasis of the cell and cause hydroxylation of DNA, denaturation of protein, and lipid peroxidation (Hoseinifar et al. 2020). Researchers observed increased lipid peroxidation in *D. labrax* treated with MPs (Barboza et al. 2018). Similar observation was found in the case of *S. aurata*, which was exposed to MPs (Solomando et al. 2020). This increased tendency of MPs to generate oxidative stress may be attributed to their large surface area, which is thought to support the absorption of oxidizing molecules (Prata et al. 2020).

Effects of MNPs on Fish Growth Rate

Researchers found the presence of polystyrene (PS) MPs (size: 5 μm) in the gut of *Trachinotus blochii*, which might harm the gut microbes. Metabolomic analysis revealed that MPs can interact with the basic building blocks of cells, viz. glucose and amino acids, thereby interfering with several signal transduction pathways, leading to multiple metabolic disorders (Yao et al. 2024). Synergistic effects have been shown between MPs and antibiotics, which significantly declined the specific growth rate and body weight gain rate in *Sebastes schlegelii* juveniles, elucidating the potential risk of their combined exposure in the environment (Tian and Liu 2024).

4.5.2 *Effects of MNPs on Other Marine Organisms*

The effects of MNPs on phytoplankton are extensive, which include retardation of cell growth, decreased chlorophyll content, and therefore, significant inhibition in photosynthesis. Study revealed that in marine microalgae (*Skeletonema costatum*), the content of chlorophyll has been found to decrease by 7% after exposure to PVC (Zhang et al. 2017; Sjollema et al. 2016). In marine environments, MPs encounter numerous chemical, physical, and biological factors that can alter the physico-chemical structures of MPs, making them more readily and differently interact with biological macro- and micro-molecules. For example, older MPs contain more scraped or abraded surfaces than the comparatively newer MP particles, and therefore function differently with their distinctive functional groups (Zhang et al. 2023; Bharath et al. 2021).

The survival and growth rate of brine shrimp (*Artemia parthenogenetica*) have been seen to decrease significantly when exposed to PS and PE MPs. Moreover, *Artemia* seemed to suffer from oxidative and immunological stress as well as problems related to gut microbes (Li et al. 2021). Apart from decreasing the chances of survival of the zooplankton, MPs can also reduce the synthesis of lipids that cause energy deficiency in marine organisms that consume zooplankton. Furthermore, zooplanktons excrete these MPs within the fecal pellets that serve not only as the vectors for MPs but also as the food source for other marine organisms, thereby causing serious problems to the organisms that belong to higher trophic levels (Cole et al. 2021).

MPs, especially PS, have been observed in coral samples collected from different sampling sites in Lombok, Indonesia. These MPs in the corals may come from fishing gear, beverage packages, and food containers (Cordova et al. 2018). A contemporary investigation demonstrated the presence of PE MPs on the surface as well as inside the coral tissues (*Goniopora columna*) and caused a reduction in the length of polyps, enzyme inhibition for detoxification, alteration in the anthozoan-algae symbiosis, and overproduction of lipid peroxide, resulting in skeletal growth inhibition, bleaching, and tissue necrosis (Reichert et al. 2018; Chen et al. 2022; Chapron et al. 2018).

PE, PP, and PVC are also observed in the esophagus, stomach, and intestine of green sea turtles (*Chelonia mydas*) and loggerhead sea turtles (*Caretta caretta*), which depend on the sea beaches for laying eggs. A recent study revealed the presence of MP particles (both PE and PS) on the surface sediments and nest bottoms of sea turtles. These MPs absorbed more pollutants and carried them to the newly hatched eggs causing hatching failure and deformities during embryonic development (Zhang et al. 2022). Not only that, but these MPs also retain more heat than the surrounding sands causing further increase in the temperature of the sand, which results in the alteration in the sex ratio of sea turtles, and thereby their reproductive success (Estrella-Jordan et al. 2023; Beckwith and Fuentes 2018).

4.6 Bibliometric Evaluation of the Environmental Impact and Ecological Risk Associated with Plastic Debris and Microplastics in Marine Ecosystems

The bibliometric analysis can be demonstrated as the quantitative examination of scientific output by taking into consideration parameters, viz. citation counts, citations per author, keyword occurrence, collaboration between countries, and institutional affiliations (Aria and Cuccurullo 2017; Donthu et al. 2020). The whole process of bibliometric analysis comprises several steps, noticeably: (i) gathering bibliographic data from Scopus database (www.scopus.com) using the search string TITLE-ABS-KEY (microplastics AND marine environment AND ecological impacts) AND (EXCLUDE (DOCTYPE, "ed") OR EXCLUDE (DOCTYPE, "le") OR EXCLUDE (DOCTYPE, "tb") OR EXCLUDE (DOCTYPE, "no") OR EXCLUDE (DOCTYPE, "cp")); (ii) reviewing and refining the collected data; (iii) analyzing the data using the Bibliometrix package (https://www.bibliometrix.org/index.html, a software developed by the researchers of University of Naples Federico II in Italy), an open source tool in R programming language; and (iv) visual presentation of analyzed data (Coulibaly et al. 2019). Network analysis of keywords was performed using VOSviewer tool (Leiden University), where keywords from publication data are visualized as nodes. Node size reflects keyword frequency, and links between nodes indicate co-occurrence (Van Eck and Waltman 2019). The following section presents a summary of bibliometric research findings on the ecological impact of microplastics in marine environments.

4.6.1 Global Publication Trends in Marine Plastics and Their Sources

The total number of articles published by different countries between 2014 and 2025 is represented in Fig. 4.2. The number of articles contributed by different countries is relatively low during the first five years (up to 2019), although there is a steep and steady growth from the beginning of 2020 till early 2025. The overall estimation shows that China has the highest number of publications with 2563 articles in the aggregate, followed by India (643 articles), Italy (436 articles), the United Kingdom (313 articles), and Brazil (269 articles). This trend has also been depicted by the global collaboration map between countries (Fig. 4.3), where the darker blue color in China indicates the highest number of publications and a gradual decrease in color intensity demonstrates the countries with publications. Figure 4.3 also puts a limelight on collaborations between countries where it is seen that countries, viz. China, India, Bangladesh, etc., are highly active in collaboration, with a frequency higher than five. The highest number of publications between 2014 and 2025 were contributed by journals like Science of The Total Environment (63), Environmental Pollution (41), Marine Pollution Bulletin (23), etc. Simultaneously, Table 4.4 highlights the most cited sources, where the article by Horton et al. 2017 has 2582, gaining 286.89 citations per year. This review article examines the occurrence, behavior, and ecological consequences of microplastics within freshwater and terrestrial ecosystems and delineates outstanding knowledge gaps and future research priorities. As mentioned by Garai et al. (2024), citation analysis plays a key role in understanding the impact of leading research and authors within a specific field. Moreover, Figs. 4.4 and 4.5 indicate the number of articles published by the most relevant authors over the last decade and the productivity of these authors within the same period, respectively. The circles represent the numerical accountability of articles published by each author, while the color denotes the density subjecting to the number of citations. Most articles were published by Wang J (22 articles), followed by Li X (10 articles), Wang Y (10 articles), Liu S (9 articles), etc. Among these, Li Z has the highest number of citations, viz. 682 in the year 2020 by gaining 113.667 citations per year. Simultaneously, a network visualization of an intellectual connection between the documents is represented via generating a bibliographic coupling (Fig. 4.5), where at least more than 3 documents cite the same previous work(s). In this diagram, each node represents an individual document, while edges between nodes denote the shared citations. The thickness of the edges often reflects the degree of coupling; documents with more shared references have thicker lines connecting them. If a cluster of nodes appears, it

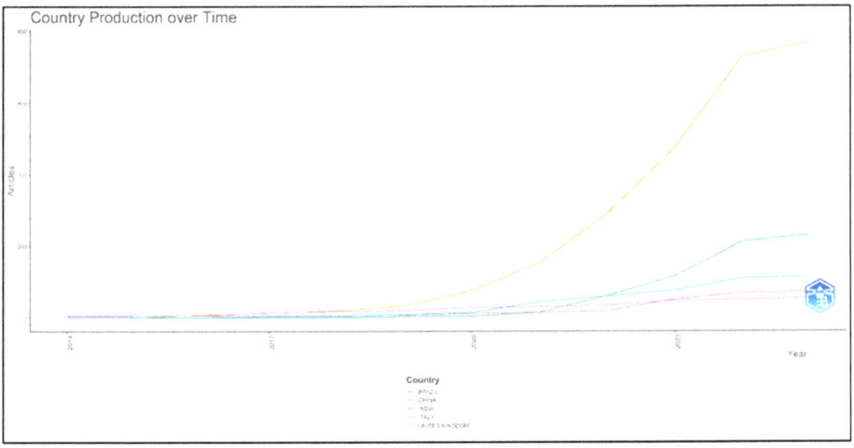

Fig. 4.2 Productivity of countries between 2014 and 2025 in publishing articles related to keywords including microplastics, marine environment, and ecological impacts

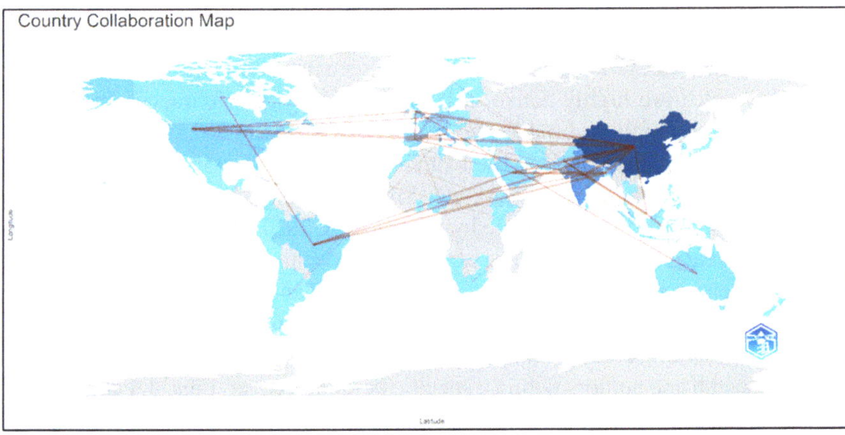

Fig. 4.3 Country collaboration map based on articles published in relevance to keywords

suggests a thematic or conceptual similarity, as these documents rely on overlapping foundational works (Fig. 4.6).

4.6.2 Most Relevant Keywords

On a general note, the keyword selected by the authors highlights the scope and significance of their article, in a research community. The analysis of authors' keywords

Table 4.4 Most relevant sources of citation in relevance to keywords

	Title and Doi of the paper	Total citations	Total citations per year
Horton et al. (2017); Science of The Total Environment	Microplastics in freshwater and terrestrial environments: evaluating the current understanding to identify the knowledge gaps and future research priorities https://doi.org/10.1016/j.scitotenv.2017.01.190	2582	286.89
do Sul and Costa (2014); Environmental Pollution	The present and future of microplastic pollution in the marine environment https://doi.org/10.1016/j.envpol.2013.10.036	1218	101.50
Avio et al. (2017); Marine Environmental Research	Plastics and microplastics in the oceans: from emerging pollutants to emerged threat https://doi.org/10.1016/j.marenvres.2016.05.012	913	101.44
Chae and An (2018); Environmental Pollution	Current research trends on plastic pollution and ecological impacts on the soil ecosystem: a review https://doi.org/10.1016/j.envpol.2018.05.008	862	107.75
Qi et al. (2020); Science of The Total Environmental	Behavior of microplastics and plastic film residues in the soil environment: a critical review https://doi.org/10.1016/j.scitotenv.2019.134722	638	106.33
Thusari and Senevirathna (2020); Heliyon	Plastic pollution in the marine environment https://doi.org/10.1016/j.heliyon. 2020.e04709	537	89.50
Anderson et al. (2016); Environmental Pollution	Microplastics in aquatic environments: implications for Canadian ecosystems https://doi.org/10.1016/j.envpol.2016.06.074	443	44.30
Harrison et al. (2014); BMC Microbiology	Rapid bacterial colonization of low-density polyethylene microplastics in coastal sediment microcosms https://doi.org/10.1186/s12866-014-0232-4	439	36.58
Henry et al. (2019); Science of The Total Environmental	Microfibres from apparel and home textiles: Prospects for including microplastics in environmental sustainability assessment https://doi.org/10.1016/j.scitotenv.2018.10.166	429	61.29
Li et al. (2019); Environmental Pollution	Using mussel as a global bioindicator of coastal microplastic pollution https://doi.org/10.1016/j.envpol.2018.10.032	422	60.29

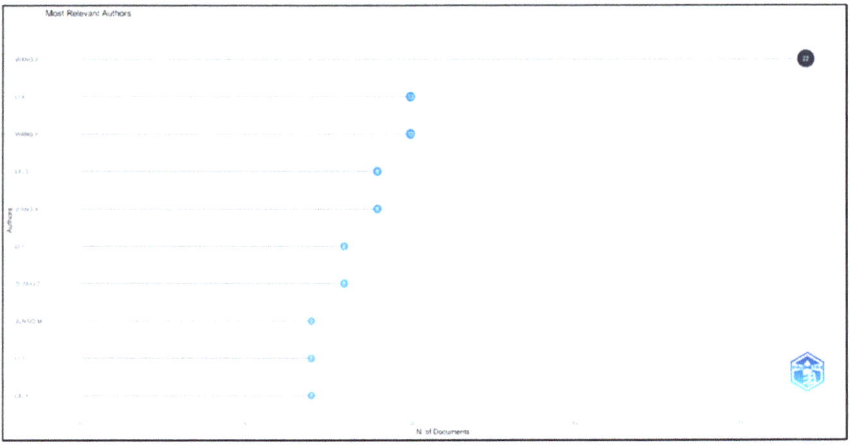

Fig. 4.4 Authors who have published the most on this topic

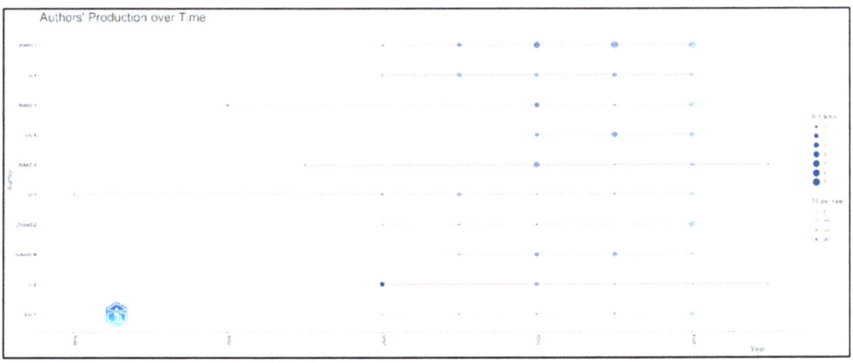

Fig. 4.5 Authors' productivity at different time

comprises gathering and examining the keywords from related research in a particular field. Thus, it helps in identifying trends and the direction of research in which it is pursued (Dixit and Jakhar 2021). After the execution of such analysis, the 50 most prominent keywords were identified that were close to the chosen theme of the topic. A word cloud in Fig. 4.7 is presented here to illustrate the most common authors' keywords from the database collected from Scopus. Here, the size of the keyword is directly proportional to its frequency across documents. More importantly, these keywords represent the key areas of research regarding the ecological impacts of microplastics in marine environments.

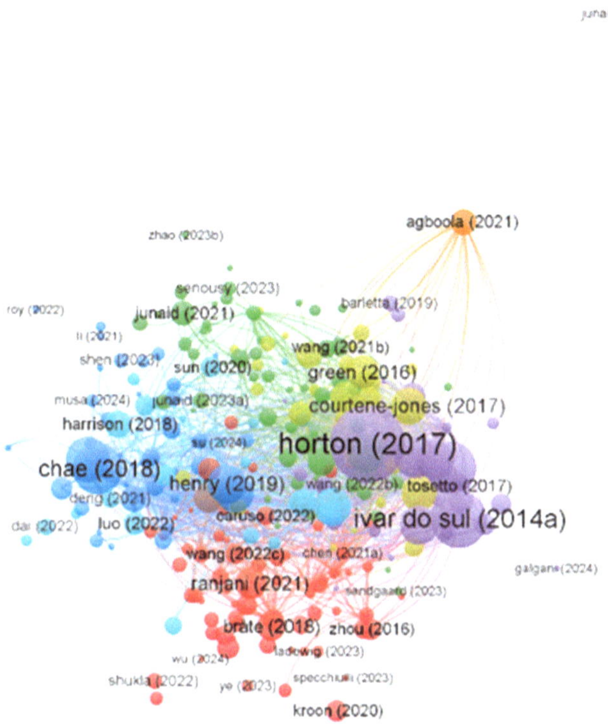

Fig. 4.6 Bibliographic coupling based on documents (> 2) owned by authors

4.6.3 Network Visualization and Overlay Visualization

Across 352 documents, a total of 1129 keywords were identified, where only 45 keywords met the requirement of appearing at least five times per document. Each keyword is represented as a node, and the size of each node indicates the frequency of each keyword occurrence. These 45 keywords were grouped into 8 clusters. The keywords "microplastics" made an appearance of 153 with a link strength of 159, followed by "plastic pollution" and "nanoplastic" with an appearance of 26, 21, and a link strength of 36 and 33, respectively. An overlay map (Figs. 4.8 and 4.9) shows the authors' keywords, where node color represents the average yearly occurrence of each keyword. Here the old keywords like "marine litter", "plastics", "marine debris", "plastic debris", "marine ecosystems", and "freshwater" are denoted by deep blue color. Similarly, the new keywords like "microplastics", "nanoplastics", "extracellular polymeric substrata", and "plastic pollution" are denoted by green color and are found to be associated with the latest terms of recent studies, viz. "ecological risks", "seawater", "environmental impact", "adsorption", "heavy metals", "human health", etc., as denoted by yellow color.

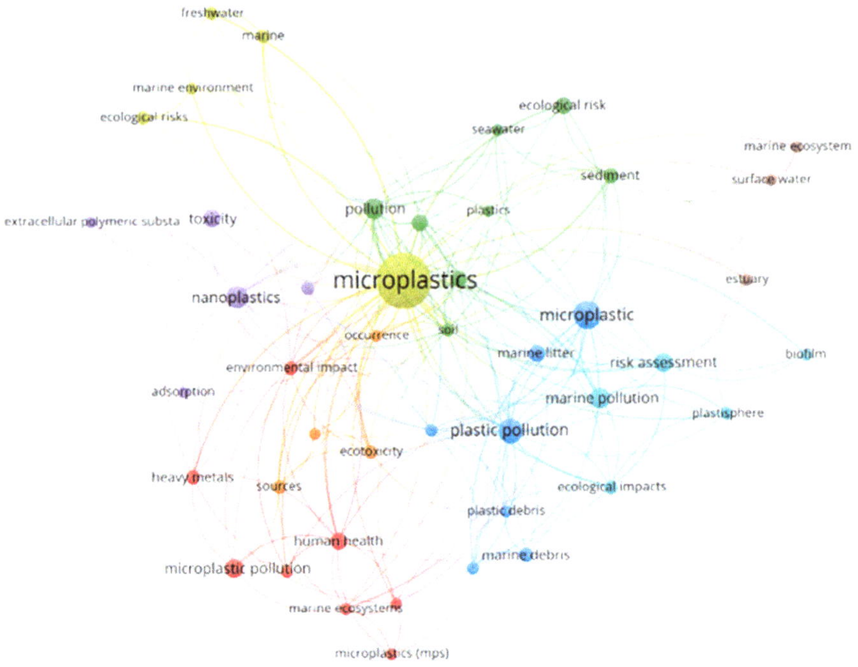

Fig. 4.7 Word cloud co-occurrence network visualization based on the authors' keywords

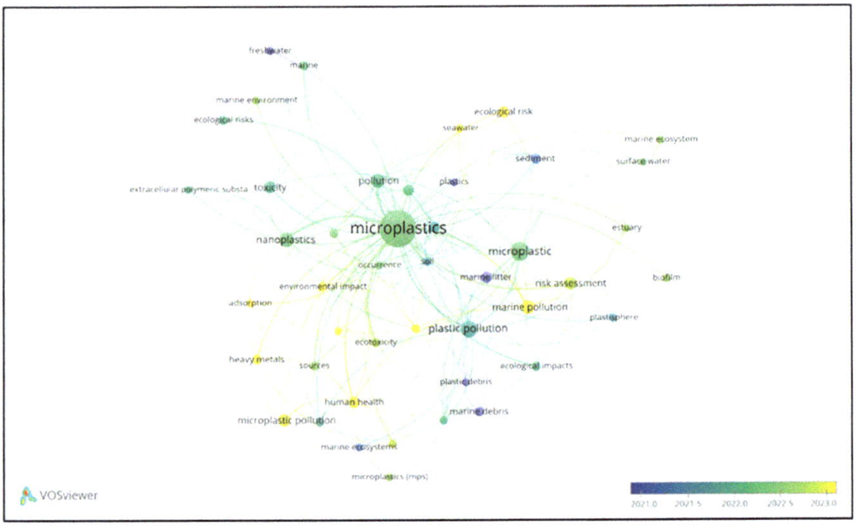

Fig. 4.8 Word cloud co-occurrence overlay visualization based on the authors' keywords

Fig. 4.9 Word cloud based on author's keywords

4.6.4 Thematic Evolution

Over the past 11 years, the analysis of keywords suggests that research themes around microplastics and their ecological impact on the marine environment have progressed significantly. Figure 4.10 representing thematic evolution successfully highlights the dominant themes and emerging trends in this field. In the first five years, researchers have predominantly nurtured the areas linking to keywords, viz. "marine litter", "plastic debris", "marine debris", "plastic pollution", "sediments", and "plastic pollution". However, the later part of the study shows that recent explorations have been done in the areas linked to "microplastic pollution", "microfiber", "climate change", "biofilm", "ecology", "risk assessment", "aquatic ecosystem", "ecotoxicology", "Antarctica", "heavy metals", "marine ecosystem", etc. The thematic evolution clearly shows the development in study areas linked to microplastics and their ecological impacts on the marine environment.

4.7 Knowledge Gaps and Future Research Directions

The substantial threat posed by plastic pollution to marine ecosystems necessitates comprehensive research to elucidate the complex challenges involved. A critical knowledge deficit exists regarding the intricate impacts of microplastics and nanoplastics on marine organisms at the cellular and molecular levels. While studies have documented the ingestion of these particles, the precise mechanisms by which they disrupt physiological processes, including cellular function, immune responses, and reproductive capabilities, remain poorly understood. Further research employing

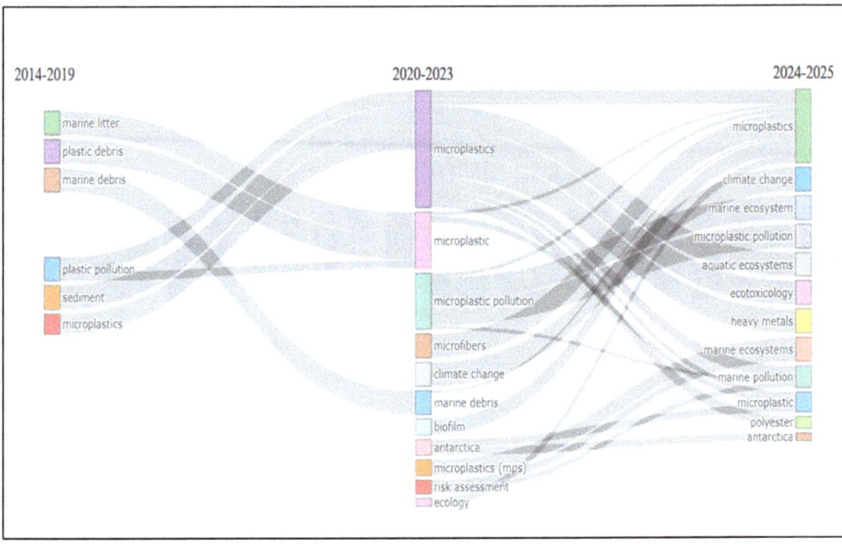

Fig. 4.10 Thematic evolution based on the authors' keywords

advanced techniques in toxicology, molecular biology, and ecophysiology is crucial to elucidate these mechanisms and assess the potential for transgenerational effects. Beyond individual organisms, a deeper understanding of the long-term, ecosystem-level consequences of plastic pollution is paramount. Current research often focuses on short-term effects, but the chronic exposure of marine communities to plastics, coupled with the cascading effects through food webs, necessitates investigations into long-term impacts on biodiversity, community structure, and ecosystem services. This requires sophisticated ecological modeling and field studies across diverse marine environments to predict the complex interplay of factors influencing ecosystem resilience. Simultaneously, the development of effective mitigation strategies is an urgent priority. Research efforts should focus on innovative approaches to plastic waste management, including the development of biodegradable alternatives, improved recycling technologies, and strategies to prevent plastic leakage into the environment. Furthermore, research into bioremediation techniques, which utilize microorganisms to break down plastics, holds promise for mitigating existing pollution. Crucially, social and economic research is needed to understand the drivers of plastic pollution and to develop policies that incentivize sustainable practices and reduce plastic consumption. Finally, improved monitoring and assessment methods are essential to track the distribution and impacts of plastic pollution in the marine environment. Current monitoring techniques often lack standardization and sensitivity, hindering accurate assessments of pollution levels and their effects on marine life. Developing advanced technologies, such as remote sensing, micro- and nano-particle detection methods, and standardized bioindicators, will enable more effective monitoring of plastic pollution across different spatial and temporal scales. Integrating these data with ecological models will improve our ability to predict future trends and inform management decisions aimed at protecting marine ecosystems from the pervasive threat of plastic pollution.

4.8 Conclusion

Plastic waste is widespread globally, so nearly all marine life has encountered plastic. However, not every interaction causes harm. The number of organisms found with plastics keeps rising, mainly due to more research – almost every species examined has plastic contact. Plastic pollution is a new problem for marine ecosystems dealing with many threats. Plastic waste in these environments causes ecological issues, affecting individual organisms, species groups, and entire ecosystems. Microplastics are readily ingested by many marine and coastal organisms because they resemble food particles. This leads to their buildup in these creatures. The accumulation can harm the affected life forms. Through food webs, microplastics amass in humans and other high-level organisms. In recent years, many global groups and charities have joined forces to tackle plastic pollution in oceans worldwide. Some countries, including in South America, have adopted regional strategies. Several countries have taken legislative action to address the plastic pollution crisis. This involves prohibiting certain plastic items and encouraging the reuse and recycling of plastics through cutting-edge technologies at both regional and national scales. In South American nations, environmental governance with pollution mitigation measures was recommended after thoroughly evaluating the unique biological and ecological characteristics of their respective ecosystems. To examine estuary pollution, focus on plastics in brackish waters. Many countries have banned some plastic products and encouraged reusing and recycling plastics with new technologies nationwide. This paper outlines effective ways to address plastic pollution in aquatic ecosystems. Key solutions include applying the 5Rs for plastic waste, i.e., refuse, reduce, reuse, repurpose, and recycle; encouraging collection of reusable plastic; holding manufacturers accountable through extended producer responsibility; launching eco-friendly initiatives through public–private partnerships; raising awareness and building capacity for environmental cleanliness; conducting research on this emerging challenge; and promoting innovations. These solutions are effective ways to reduce and control plastic pollution in important water environments. The growth of plastic pollution will continue, but how much it increases depends on what governments, businesses, suppliers, and communities do.

References

Abalansa S, Mahrad B, Vondolia G, Icely J, Newton A (2020) The marine plastic litter issue: a social-economic analysis. Sustainability 12(20):8677. https://doi.org/10.3390/su12208677

Acarer S (2023) Microplastics in wastewater treatment plants: sources, properties, removal efficiency, removal mechanisms, and interactions with pollutants. Water Sci Techno 87:685–710 Https://doi.org/10.2166/wst.2023.022

Alexiadou P, Foskolos I, Frantzis A (2019) Ingestion of macroplastics by odontocetes of the Greek Seas, Eastern Mediterranean: Often deadly! Mar Pollut Bull 146:67–75. https://doi.org/10.1016/j.marpolbul.2019.05.055

Allyn EM, Scordino JJ (2020) Entanglement rates and haulout abundance trends of Steller (*Eumetopias jubatus*) and California (*Zalophus californianus*) sea lions on the north coast of Washington state. PloS One 15(8):e0237178. https://doi.org/10.1371/journal.pone.0237178

Almroth BC, Eggert H (2019) Marine plastic pollution: sources, impacts, and policy issues. Review of Environmental Economics and Policy 13(2):317–326 Https://doi.org/10.1093/reep/rez012

Amaneesh C, Anna Balan S, Silpa PS, Kim JW, Greeshma K, Aswathi Mohan A, Robert Antony A, Grossart HP, Kim HS, Ramanan R (2023) Gross negligence: impacts of microplastics and plastic leachates on phytoplankton community and ecosystem dynamics. Environ Sci Technol 57(1):5–24. https://doi.org/10.1021/acs.est.2c05817

Anderson JC, Park BJ, Palace VP (2016) Microplastics in aquatic environments: implications for Canadian ecosystems. Environ Pollut 218:269–280. https://doi.org/10.1016/j.envpol.2016.06.074

Aria M, Cuccurullo C (2017) Bibliometrix: An R-tool for comprehensive science mapping analysis. J Informet 11(4):959–975. https://doi.org/10.1016/j.joi.2017.08.007

Arias AH, Ronda AC, Oliva AL, Marcovecchio JE (2019) Evidence of microplastic ingestion by fish from the Bahía Blanca estuary in Argentina, South America. Bull Environ Contam Toxicol 102(6):750–756. https://doi.org/10.1007/s00128-019-02604-2

Avio CG, Gorbi S, Regoli F (2017) Plastics and microplastics in the oceans: from emerging pollutants to emerged threat. Mar Environ Res 128:2–11

Baak JE, Provencher JF, Mallory ML (2020) Plastic ingestion by four seabird species in the Canadian Arctic: comparisons across species and time. Mar Pollut Bull 158:111386

Banaee M, Soltanian S, Sureda A, Gholamahosseini A, Haghi BN, Akhlaghi M, Derikvandy A (2019) Evaluation of single and combined effects of cadmium and microplastic particles on biochemical and immunological parameters of common carp (*Cyprinus carpio*). Chemosphere 236:124335. https://doi.org/10.1016/j.chemosphere.2019.07.066.10.1016/j.marenvres.2016.05.012

Bancone CE, Turner SD, Ivar do Sul JA, Rose NL (2020) The paleoecology of microplastic contamination. Front Environ Sci 8:574008. https://doi.org/10.3389/fenvs.2020.574008

Bansal M, Santhiya D, Sharma JG (2021) Behavioural mechanisms of microplastic pollutants in marine ecosystem: challenges and remediation measurements. Water Air Soil Pollut 232(9):372. https://doi.org/10.1007/s11270-021-05301-1

Barboza LGA, Vieira LR, Branco V, Figueiredo N, Carvalho F, Carvalho C, Guilhermino L (2018) Microplastics cause neurotoxicity, oxidative damage and energy-related changes and interact with the bioaccumulation of mercury in the European seabass, *Dicentrarchus labrax* (Linnaeus, 1758). Aquat Toxicol 195:49–57. https://doi.org/10.1016/j.aquatox.2017.12.008

Barnes DK, Galgani F, Thompson RC, Barlaz M (2009) Accumulation and fragmentation of plastic debris in global environments. Philos Trans R Soc Lond Ser B Biol Sci 364(1526):1985–1998. https://doi.org/10.1098/rstb.2008.0205

Basto MN, Nicastro KR, Tavares AI, McQuaid CD, Casero M, Azevedo F, Zardi GI (2019) Plastic ingestion in aquatic birds in Portugal. Mar Pollut Bull 138:19–24. https://doi.org/10.1016/j.marpolbul.2018.11.024

Beckwith VK, Fuentes MM (2018) Microplastic at nesting grounds used by the northern Gulf of Mexico loggerhead recovery unit. Mar Pollut Bull 131:32–37 Https://doi.org/10.1016/j.marpolbul.2018.04.001
Bessa F, Barría P, Neto JM, Frias JP, Otero V, Sobral P, Marques JC (2018) Occurrence of microplastics in commercial fish from a natural estuarine environment. Mar Pollut Bull 128:575–584. https://doi.org/10.1016/j.marpolbul.2018.01.044
Bharath KM, Natesan U, Vaikunth R, Kumar PKR, Ruthra R, Srinivasalu S (2021) Spatial distribution of microplastic concentration around landfill sites and its potential risk on groundwater. Chemosphere 277:130263. https://doi.org/10.1016/j.chemosphere.2021.130263
Blettler MC, Mitchell C (2021) Dangerous traps: Macroplastic encounters affecting freshwater and terrestrial wildlife. Sci Total Environ 798:149317. https://doi.org/10.1016/j.scitotenv.2021.149317
Bond AL, Montevecchi WA, Guse N, Regular PM, Garthe S, Rail JF (2012) Prevalence and composition of fishing gear debris in the nests of northern gannets (*Morus bassanus*) are related to fishing effort. Mar Pollut Bull 64(5):907–911. https://doi.org/10.1016/j.marpolbul.2012.03.011
Bordbar L, Kapiris K, Kalogirou S, Anastasopoulou A (2018) First evidence of ingested plastics by a high commercial shrimp species (*Plesionika narval*) in the eastern Mediterranean. Mar Pollut Bull 136:472–476. https://doi.org/10.1016/j.marpolbul.2018.09.030
Boren LJ, Morrissey M, Muller CG, Gemmell NJ (2006) Entanglement of New Zealand fur seals in man-made debris at Kaikoura, New Zealand. Marine Pollution Bulletin 52(4):442–446. https://doi.org/10.1016/j.marpolbul.2005.12.003
Borges-Ramírez MM, Mendoza-Franco EF, Escalona-Segura G, Osten JR (2020) Plastic density as a key factor in the presence of microplastic in the gastrointestinal tract of commercial fishes from Campeche Bay, Mexico. Environ Pollut (Barking, Essex: 1987) 267:115659. https://doi.org/10.1016/j.envpol.2020.115659
Brandts I, Barría C, Martins MA, Franco-Martínez L, Barreto A, Tvarijonaviciute A, Tort L, Oliveira M, Teles M (2021) Waterborne exposure of gilthead seabream (*Sparus aurata*) to polymethylmethacrylate nanoplastics causes effects at cellular and molecular levels. J Hazard Mater 403:123590. https://doi.org/10.1016/j.jhazmat.2020.123590
Brandts I, Teles M, Tvarijonaviciute A, Pereira ML, Martins MA, Tort L, Oliveira M (2018) Effects of polymethylmethacrylate nanoplastics on *Dicentrarchus labrax*. Genomics 110(6):435–441. https://doi.org/10.1016/j.ygeno.2018.10.006
Brooks, A., Jambeck, J., & Mozo-Reyes, E. (2020). Plastic waste management and leakage in Latin America and the Caribbean..
Butterworth A (2016) A review of the welfare impact on pinnipeds of plastic marine debris. Front Mar Sci 3:149. https://doi.org/10.3389/fmars.2016.00149
Canning-Clode J, Sepulveda P, Almeida S, Monteiro J (2020) Will covid-19 containment and treatment measures drive shifts in marine litter pollution? Front Mar Sci 7.https://doi.org/10.3389/fmars.2020.00691
Carretta, J. V., Muto, M., Greenman, J., Wilkinson, K. M., Lawson, D. D., Viezbicke, J., & Jannot, J. E. (2017). Sources of human-related injury and mortality for US Pacific West Coast marine mammal stock assessments, 2011–2015.https://doi.org/10.7289/V5/TM-SWFSC-579
Carson HS, Lamson MR, Nakashima D, Toloumu D, Hafner J, Maximenko N, McDermid KJ (2013) Tracking the sources and sinks of local marine debris in Hawai'i. Mar Environ Res 84:76–83. https://doi.org/10.1016/j.marenvres.2012.12.002
Cau A, Sbrana A, Franceschini S, Fiorentino F, Follesa MC, Galgani F, Garofalo G, Gerigny O, Profeta A, Rinelli P, Sbrana M, Russo T (2024) What, where, and when: Spatial-temporal distribution of macro-litter on the seafloor of the western and central Mediterranean sea. Environ Pollut (Barking, Essex: 1987) 342:123028. https://doi.org/10.1016/j.envpol.2023.123028
Cawthorn MW (1984, November) Entanglement in, and ingestion of, plastic litter by marine mammals, sharks, and turtles in New Zealand waters. In: Proceedings of the workshop on the fate and impact of marine debris, vol 16, p 29

Cesarini G, Secco S, Battisti C, Questino B, Marcello L, Scalici M (2022) Temporal changes of plastic litter and associated encrusting biota: Evidence from Central Italy (Mediterranean Sea). Mar Pollut Bull 181:113890. https://doi.org/10.1016/j.marpolbul.2022.113890

Chae Y, An YJ (2018) Current research trends on plastic pollution and ecological impacts on the soil ecosystem: a review. Environ Pollut (Barking, Essex: 1987) 240:387–395. https://doi.org/10.1016/j.envpol.2018.05.008

Chapron L, Peru E, Engler A, Ghiglione JF, Meistertzheim AL, Pruski AM, Purser A, Vétion G, Galand PE, Lartaud F (2018) Macro- and microplastics affect cold-water corals growth, feeding and behaviour. Sci Rep 8(1):15299. https://doi.org/10.1038/s41598-018-33683-6

Chen YT, Ding DS, Lim YC, Singhania RR, Hsieh S, Chen CW, Hsieh SL, Dong CD (2022) Impact of polyethylene microplastics on coral *Goniopora columna* causing oxidative stress and histopathology damages. Sci Total Environ 828:154234. https://doi.org/10.1016/j.scitotenv.2022.154234

Chen M, Du M, Jin A, Chen S, Dasgupta S, Li J, Xu H, Ta K, Peng X (2020) Forty-year pollution history of microplastics in the largest marginal sea of the western Pacific. Geochem Perspect Lett 13:42–47. https://doi.org/10.7185/geochemlet.2012

Chen Y, Awasthi AK, Wei F, Tan Q, Li J (2021) Single-use plastics: Production, usage, disposal, and adverse impacts. The Science of the Total Environment 752:141772. https://doi.org/10.1016/j.scitotenv.2020.141772

Chukwuone NA, Amaechina EC, Ifelunini IA (2022) Determinants of household's waste disposal practices and willingness to participate in reducing the flow of plastics into the ocean: Evidence from coastal city of Lagos Nigeria. PLoS One 17(4):e0267739. https://doi.org/10.1371/journal.pone.0267739

Clark BL, Carneiro APB, Pearmain EJ, Rouyer MM, Clay TA, Cowger W, Phillips RA, Manica A, Hazin C, Eriksen M, González-Solís J, Adams J, Albores-Barajas YV, Alfaro-Shigueto J, Alho MS, Araujo DT, Arcos JM, Arnould JPY, Barbosa NJP, Barbraud C, Dias MP (2023) Global assessment of marine plastic exposure risk for oceanic birds. Nat Commun 14(1):3665. https://doi.org/10.1038/s41467-023-38900-z

Cole M, Lindeque PK, Fileman E, Clark J, Lewis C, Halsband C, Galloway TS (2016) Microplastics alter the properties and sinking rates of zooplankton faecal pellets. Environ Sci Technol 50(6):3239–3246. https://doi.org/10.1021/acs.est.5b05905

Cole M, Lindeque P, Fileman E, Halsband C, Goodhead R, Moger J, Galloway TS (2013) Microplastic ingestion by zooplankton. Environ Sci Technol 47(12):6646–6655. https://doi.org/10.1021/es400663f

Cole M, Lindeque P, Halsband C, Galloway TS (2011) Microplastics as contaminants in the marine environment: a review. Mar Pollut Bull 62(12):2588–2597. https://doi.org/10.1016/j.marpolbul.2011.09.025

Colmenero AI, Barría C, Broglio E, García-Barcelona S (2017) Plastic debris straps on threatened blue shark *Prionace glauca*. Mar Pollut Bull 115(1–2):436–438. https://doi.org/10.1016/j.marpolbul.2017.01.011

Cordova MR, Hadi TA, Prayudha B (2018) Occurrence and abundance of microplastics in coral reef sediment: a case study in Sekotong, Lombok-Indonesia. AES Bioflux 10(1):23–29

Costa, R. A., Sá, S., Pereira, A. T., Ângelo, A. R., Vaqueiro, J., Ferreira, M., & Eira, C. (2020). Prevalence of entanglements of seabirds in marine debris in the central Portuguese coast. *Marine pollution bulletin*, *161*(Pt A), 111746. https://doi.org/10.1016/j.marpolbul.2020.111746

Coulibaly S, Kamsu-Foguem B, Kamissoko D, Traore D (2019) Deep neural networks with transfer learning in millet crop images. Comput Ind 108:115–120. https://doi.org/10.1016/j.compind.2019.02.003

Covernton GA, Davies HL, Cox KD, El-Sabaawi R, Juanes F, Dudas SE, Dower JF (2021) A Bayesian analysis of the factors determining microplastics ingestion in fishes. J Hazard Mater 413:125405. https://doi.org/10.1016/j.jhazmat.2021.125405

Critchell K, Hoogenboom MO (2018) Effects of microplastic exposure on the body condition and behaviour of planktivorous reef fish (*Acanthochromis polyacanthus*). PLoS ONE 13(3):e0193308. https://doi.org/10.1371/journal.pone.0193308

Curtis S, Elwen SH, Dreyer N, Gridley T (2021) Entanglement of Cape fur seals (*Arctocephalus pusillus pusillus*) at colonies in central Namibia. Mar Pollut Bull 171:112759. https://doi.org/10.1016/j.marpolbul.2021.112759

da Silva Mendes S, de Carvalho RH, de Faria AF, de Sousa BM (2015) Marine debris ingestion by Chelonia mydas (Testudines: Cheloniidae) on the Brazilian coast. Mar Pollut Bull 92(1–2):8–10. https://doi.org/10.1016/j.marpolbul.2015.01.010

Dąbrowska A, Stachowicz M, Szymiczek M (2024) Glitters in fishing ground baits - A direct source of primary microplastics in soil and freshwater ecosystems. Chemosphere 369:143842. https://doi.org/10.1016/j.chemosphere.2024.143842

Dar MA, Dhole NP, Pawar KD, Xie R, Shahnawaz M, Pandit RS, Sun J (2022) Ecotoxic effects of the plastic waste on marine fauna: an overview. In: Impact of plastic waste on the marine biota, pp 287–300. https://doi.org/10.1007/978-981-16-5403-9_15

Das A (2023) The emerging role of microplastics in systemic toxicity: Involvement of reactive oxygen species (ROS). The Science of the Total Environment 895:165076. https://doi.org/10.1016/j.scitotenv.2023.165076

Dasgupta S, Sarraf M, Wheeler D (2022) Plastic waste cleanup priorities to reduce marine pollution: A spatiotemporal analysis for Accra and Lagos with satellite data. The Science of the Total Environment 839:156319. https://doi.org/10.1016/j.scitotenv.2022.156319

Dasgupta S, Peng X, Xu H, Ta K, Chen S, Li J, Du M (2021) Deep seafloor plastics as the source and sink of organic pollutants in the northern South China Sea. The Science of the Total Environment 765:144228. https://doi.org/10.1016/j.scitotenv.2020.144228

Daudt NW, Bugoni L, Nunes GT (2023) Plastics and waterbirds in Brazil: a review of ingestion, nest materials and entanglement reveals substantial knowledge gaps and opportunities for research. Environ Pollut (Barking, Essex: 1987) 316(Pt 2):120615. https://doi.org/10.1016/j.envpol.2022.120615

de Stephanis R, Giménez J, Carpinelli E, Gutierrez-Exposito C, Cañadas A (2013) As main meal for sperm whales: plastics debris. Mar Pollut Bull 69(1–2):206–214. https://doi.org/10.1016/j.marpolbul.2013.01.033

Derraik JG (2002) The pollution of the marine environment by plastic debris: a review. Mar Pollut Bull 44(9):842–852. https://doi.org/10.1016/s0025-326x(02)00220-5

Deshpande PC, Philis G, Brattebø H, Fet AM (2019) Using Material Flow Analysis (MFA) to generate the evidence on plastic waste management from commercial fishing gears in Norway. Resources, Conservation & Recycling: X 5:100024. https://doi.org/10.1016/j.rcrx.2019.100024

Dey U, Chell S, Mondal M, Das K, Raj D, Pandey G, Meraj PK, Almazroui M, Verma S (2024) Potential threat of microplastic pollution on coastal-marine ecosystem—an emerging economic setback and question to blue Economy? Earth Syst Environ 1–29. https://doi.org/10.1007/s41748-024-00485-y

Digka N, Patsiou D, Hatzonikolakis Y, Raitsos DE, Skia G, Koutsoubas D, Dimitriadis C, Tsangaris C (2024) Microplastic ingestion in mussels from the East Mediterranean Sea: Exploring its impacts in nature and controlled conditions. The Science of the Total Environment 946:174268. https://doi.org/10.1016/j.scitotenv.2024.174268

Ding J, Zhang S, Razanajatovo RM, Zou H, Zhu W (2018) Accumulation, tissue distribution, and biochemical effects of polystyrene microplastics in the freshwater fish red tilapia (*Oreochromis niloticus*). Environ Pollut 238:1–9. https://doi.org/10.1016/j.envpol.2018.03.001

Dixit A, Jakhar SK (2021) Airport capacity management: A review and bibliometric analysis. J Air Transp Manag 91:102010. https://doi.org/10.1016/j.jairtraman.2020.102010

Ivar do Sul, J. A., & Costa, M. F. (2014). The present and future of microplastic pollution in the marine environment. *Environmental pollution (Barking, Essex: 1987)*, *185*, 352–364. https://doi.org/10.1016/j.envpol.2013.10.036

Do H, Armstrong CW (2023) Ghost fishing gear and their effect on ecosystem services – Identification and knowledge gaps. Mar Policy 150:105528. https://doi.org/10.1016/j.marpol.2023.105528

Dong Y, Gao M, Song Z, Qiu W (2020) As(III) adsorption onto different-sized polystyrene microplastic particles and its mechanism. Chemosphere 239:124792. https://doi.org/10.1016/j.chemosphere.2019.124792

Donnelly-Greenan EL, Nevins HM, Harvey JT (2019) Entangled seabird and marine mammal reports from citizen science surveys from coastal California (1997–2017). Mar Pollut Bull 149:110557. https://doi.org/10.1016/j.marpolbul.2019.110557

Donthu N, Kumar S, Mukherjee D, Pandey N, Lim WM (2021) How to conduct a bibliometric analysis: An overview and guidelines. J Bus Res 133:285–296. https://doi.org/10.1016/j.jbusres.2021.04.070

Duncan EM, Broderick AC, Fuller WJ, Galloway TS, Godfrey MH, Hamann M, Limpus CJ, Lindeque PK, Mayes AG, Omeyer LCM, Santillo D, Snape RTE, Godley BJ (2019) Microplastic ingestion ubiquitous in marine turtles. Glob Change Biol 25(2):744–752. https://doi.org/10.1111/gcb.14519

Duncan EM, Botterell ZL, Broderick AC, Galloway TS, Lindeque PK, Nuno A, Godley BJ (2017) A global review of marine turtle entanglement in anthropogenic debris: a baseline for further action. Endangered Species Research 34:431–448. https://doi.org/10.3354/esr00865

Elhacham E, Ben-Uri L, Grozovski J, Bar-On YM, Milo R (2020) Global human-made mass exceeds all living biomass. Nature 588(7838):442–444. https://doi.org/10.1038/s41586-020-3010-5

Emmerik T, Loozen M, Oeveren K, Buschman F, Prinsen G (2019) Riverine plastic emission from Jakarta into the ocean. Environ Res Lett 14(8):084033 Https://doi.org/10.1088/1748-9326/ab30e8

Ennen JR, Cecala KK, Gould P, Colvin R, Denison J, Garig DF, Hyder S, Recker L, Davenport JM (2021) Size matters: the influence of trap and mesh size on turtle captures. Wildl Soc Bull 45(1):130–137https://doi.org/10.1002/wsb.1163

Eriksen M, Lebreton LC, Carson HS, Thiel M, Moore CJ, Borerro JC, Galgani F, Ryan PG, Reisser J (2014) Plastic pollution in the world's oceans: more than 5 trillion plastic pieces weighing over 250,000 tons afloat at sea. PLoS ONE 9(12):e111913. https://doi.org/10.1371/journal.pone.0111913

Erkan HS, Turan NB, Albay M, Engin GO (2021) A preliminary study on the distribution and morphology of microplastics in the coastal areas of Istanbul, the metropolitan city of Turkey: The effect of location differences. J Clean Prod 307:127320. https://doi.org/10.1016/j.jclepro.2021.127320

Espinosa C, Cuesta A, Esteban MÁ (2017) Effects of dietary polyvinylchloride microparticles on general health, immune status and expression of several genes related to stress in gilthead seabream (*Sparus aurata* L.). Fish Shellfish Immunol 68:251–259. https://doi.org/10.1016/j.fsi.2017.07.006

Espinosa C, Esteban MÁ, Cuesta A (2019) Dietary administration of PVC and PE microplastics produces histological damage, oxidative stress and immunoregulation in European sea bass (*Dicentrarchus labrax* L.). Fish Shellfish Immunol 95:574–583. https://doi.org/10.1016/j.fsi.2019.10.072

Estrella-Jordan BA, Lango-Reynoso F, Castañeda-Chávez MDR, Montoya-Mendoza J, Reynier-Valdes D (2023) Microplastic pollution in sea turtle nests on the beaches of Nautla and Vega de Alatorre, Veracruz. Microplastics 2(2):182–191. https://doi.org/10.3390/microplastics2020014

Expósito N, Rovira J, Sierra J, Folch J, Schuhmacher M (2021) Microplastics levels, size, morphology and composition in marine water, sediments and sand beaches. Case study of Tarragona coast (western Mediterranean). Sci Total Environ 786:147453. https://doi.org/10.1016/j.scitotenv.2021.147453

Fife DT, Robertson GJ, Shutler D, Braune BM, Mallory ML (2015) Trace elements and ingested plastic debris in wintering dovekies (Alle alle). Mar Pollut Bull 91(1):368–371. https://doi.org/10.1016/j.marpolbul.2014.11.029

Focardi, A., Moore, L., Raina, J., Seymour, J., Paulsen, I., & Tetu, S. (2022). Plastic leachates impair picophytoplankton and dramatically reshape the marine microbiome. *Microbiome*, 10(1). Https://doi.org/https://doi.org/10.1186/s40168-022-01369-x

Franco-Trecu V, Drago M, Katz H, Machín E, Marín Y (2017) With the noose around the neck: marine debris entangling otariid species. Environ Pollut (Barking, Essex: 1987) 220(Pt B):985–989. https://doi.org/10.1016/j.envpol.2016.11.057

Gall SC, Thompson RC (2015) The impact of debris on marine life. Mar Pollut Bull 92(1–2):170–179. https://doi.org/10.1016/j.marpolbul.2014.12.041

Garai S, Bhattacharjee C, Sarkar S, Moulick D, Dey S, Jana S, Dhar A, Roy A, Mondal K, Mondal M, Mukherjee S, Ghosh S, Singh P, Ramteke P, Manna D, Hazra S, Malakar P, Banerjee H, Brahmachari K, Hossain A (2024) Microplastics in the soil-water-food nexus: Inclusive insight into global research findings. The Science of the Total Environment 946:173891. https://doi.org/10.1016/j.scitotenv.2024.173891

Garcés-Ordóñez O, Mejía-Esquivia KA, Sierra-Labastidas T, Patiño A, Blandón LM, Espinosa Díaz LF (2020) Prevalence of microplastic contamination in the digestive tract of fishes from mangrove ecosystem in Cispata, Colombian Caribbean. Mar Pollut Bull 154:111085. https://doi.org/10.1016/j.marpolbul.2020.111085

Germanov ES, Marshall AD, Hendrawan IG, Admiraal R, Rohner CA, Argeswara J, Wulandari R, Himawan MR, Loneragan NR (2019) Microplastics on the menu: plastics pollute Indonesian manta ray and whale shark feeding grounds. Front Mar Sci 6:487857. https://doi.org/10.3389/fmars.2019.00679

Geyer R, Jambeck JR, Law KL (2017) Production, use, and fate of all plastics ever made. Sci Adv 3(7):e1700782. https://doi.org/10.1126/sciadv.1700782

Ghaouar H, Boussellaa W, Jribi I (2024) Ghost gears in the Gulf of Gabès: alarming situation and sustainable solution perspectives. Sustainability 16(7):2632. https://doi.org/10.3390/su16072632

Gilardi KV, Carlson-Bremer D, June JA, Antonelis K, Broadhurst G, Cowan T (2010) Marine species mortality in derelict fishing nets in Puget Sound, WA and the cost/benefits of derelict net removal. Mar Pollut Bull 60(3):376–382. https://doi.org/10.1016/j.marpolbul.2009.10.016

Gilman E (2015) Status of international monitoring and management of abandoned, lost and discarded fishing gear and ghost fishing. Mar Policy 60:225–239. https://doi.org/10.1016/j.marpol.2015.06.016

Gilman E, Antonelis K, Drinkwin J, Gorgin S, Suuronen P, Thomas SN, Wilson J (2023) Introduction to the Marine Policy special issue on abandoned, lost and discarded fishing gear: Causes, magnitude, impacts, mitigation methods and priorities for monitoring and evidence-informed management. Mar Policy 155:105738. https://doi.org/10.1016/j.marpol.2023.105738

Grundlehner A, Diepens NJ, Linders T, Peeters ETHM, Koelmans AA (2023) Towards continuous mass and size distributions for beach plastic litter: Spatiotemporal analyses of abundance and composition. J Hazard Mater 458:131984. https://doi.org/10.1016/j.jhazmat.2023.131984

Hamed M, Soliman HAM, Osman AGM, Sayed AEH (2019) Assessment the effect of exposure to microplastics in Nile Tilapia (*Oreochromis niloticus*) early juvenile: I. blood biomarkers. Chemosphere 228:345–350. https://doi.org/10.1016/j.chemosphere.2019.04.153

Hankins C, Duffy A, Drisco K (2018) Scleractinian coral microplastic ingestion: Potential calcification effects, size limits, and retention. Mar Pollut Bull 135:587–593. https://doi.org/10.1016/j.marpolbul.2018.07.067

Hardesty BD, Roman L, Duke NC, Mackenzie JR, Wilcox C (2021) Abandoned, lost and discarded fishing gear 'ghost nets' are increasing through time in Northern Australia. Mar Pollut Bull 173(Pt A):112959. https://doi.org/10.1016/j.marpolbul.2021.112959

Harrison JP, Schratzberger M, Sapp M, Osborn AM (2014) Rapid bacterial colonization of low-density polyethylene microplastics in coastal sediment microcosms. BMC Microbiol 14:232. https://doi.org/10.1186/s12866-014-0232-4

Hartmann NB, Hüffer T, Thompson RC, Hassellöv M, Verschoor A, Daugaard AE, Rist S, Karlsson T, Brennholt N, Cole M, Herrling MP, Hess MC, Ivleva NP, Lusher AL, Wagner M (2019) Are we speaking the same language? Recommendations for a definition and categorization framework for plastic debris. Environ Sci Technol 53(3):1039–1047. https://doi.org/10.1021/acs.est.8b05297

Henry B, Laitala K, Klepp IG (2019) Microfibres from apparel and home textiles: Prospects for including microplastics in environmental sustainability assessment. Sci Total Environ 652:483–494. https://doi.org/10.1016/j.scitotenv.2018.10.166

Horton AA, Walton A, Spurgeon DJ, Lahive E, Svendsen C (2017) Microplastics in freshwater and terrestrial environments: Evaluating the current understanding to identify the knowledge gaps and future research priorities. The Science of the Total Environment 586:127–141. https://doi.org/10.1016/j.scitotenv.2017.01.190

Hoseinifar SH, Yousefi S, Van Doan H, Ashouri G, Gioacchini G, Maradonna F, Carnevali O (2020) Oxidative stress and antioxidant defense in fish: the implications of probiotic, prebiotic, and synbiotics. Reviews in Fisheries Science & Aquaculture 29(2):198–217. https://doi.org/10.1080/23308249.2020.1795616

Hu M, Palić D (2020) Micro- and nano-plastics activation of oxidative and inflammatory adverse outcome pathways. Redox Biol 37:101620. https://doi.org/10.1016/j.redox.2020.101620

Hutchinson MR, Itano DG, Muir JA, Holland KN (2015) Post-release survival of juvenile silky sharks captured in a tropical tuna purse seine fishery. Mar Ecol Prog Ser 521:143–154. https://doi.org/10.3354/meps11073

Isangedighi IA, David GS, Obot OI (2020) Plastic waste in the aquatic environment: impacts and management. In: Analysis of nanoplastics and microplastics in food. CRC Press, pp 15–43. https://doi.org/10.31058/j.envi.2018.21001

Jepsen EM, de Bruyn PJN (2019) Pinniped entanglement in oceanic plastic pollution: A global review. Mar Pollut Bull 145:295–305. https://doi.org/10.1016/j.marpolbul.2019.05.042

Jonathan MP, Sujitha SB, Rodriguez-Gonzalez F, Villegas LEC, Hernández-Camacho CJ, Sarkar SK (2021) Evidences of microplastics in diverse fish species off the Western Coast of Pacific Ocean, Mexico. Ocean Coast Manag. https://doi.org/10.1016/j.ocecoaman.2021.105544204, 105544.

Jovanović B (2017) Ingestion of microplastics by fish and its potential consequences from a physical perspective. Integr Environ Assess Manag 13(3):510–515. https://doi.org/10.1002/ieam.1913

Jovanović B, Gökdağ K, Güven O, Emre Y, Whitley EM, Kideys AE (2018) Virgin microplastics are not causing imminent harm to fish after dietary exposure. Mar Pollut Bull 130:123–131. https://doi.org/10.1016/j.marpolbul.2018.03.016

Karamanlidis AA, Androukaki E, Adamantopoulou S, Chatzispyrou A, Johnson WM, Kotomatas S, Dendrinos P (2008) Assessing accidental entanglement as a threat to the Mediterranean monk seal *Monachus monachus*. Endangered Species Research 5(2–3):205–213. https://doi.org/10.3354/esr00092

Katsanevakis S (2008) Marine debris, a growing problem: Sources, distribution, composition, and impacts. New Research. Nova Science Publishers, New York, Marine Pollution, pp 53–100

Kim JH, Yu YB, Choi JH (2021) Toxic effects on bioaccumulation, hematological parameters, oxidative stress, immune responses and neurotoxicity in fish exposed to microplastics: A review. J Hazard Mater 413:125423. https://doi.org/10.1016/j.jhazmat.2021.125423

Koelmans A, Bakir A, Burton G, Janssen C (2016) Microplastic as a vector for chemicals in the aquatic environment: critical review and model-supported reinterpretation of empirical studies. Environ Sci Technol 50(7):3315–3326. https://doi.org/10.1021/acs.est.5b06069

Kruse, K., Knickmeier, K., Brennecke, D., Unger, B., & Siebert, U. (2023). Plastic debris and its impacts on marine mammals. In *Marine mammals: a deep dive into the world of science* (pp. 49–62). Cham: Springer International Publishing. https://doi.org/10.1007/978-3-031-06836-2_4

Kurtela A, Antolović N (2019) The problem of plastic waste and microplastics in the seas and oceans: impact on marine organisms. Croatian Journal of Fisheries 77(1):51–56. https://doi.org/10.2478/cjf-2019-0005

Lam TWL, Tsui YCJ, Cheng YL, Ma ATH, Fok L (2023) Microplastic contamination in edible clams from popular recreational clam-digging sites in Hong Kong and implications for human

health. The Science of the Total Environment 875:162576. https://doi.org/10.1016/j.scitotenv.2023.162576
Landrigan, P. J., Stegeman, J. J., Fleming, L. E., Allemand, D., Anderson, D. M., Backer, L. C., Brucker-Davis, F., Chevalier, N., Corra, L., Czerucka, D., Bottein, M. D., Demeneix, B., Depledge, M., Deheyn, D. D., Dorman, C. J., Fénichel, P., Fisher, S., Gaill, F., Galgani, F., Gaze, W. H., … Rampal, P. (2020). Human Health and Ocean Pollution. *nnals of global health*, *86*(1), 151. https://doi.org/10.5334/aogh.2831
Lawson TJ, Wilcox C, Johns K, Dann P, Hardesty BD (2015) Characteristics of marine debris that entangle Australian fur seals (*Arctocephalus pusillus doriferus*) in southern Australia. Mar Pollut Bull 98(1–2):354–357. https://doi.org/10.1016/j.marpolbul.2015.05.053
Lebreton LCM, van der Zwet J, Damsteeg JW, Slat B, Andrady A, Reisser J (2017) River plastic emissions to the world's oceans. Nat Commun 8:15611. https://doi.org/10.1038/ncomms15611
Lei K, Qiao F, Liu Q, Wei Z, Qi H, Cui S, Yue X, Deng Y, An L (2017) Microplastics releasing from personal care and cosmetic products in China. Mar Pollut Bull 123(1–2):122–126. https://doi.org/10.1016/j.marpolbul.2017.09.016
Lei L, Wu S, Lu S, Liu M, Song Y, Fu Z, Shi H, Raley-Susman KM, He D (2018) Microplastic particles cause intestinal damage and other adverse effects in zebrafish *Danio rerio* and nematode Caenorhabditis elegans. The Science of the Total Environment 619–620:1–8. https://doi.org/10.1016/j.scitotenv.2017.11.103
Li H, Chen H, Wang J, Li J, Liu S, Tu J, Chen Y, Zong Y, Zhang P, Wang Z, Liu X (2021) Influence of microplastics on the growth and the intestinal microbiota composition of Brine Shrimp. Front Microbiol 12:717272. https://doi.org/10.3389/fmicb.2021.717272
Li H, Liu H, Bi L, Liu Y, Jin L, Peng R (2024) Immunotoxicity of microplastics in fish. Fish Shellfish Immunol 150:109619. https://doi.org/10.1016/j.fsi.2024.109619
Li J, Lusher AL, Rotchell JM, Deudero S, Turra A, Bråte ILN, Sun C, Shahadat Hossain M, Li Q, Kolandhasamy P, Shi H (2019) Using mussel as a global bioindicator of coastal microplastic pollution. Environ Pollut (Barking, Essex: 1987) 244:522–533. https://doi.org/10.1016/j.envpol.2018.10.032
Liu K, Wu T, Wang X, Song Z, Zong C, Wei N, Li D (2019) Consistent transport of terrestrial microplastics to the ocean through atmosphere. Environ Sci Technol 53(18):10612–10619. https://doi.org/10.1021/acs.est.9b03427
Ma J, Zhao J, Zhu Z, Li L, Yu F (2019) Effect of microplastic size on the adsorption behavior and mechanism of triclosan on polyvinyl chloride. Environ Pollut 254:113104. https://doi.org/10.1016/j.envpol.2019.113104
Macleod M, Arp HPH, Tekman MB, Jahnke A (2021) The global threat from plastic pollution. Science 373(6550):61–65. https://doi.org/10.1126/science.aba7799
Maes T, Barry J, Leslie H, Vethaak A, Nicolaus E, Law R, Lyons B, Martinez R, Harley B, Thain J (2018) Below the surface: Twenty-five years of seafloor litter monitoring in coastal seas of North West Europe (1992–2017). Sci Total Environ 630:790–798. https://doi.org/10.1016/j.scitotenv.2018.02.247
Markic A, Gaertner J-C, Gaertner-Mazouni N, Koelmans AA (2020) Plastic ingestion by marine fish in the wild. Crit Rev Environ Sci Technol 50(7):657–697. https://doi.org/10.1080/10643389.2019.1627993
Martin C, Baalkhuyur F, Valluzzi L, Saderne V, Cusack M, Almahasheer H, Krishnakumar P, Rabaoui L, Qurban M, Arias-Ortiz A (2020) Exponential increase of plastic burial in mangrove sediments as a major plastic sink. Sci Adv 6(44):eaaz5593. https://doi.org/10.1126/sciadv.aaz5593)
Ma'ruf M (2019) Indonesia response and recent development of law and policy in addressing marine plastic litter. J Indonesian Legal Stud 4(2):167–188. https://doi.org/10.15294/jils.v4i2.34757)
Meaza I, Toyoda J, Wise J (2021) Microplastics in sea turtles, marine mammals and humans: a one environmental health perspective. Front Environ Sci 8.https://doi.org/10.3389/fenvs.2020.575614

Meem RA, Ahmed A, Maraz KM, Md Shamim Hossain Khan RA (2021) A review on the impact of plastic debris on marine environment. Mod Concepts Mater Sci 4:1–7. https://doi.org/10.33552/MCMS.2021.04.000591

Meijer L, Emmerik T, Ent R, Schmidt C, Lebreton L (2021) More than 1000 rivers account for 80% of global riverine plastic emissions into the ocean. Sci Adv 7(18):eaba5803. https://doi.org/10.1126/sciadv.aaz5803

Moore E, Lyday S, Roletto J, Litle K, Parrish JK, Nevins H, Harvey J, Mortenson J, Greig D, Piazza M (2009) Entanglements of marine mammals and seabirds in central California and the north-west coast of the United States 2001–2005. Mar Pollut Bull 58(7):1045–1051. https://doi.org/10.1016/j.marpolbul.2009.02.016

Morales-Caselles C, Viejo J, Martí E, González-Fernández D, Pragnell-Raasch H, González-Gordillo JI, Montero E, Arroyo GM, Hanke G, Salvo VS (2021) An inshore–offshore sorting system revealed from global classification of ocean litter. Nat Sustain 4(6):484–493. https://doi.org/10.1038/s41893-021-00694-z

Morgan E, Hutchinson D, Gaion A (2021) Plastic ingestion by the small-spotted catshark (*Scyliorhinus canicula*) from the south west coast of the United Kingdom. Bull Environ Contam Toxicol 106(6):910–915. https://doi.org/10.1007/s00128-021-03144-4

Munoz M, Ortiz D, Nieto-sandoval J, De Pedro ZM, Casas JA (2021) Adsorption of micropollutants onto realistic microplastics: Role of microplastic nature, size, age, and NOM fouling. Chemosphere 283:131085. https://doi.org/10.1016/j.chemosphere.2021.131085

Naidoo T, Glassom D (2019) Decreased growth and survival in small juvenile fish, after chronic exposure to environmentally relevant concentrations of microplastic. Mar Pollut Bull 145:254–259. https://doi.org/10.1016/j.marpolbul.2019.05.050

Oliveira M, Ribeiro A, Hylland K, Guilhermino L (2013) Single and combined effects of microplastics and pyrene on juveniles (0+ group) of the common goby *Pomatoschistus microps* (Teleostei, Gobiidae). Ecol Ind 34:641–647. https://doi.org/10.1016/j.ecolind.2013.06.019

Padula AD, Ronda AC, Rodríguez Pirani LS, Picone AL, Romano RM, Giardino GV, Seco Pon JP, Machado R, Gerpe M, Rodriguez D, Denuncio P (2025) Microplastics in the digestive tract of an endangered cetacean of the Southwest Atlantic Ocean: The franciscana dolphin. Mar Pollut Bull 210:117348. https://doi.org/10.1016/j.marpolbul.2024.117348

Parga Martínez KB, Tekman MB, Bergmann M (2020) Temporal trends in marine litter at three stations of the HAUSGARTEN observatory in the Arctic deep sea. Front Mar Sci 7:321. https://doi.org/10.3389/fmars.2020.00321

Parton KJ, Galloway TS, Godley BJ (2019) Global review of shark and ray entanglement in anthropogenic marine debris. Endangered Species Research 39:173–190. https://doi.org/10.3354/esr00976

Perez-Venegas DJ, Valenzuela-Sánchez A, Montalva F, Pavés H, Seguel M, Wilcox C, Galbán-Malagón C (2021) Towards understanding the effects of oceanic plastic pollution on population growth for a South American fur seal (*Arctocephalus australis australis*) colony in Chile. Environ Pollut 279:116881. https://doi.org/10.1016/j.envpol.2021.116881

Pestana, G. (2023). Digital chain of custody for a blue circular economy. *European Conference on Knowledge Management, 24*(2), 1034–1041. https://doi.org/10.34190/eckm.24.2.1395

Pierce KE, Harris RJ, Larned LS, Pokras MA (2004) Obstruction and starvation associated with plastic ingestion in a Northern Gannet *Morus bassanus* and a Greater Shearwater *Puffinus gravis*. Mar Ornithol 32:187–189

Pojar I, Kochleus C, Dierkes G, Ehlers SM, Reifferscheid G, Stock F (2021) Quantitative and qualitative evaluation of plastic particles in surface waters of the Western Black Sea. Environ Pollut 268:115724. https://doi.org/10.1016/j.envpol.2020.115724

Prampramote J, Boonhoh W, Intongead S, Sakornwimol W, Prachamkhai P, Sansamur C, Hayakijkosol O, Wongtawan T (2022) Association of ocean macroplastic debris with stranded sea turtles in the Central Gulf of Thailand. Endangered Species Research 47:333–343. https://doi.org/10.3354/esr01155

Prata JC, da Costa JP, Lopes I, Duarte AC, Rocha-Santos T (2020) Environmental exposure to microplastics: An overview on possible human health effects. Sci Total Environ 702:134455. https://doi.org/10.1016/j.scitotenv.2019.134455

Prokić MD, Gavrilović BR, Radovanović TB, Gavrić JP, Petrović TG, Despotović SG, Faggio C (2021) Studying microplastics: Lessons from evaluated literature on animal model organisms and experimental approaches. J Hazard Mater 414:125476. https://doi.org/10.1016/j.jhazmat.2021.125476

Provencher JF, Bond AL, Avery-Gomm S, Borrelle SB, Rebolledo ELB, Hammer S, Kühn S, Lavers JL, Mallory ML, Trevail A (2017) Quantifying ingested debris in marine megafauna: A review and recommendations for standardization. Anal Methods 9(9):1454–1469. https://doi.org/10.1039/C7AY00735E

Provencher JF, Borrelle SB, Bond AL, Lavers JL, Van Franeker JA, Kühn S, Hammer S, Avery-Gomm S, Mallory ML (2019) Recommended best practices for plastic and litter ingestion studies in marine birds: Collection, processing, and reporting. Facets 4(1):111–130. https://doi.org/10.1139/facets-2018-0044

Qi R, Jones DL, Li Z, Liu Q, Yan C (2020) Behavior of microplastics and plastic film residues in the soil environment: A critical review. Sci Total Environ 703:134722. https://doi.org/10.1016/j.scitotenv.2019.134722

Rangel-Buitrago N, Neal WJ, Galgani F (2023) Plastics in the Anthropocene: A multifaceted approach to marine pollution management. Mar Pollut Bull 194:115359. https://doi.org/10.1016/j.marpolbul.2023.115359

Râpă M (2024) An overview of the current trends in marine plastic litter management for a sustainable development. Recycling 9(2):30. https://doi.org/10.3390/recycling9020030

Reichert J, Schellenberg J, Schubert P, Wilke T (2018) Responses of reef building corals to microplastic exposure. Environ Pollut 237:955–960. https://doi.org/10.1016/j.envpol.2017.11.006

Richardson K, Hardesty BD, Vince J, Wilcox C (2022) Global estimates of fishing gear lost to the ocean each year. Sci Adv 8(41):eabq0135. https://doi.org/10.1126/sciadv.abq0135

Richardson K, Hardesty BD, Vince JZ, Wilcox C (2021) Global causes, drivers, and prevention measures for lost fishing gear. Front Mar Sci 8:690447. https://doi.org/10.3389/fmars.2021.690447

Robuck AR, Hudak CA, Agvent L, Emery G, Ryan PG, Perold V, Powers KD, Pedersen J, Thompson MA, Suca JJ (2022) Birds of a feather eat plastic together: High levels of plastic ingestion in Great Shearwater adults and juveniles across their annual migratory cycle. Front Mar Sci 8:719721. https://doi.org/10.3389/fmars.2021.719721

Rochman C (2015) The complex mixture, fate and toxicity of chemicals associated with plastic debris in the marine environment, pp 117–140. https://doi.org/10.1007/978-3-319-16510-3_5

Rochman CM, Hoh E, Kurobe T, Teh SJ (2013) Ingested plastic transfers hazardous chemicals to fish and induces hepatic stress. Sci Rep 3:3263. https://doi.org/10.1038/srep03263

Rodríguez B, Bécares J, Rodríguez A, Arcos JM (2013) Incidence of entanglements with marine debris by northern gannets (*Morus bassanus*) in the non-breeding grounds. Mar Pollut Bull 75(1–2):259–263. https://doi.org/10.1016/j.marpolbul.2013.07.003

Roman L, Hardesty BD, Hindell MA, Wilcox C (2019) A quantitative analysis linking seabird mortality and marine debris ingestion. Sci Rep 9(1):3202. https://doi.org/10.1038/s41598-018-36585-9

Roman L, Butcher RG, Stewart D, Hunter S, Jolly M, Kowalski P, Hardesty BD, Lenting B (2021) Plastic ingestion is an underestimated cause of death for southern hemisphere albatrosses. Conserv Lett 14(3):e12785. https://doi.org/10.1111/conl.12785

Saavedra F (2024) Sea of plastic: representations of the sea and pro-environmental attitudes and behaviors regarding marine plastic pollution in Peru and Chile. Front Psychol 14. https://doi.org/10.3389/fpsyg.2023.1308796

Salazar-Casals A, de Reus K, Greskewitz N, Havermans J, Geut M, Villanueva S, Rubio-Garcia A (2022, August) Increased incidence of entanglements and ingested marine debris in Dutch seals from 2010 to 2020. Oceans 3(3):389–400. https://doi.org/10.3390/oceans3030026

Schmidt C, Kühnel D, Materić D, Stubenrauch J, Schubert K, Luo A, Wendt-Potthoff K, Jahnke A (2024) A multidisciplinary perspective on the role of plastic pollution in the triple planetary crisis. Environ Int 193:109059. https://doi.org/10.1016/j.envint.2024.109059

Schmidt C, Krauth T, Wagner S (2017) Export of plastic debris by rivers into the sea. Environ Sci Technol 51(21):12246–12253. https://doi.org/10.1021/acs.est.7b02368

Setälä O, Lehtiniemi M, Coppock R, Cole M (2018) Microplastics in marine food webs. In: Microplastic contamination in aquatic environments. Elsevier, pp 339–363. https://doi.org/10.1021/acs.est.7b02368

Shi Y (2023) Analysis of marine plastic pollution and environmental problems. In Highlights in Science engineering and technology, vol 59, pp 56–61. https://doi.org/10.54097/hset.v59i.10049

Simeonova A, Chuturkova R (2020) Macroplastic distribution (Single-use plastics and some Fishing gear) from the northern to the southern Bulgarian Black Sea coast. Regional Studies in Marine Science 37:101329. https://doi.org/10.1016/j.rsma.2020.101329

Sjollema SB, Redondo-Hasselerharm P, Leslie HA, Kraak MHS, Vethaak AD (2016) Do plastic particles affect microalgal photosynthesis and growth? Aquatic Toxicology (Amsterdam, Netherlands) 170:259–261. https://doi.org/10.1016/j.aquatox.2015.12.002

Solomando A, Capó X, Alomar C, Álvarez E, Compa M, Valencia JM, Pinya S, Deudero S, Sureda A (2020) Long-term exposure to microplastics induces oxidative stress and a pro-inflammatory response in the gut of *Sparus aurata* Linnaeus, 1758. Environ Pollut (Barking, Essex: 1987) 266(Pt 1):115295. https://doi.org/10.1016/j.envpol.2020.115295

Sonke J, Koenig A, Segur T, Yakovenko N (2024) Global environmental plastics dispersal under OECD policy scenarios towards 2060. https://doi.org/10.31223/X5W990

Stelfox M, Hudgins J, Sweet M (2016) A review of ghost gear entanglement amongst marine mammals, reptiles and elasmobranchs. Mar Pollut Bull 111(1–2):6–17. https://doi.org/10.1016/j.marpolbul.2016.06.034

Su L, Deng H, Li B, Chen Q, Pettigrove V, Wu C, Shi H (2019) The occurrence of microplastic in specific organs in commercially caught fishes from coast and estuary area of east China. J Hazard Mater 365:716–724. https://doi.org/10.1016/j.jhazmat.2018.11.024

Sun Y, Xu W, Gu Q, Chen Y, Zhou Q, Zhang L, Gu L, Huang Y, Lyu K, Yang Z (2019) Small-Sized Microplastics Negatively Affect Rotifers: Changes in the Key Life-History Traits and Rotifer-*Phaeocystis* Population Dynamics. Environ Sci Technol 53(15):9241–9251. https://doi.org/10.1021/acs.est.9b02893

Suwaki CH, De-La-Cruz LT, Lopes RM (2020) Impacts of microplastics on the swimming behavior of the copepod *Temora turbinata* (Dana, 1849). Fluids 5(3):103. https://doi.org/10.3390/fluids5030103

Tan E, Jaafar N, Tan S, Zanuri N (2022) A review of plastic and microplastic pollution towards the malaysian marine environment. IOP Conf Ser Earth Environ Sci 1013(1):012012. https://doi.org/10.1088/1755-1315/1013/1/012012

Tetu SG, Sarker I, Schrameyer V, Pickford R, Elbourne LDH, Moore LR, Paulsen IT (2019) Plastic leachates impair growth and oxygen production in *Prochlorococcus*, the ocean's most abundant photosynthetic bacteria. Commun Biol 184. https://doi.org/10.1038/s42003-019-0410-x

Thiel M, Luna-Jorquera G, Álvarez-Varas R, Gallardo C, Hinojosa IA, Luna N, Miranda-Urbina D, Morales N, Ory N, Pacheco AS, Portflitt-Toro M (2018) Impacts of marine plastic pollution from continental coasts to subtropical gyres—fish, seabirds, and other vertebrates in the SE Pacific. Front Mar Sci 5:238. https://doi.org/10.3389/fmars.2018.00238

Thushari GGN, Senevirathna JDM (2020) Plastic pollution in the marine environment. Heliyon 6(8):e04709. https://doi.org/10.1016/j.heliyon.2020.e04709

Tian H, Liu L (2024) Effects of antibiotics and microplastics on the growth of Sebastes schlegelii. In: E3S web of conferences, vol 536, p 02008. EDP Sciences. https://doi.org/10.1051/e3sconf/202453602008

Turner A, Amos SL, Williams T (2021) Coastal dunes as a sink and secondary source of marine plastics: a study at Perran Beach, southwest England. Mar Pollut Bull 173(Pt B):113133. https://doi.org/10.1016/j.marpolbul.2021.113133

Tuuri EM, Leterme SC (2023) How plastic debris and associated chemicals impact the marine food web: a review. Environ Pollut (Barking, Essex: 1987) 321:121156. https://doi.org/10.1016/j.envpol.2023.121156

Tyllianakis E, Ferrini S (2021) Personal attitudes and beliefs and willingness to pay to reduce marine plastic pollution in indonesia. Mar Pollut Bull 173:113120. https://doi.org/10.1016/j.marpolbul.2021.113120

Urban-Malinga B, Zalewski M, Jakubowska A, Wodzinowski T, Malinga M, Pałys B, Dąbrowska A (2020) Microplastics on sandy beaches of the southern Baltic Sea. Mar Pollut Bull 155:111170. https://doi.org/10.1016/j.marpolbul.2020.111170

Van Dyck IP, Nunoo FK, Lawson ET (2016) An empirical assessment of marine debris, seawater quality and littering in Ghana. J Geosci Environ Protect 4(5):21–36. https://doi.org/10.4236/gep.2016.45003

Van Eck NJ, Waltman L (2019) Vosviewer Manual. Version 1.6.13. Universiteit Leiden. https://www.vosviewer.com/documentation/Manual_vosviewer_1.6.13.pdf

Vega-Herrera A, Llorca M, Savva K, León VM, Abad E, Farré M (2021) Screening and quantification of micro (nano) plastics and plastic additives in the seawater of Mar Menor Lagoon. Front Mar Sci 8:697424. https://doi.org/10.3389/fmars.2021.697424

Votier SC, Archibald K, Morgan G, Morgan L (2011) The use of plastic debris as nesting material by a colonial seabird and associated entanglement mortality. Mar Pollut Bull 62(1):168–172. https://doi.org/10.1016/j.marpolbul.2010.11.009

Watson AR, Blount C, McPhee DP, Zhang D, Smith MPL, Reeds K, Williamson JE (2022) Source, fate and management of recreational fishing marine debris. Mar Pollut Bull 178:113500. https://doi.org/10.1016/j.marpolbul.2022.113500

Wedemeyer-Strombel KR, Balazs GH, Johnson JB, Peterson TD, Wicksten MK, Plotkin PT (2015) High frequency of occurrence of anthropogenic debris ingestion by sea turtles in the North Pacific Ocean. Mar Biol 162:2079–2091. https://doi.org/10.1007/s00227-015-2738-1

Welden NA, Cowie PR (2017) Degradation of common polymer ropes in a sublittoral marine environment. Mar Pollut Bull 118(1–2):248–253. https://doi.org/10.1016/j.marpolbul.2017.02.072

Welden NAC, Cowie PR (2016) Environment and gut morphology influence microplastic retention in langoustine, Nephrops norvegicus. Environ pollution (Barking, Essex: 1987) 214:859–865. https://doi.org/10.1016/j.envpol.2016.03.067

Wen B, Zhang N, Jin SR, Chen ZZ, Gao JZ, Liu Y, Liu HP, Xu Z (2018) Microplastics have a more profound impact than elevated temperatures on the predatory performance, digestion and energy metabolism of an Amazonian cichlid. Aquatic Toxicology (Amsterdam, Netherlands) 195:67–76. https://doi.org/10.1016/j.aquatox.2017.12.010

Wilcox C, Heathcote G, Goldberg J, Gunn R, Peel D, Hardesty BD (2015) Understanding the sources and effects of abandoned, lost, and discarded fishing gear on marine turtles in northern Australia. Conservation Biology: the Journal of the Society for Conservation Biology 29(1):198–206. https://doi.org/10.1111/cobi.12355

Yao FC, Jin CX, Liang H, Zhang Y, Gu Y, Song FB, Zhou Z, Sun JL, Luo J (2024) Microplastics weaken the digestion and absorption functions in the golden pompano (*Trachinotus blochii*) by affecting the intestinal structure, bacteria and metabolites. Chemosphere 362:142415. https://doi.org/10.1016/j.chemosphere.2024.142415

Yap H, Shukor M, Yasid N (2022) The microplastics occurrence and toxic effects in marine environment. Journal of Environmental Microbiology and Toxicology 10(2):1–6 Https://doi.org/10.54987/jemat.v10i2.733

Yin L, Chen B, Xia B, Shi X, Qu K (2018) Polystyrene microplastics alter the behavior, energy reserve and nutritional composition of marine jacopever (*Sebastes schlegelii*). J Hazard Mater 360:97–105. https://doi.org/10.1016/j.jhazmat.2018.07.110

Yong CQY, Valiyaveettil S, Tang BL (2020) Toxicity of microplastics and nanoplastics in mammalian systems. Int J Environ Res Public Health 17(5):1509. https://doi.org/10.3390/ijerph17051509

Yorio P, Marinao C, Kasinsky T, Ibarra C, Suárez N (2020) Patterns of plastic ingestion in Kelp Gull (*Larus dominicanus*) populations breeding in northern Patagonia, Argentina. Mar Pollut Bull 156:111240. https://doi.org/10.1016/j.marpolbul.2020.111240

Zhang C, Chen X, Wang J, Tan L (2017) Toxic effects of microplastic on marine microalgae Skeletonema costatum: interactions between microplastic and algae. Environ Pollut (Barking, Essex: 1987) 220(Pt B):1282–1288. https://doi.org/10.1016/j.envpol.2016.11.005

Zhang T, Lin L, Li D, Wang J, Liu Y, Li R, Wu S, Shi H (2022) Microplastic pollution at Qilianyu, the largest green sea turtle nesting grounds in the northern South China Sea. PeerJ 10:e13536. https://doi.org/10.7717/peerj.13536

Zhang X, Liu L, Chen X, Li J, Chen J, Wang H (2023) The fate and risk of microplastic and antibiotic sulfamethoxazole coexisting in the environment. Environ Geochem Health 45(6):2905–2915. https://doi.org/10.1007/s10653-022-01385-8

Zhang M, Haward M, McGee J (2020) Marine plastic pollution in the polar south: Responses from Antarctic Treaty System. Polar Rec 56:e36. https://doi.org/10.1017/S0032247420000388

Zhao C (2023) A study of marine plastic pollution abatement: a bibliometric analysis on status and development trend. Acad J Environ Earth Sci 5(10), 99–104. https://doi.org/10.25236/AJEE.2023.051013

Zhou J, Luo D (2024) The global governance of marine plastic pollution: rethinking the extended producer responsibility system. Front Mar Sci 11:1363269. https://doi.org/10.3389/fmars.2024.1363269

Zwollo P, Quddos F, Bagdassarian C, Seeley ME, Hale RC, Abderhalden L (2021) Polystyrene microplastics reduce abundance of developing B cells in rainbow trout (Oncorhynchus mykiss) primary cultures. Fish Shellfish Immunol 114:102–111. https://doi.org/10.1016/j.fsi.2021.04.014

Chapter 5
The Interaction of Toxic Chemicals with Micro/Nanoplastics

Manuel Palencia, Angélica García-Quintero, Ranjita Saikia, and Kevin H. Libreros-Castillo

Abstract One of the environmental problems of most concern to citizens, governments, and scientists is the increasing contamination caused by synthetic plastics; in recent years, there has been a rising awareness concerning microplastics and nanoplastics manufactured and generated through physicochemical and biological degradation processes. This dimension promotes interaction with biological entities and enhances their mobility through air, soil, and water bodies, where they can interact with several pollutants. These pollutants have the potential to adsorb, transport, concentrate, and desorb from the surface of micro/nanoplastics, generating co-contamination processes more widespread than their macroscopic counterparts, which have exhibited synergistic and antagonistic effects on the organisms with which they can interact and/or internalize, being this a research field of great interest and development at present. To provide a comprehensive understanding of the relationship between chemical pollutants and micro/nanoplastics, this chapter is organized around three main themes: the sorption behavior of micro/nanoplastics and the factors that influence this interaction, the diverse types of chemical pollutants that can be adsorbed in micro/nanoplastics, and the environmental impact of these interactions in soil, air, and water bodies, with a particular focus on marine and coastal ecosystems.

Keywords Co-contamination · Adsorption · Molecular interactions · Aging processes · Pollutant vectors

M. Palencia (✉) · A. García-Quintero · K. H. Libreros-Castillo
Research Group in Science with Technological Applications (GI–CAT), Department of Chemistry, Faculty of Natural and Exact Sciences, Universidad del Valle, Cali, Colombia
e-mail: manuel.palencia@correounivalle.edu.co

A. García-Quintero · R. Saikia · K. H. Libreros-Castillo
Mindtech Research Group (Mindtech-RG), Mindtech s.a.s., Montería/Cali, Colombia

A. García-Quintero
Research Group in Sustainable Development and Innovation (GIDSI), Institute of Analytical Science and Technology Golden–Hammer, Montería/Cali, Colombia

S. K. Panda and V. P.M. (eds.), *Bionanocomposites from Marine Origin*, Advanced Structured Materials 247, https://doi.org/10.1007/978-3-032-10749-7_5

5.1 Introduction

Polymeric materials are macromolecules formed by repetitive units denominated monomers; covalent bonds connect these monomers, enabling the formation of a diverse range of natural, semi-synthetic, and synthetic materials relevant to human societies (Agboola and Benson 2021). These materials allow for the storage and transmission of genetic information to the packaging and preservation of food products, as in the case of DNA and polyethylene, respectively (Agboola and Benson 2021; Desidery and Lanotte 2022). Polymers can be classified according to their thermal behavior as thermosetting or thermoplastic; the latter category refers to the variation in the mechanical properties that make polymers moldable when subjected to heating or pressure (Desidery and Lanotte 2022). Among the diversity of these polymeric materials, synthetic plastics, i.e., those that are artificially obtained and can be molded, have emerged as essential materials for economic and social welfare since the beginning of the twentieth century; in turn, these materials allow adding value to a broad spectrum of petroleum by-products (Ajithkumar et al. 2024; Palencia et al. 2021).

Although there is a wide variety of plastics, with a notorious spectrum of physical and chemical properties and the presence of various additives, these materials commonly have low density, modulation of their physicochemical properties, as well as mechanical, chemical, and biological resistance; which allows for a broad versatility that results in their usage in various everyday applications, including storage systems, textile materials, construction elements, medical supplies, and more (see Fig. 5.1) (Komorowska-Kaufman and Marciniak 2024). As a consequence of these characteristics, over 320 million tons of plastic are produced globally annually, approximately 40% of which is used for single-use packaging, and about 8 million tons are deposited in the environment (Ajithkumar et al. 2024; Komorowska-Kaufman and Marciniak 2024; Palencia et al. 2016; Thompson et al. 2009; Sudarsan et al. 2024). The latter is an aspect of considerable concern because, despite the numerous benefits of these materials, there are serious environmental considerations related to every stage of their life cycle. Regarding their production, synthetic plastics are typically generated from raw materials such as fossil fuels, petroleum, and natural gas, whose derivatives generate monomers such as ethylene and styrene; in this conversion, pollutants and greenhouse gases such as carbon dioxide and nitrogen oxides are formed (Kim et al. 2023). Regarding their final disposal, synthetic plastics can be reused, recycled, and predominantly deposited in landfills; nevertheless, due to inadequate management practices, a significant amount ends up in the environment, leading to contamination processes. A typical example of this issue is the presence of low-density plastics, such as polystyrene (PS) and polypropylene (PP), in marine and coastal ecosystems (Agboola and Benson 2021; Bodus et al. 2024; Liu et al. 2021; Um et al. 2023).

On the other hand, most synthetic plastics are not biodegradable because their chemical structure is not optimally susceptible to biological action; in fact, their degradation is conditioned on various factors, including molecular weight, degree

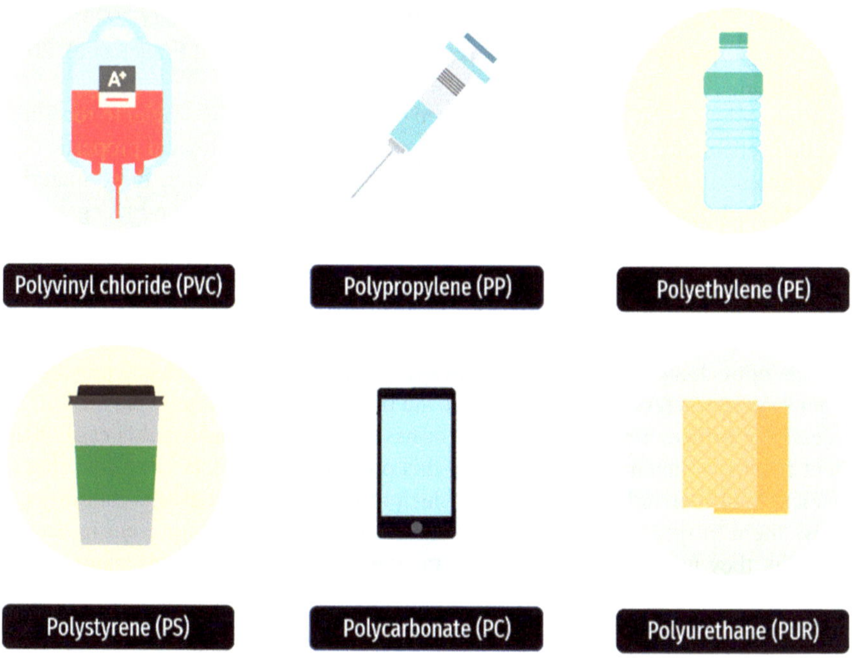

Fig. 5.1 Typical applications of synthetic plastic polymers

of crystallinity, hydrophobicity, solubility, and environmental conditions, which, in extreme cases, can lead to the degradation of the material after centuries (Talukdar et al. 2024). During this period, plastic is subjected to pressures such as mechanical abrasion, thermal degradation, ultraviolet light, and organism activity, factors that contribute to the formation of plastic particles in micrometric and nanometric sizes, classifying them as secondary micro and nanoplastics (MNPs) (Agboola and Benson 2021; Komorowska-Kaufman and Marciniak 2024; Talukdar et al. 2024; Chellasamy et al. 2023; Hajji et al. 2024; Zhu and Wang 2020; Monira et al. 2023; Permana et al. 2025). Meanwhile, primary MNPs are intentionally generated particles for applications such as microbeads and exfoliating particles in personal and beauty care products, cleaning supplies, and drug delivery vectors (Agboola and Benson 2021; Al Harraq and Bharti 2021; Witczak and Pokorska-Niewiada 2023). Entities such as the European Food Safety Authority (EFSA) have formally defined microplastics (MPs) as plastic particles ranging from 0.1 to 5000 μm and nanoplastics (NPs) as plastic particles in a range of approximately 1–100 nm (although these definitions are still under discussion) (Komorowska-Kaufman and Marciniak 2024; Hajji et al. 2024; Witczak and Pokorska-Niewiada 2023; Sheriff et al. 2023; Upadhyay et al. 2024; Arregocés-Garcés et al. 2024; Franco et al. 2020; Huang et al. 2023; Shukla et al. 2024; Yu et al. 2021). It is important to note that nanomaterials must have at least one dimension within this scale; for instance, a plastic sheet with one dimension measured in nanometers is classified as nanoplastic (Komorowska-Kaufman and

Marciniak 2024; Hajji et al. 2024; Witczak and Pokorska-Niewiada 2023; Sheriff et al. 2023; Upadhyay et al. 2024; Arregocés-Garcés et al. 2024; Franco et al. 2020; Huang et al. 2023; Shukla et al. 2024; Yu et al. 2021).

While MPs and NPs are classified based on particle size, it is crucial to recognize that their chemical composition and physicochemical and biological properties can be significantly diverse. Although they are all insoluble in water, these materials can exhibit different behaviors, such as being biodegradable or non-biodegradable, lipophilic or hydrophilic, and spherical or lamellar, among others; consequently, the environmental mobility of polyvinylchloride (PVC) MPs should not be expected to be analogous to that of polyethylene (PE) MPs (Chellasamy et al. 2023). At the same time, it should be considered that NPs are at a scale of matter in which their surface, optical, and electronic properties differ from those of the same materials at larger scales; as a result, MPs and NPs made from the same synthetic plastic will not necessarily behave in the same manner or have similar impacts (Shukla et al. 2024; Yu et al. 2021; Santana et al. 2024). In this order of ideas, MNPs present inherent characteristics derived from their size, which is associated with greater mobility that allows them to spread over vast areas, making them nearly omnipresent in ecosystems. So, they have been detected in water samples, soils, food, and water bodies. In some instances, their presence has led to adverse effects, prompting them to be classified as emerging contaminants (Ajithkumar et al. 2024; Palencia et al. 2021; Komorowska-Kaufman and Marciniak 2024; Um et al. 2023; Talukdar et al. 2024; Chellasamy et al. 2023; Hajji et al. 2024; Sheriff et al. 2023). Due to their dimensions and interactions, they can trigger biological responses such as oxidative stress and, in turn, allow the release of certain additives through desorption or internalization, potentially impacting public and environmental health (Ajithkumar et al. 2024; Palencia et al. 2021; Komorowska-Kaufman and Marciniak 2024; Um et al. 2023; Sheriff et al. 2023; Jiang et al. 2022; Fan et al. 2023). These additives are chemical substances that enhance polymer properties; they include plasticizers such as Bisphenol A (BPA), which is recognized as an endocrine disruptor and potential carcinogen, and flame retardants such as polybrominated diphenyl ethers, which are associated with reproduction disorders (Agboola and Benson 2021; Yu et al. 2021; Forero et al. 2022; Adegoke et al. 2023).

Additionally, MNPs exposed to environmental conditions undergo aging processes, such as oxidation, photodegradation, and erosion, which alter their crystalline structure, thermal stability, mechanical strength, and surface composition (Ruggieri et al. 2024). The latter is one of the crucial factors in the modulation of interactions between MNPs with other entities, so that both pristine and aged MNPs can interact with different substances in the environment, including natural and anthropogenic pollutants, an interaction that can lead to the formation of composite pollutants, which may have a synergistic effect on toxicity and environmental impact (Al Harraq and Bharti 2021; Shukla et al. 2024; Xie et al. 2025). Therefore, MNPs can absorb, accumulate, transport, and desorb chemical, physical, and biological contaminants through environmental compartments, trophic levels, and tissues. This process increases the mobility of these contaminants in the environment and enhances the likelihood of their entry into organisms and natural systems at high concentrations, leading to problems such as soil structure alteration, bioaccumulation in aquatic and terrestrial organisms, and severe and chronic diseases in humans, such as tissue damage and carcinogenesis (Agboola and Benson 2021; Thompson et al. 2009; Shukla et al. 2024; Forero et al. 2022; Costigan et al. 2022; Porta 2021; Thomas 2022; Liang et al. 2024; Zhi et al. 2025; Beggel et al. 2024). Scenarios of notable concern because it is expected that with time, a rise in the concentration of MNPs will be found in the environment due to the growing presence of synthetic plastic waste, as well as the accumulation of organic and inorganic pollutants that may interact with and be transported by these particles. In this context, this chapter aims to provide a comprehensive overview of the interactions between MNPs and chemical pollutants. It will begin by examining the sorption behavior of MNPs and the factors that influence these interactions; next, it will focus on the various chemical pollutants that MNPs can adsorb; finally, the chapter will address the environmental impacts of the interactions between MNPs and pollutants.

5.2 Sorption Behavior of Micro/Nanoplastics and Influencing Factors

A surface sorption process refers to the phenomenon by which atoms, molecules, and/or ions are transferred from a fluid phase, gaseous or liquid in nature, to a solid surface, described through adsorption and absorption processes (Yu et al. 2021). Specifically, adsorption is the process where entities adhere to a surface through reversible physical interactions (physisorption) and/or irreversible chemical interactions (chemisorption). Hydrophilic and hydrophobic intermolecular forces primarily drive physisorption, while chemisorption involves the formation of chemical bonds between the entity, which is located at the interface of the solid and fluid phases, and the surface itself; in contrast, absorption involves the entity diffusing into the solid material (Yu et al. 2021; Xie et al. 2025). In the case of MNPs, adsorption processes tend to prevail with the consequent decrease in interfacial free energy, which is influenced by the pore-filling mechanism governed by molecular diffusion in microporous particles. In turn, depending on the porosity, particle size, and environmental conditions, absorption phenomena can play a significant role; for instance, studies have shown that pollutants like fluoranthene and phenanthrene mainly undergo adsorption phenomena in MPs with dimensions smaller than 10 μm, while MPs around 100 μm also exhibit relevant absorption behavior (Agboola and Benson 2021; Al Harraq and Bharti 2021).

The principal interactions that can coexist in the same particle and enable, mediate, and modulate adsorption can be categorized as hydrophobic, hydrophilic, and electron donor–acceptor interactions (Yu et al. 2021; Sun et al. 2022). *Hydrophobic interactions* refer to the forces that occur between molecules, segments of molecules, and/or materials that have hydrophobic characteristics, i.e., those that repel water (a hydrophilic substance). These interactions are driven by the tendency of non-polar molecules to group together to minimize their contact with water. Additionally, these are related to dispersive forces, which arise from the electric fields created by the movement of electrons; this movement leads to variations in the dipole between the nucleus and the electron clouds, polarizing the electron cloud and inducing a dipole (Rafa et al. 2023; Li et al. 2025a). It should be noted that most MNPs released into the environment are hydrophobic, derived from PE, PS, and PP; therefore, hydrophobic interactions are one of the most relevant sorption mechanisms (Komorowska-Kaufman and Marciniak 2024; Al Harraq and Bharti 2021; Rafa et al. 2023). This sorption capacity of hydrophobic organic compounds on hydrophobic MNPs strongly correlates with descriptors such as the octanol–water partition coefficient (K_{ow}) and the hexadecane-water partition coefficient (K_{hw}) (Agboola and Benson 2021; Mohana et al. 2022).

In turn, one important type of hydrophobic interaction is π-π stacking, which results from the attraction between the electronic clouds of the aromatic rings present in polymers such as PS, widely used as an outdoor material, and aromatic organic substances such as naphthalene and its derivatives, molecules of great relevance in the pharmaceutical and dye industries (Um et al. 2023; Rafa et al. 2023; Li et al. 2025a; Mohana et al. 2022; Zhang et al. 2024a). This interaction influences the differential affinity between aromatic organic pollutants and MNPs made from polymers with aromatic rings; it has been demonstrated that π-π stacking is primarily responsible for the preferential adsorption of aromatic molecules on PS (over PP and PE particles) of aromatic molecules, including various compounds such as atorvastatin (a pharmaceutical), aniline (a dye), phenanthrene (an industrial precursor), 17 β-estradiol (a hormone), and polychlorinated biphenyls (PCBs) (a plasticizer) (Yu et al. 2021; Chen et al. 2023; Zhou et al. 2021). In the latter case, it has been reported that PCBs' adsorption on PS NPs can be one to two orders of magnitude higher compared to PE MPs; however, these results are influenced not only by π-π stacking interactions but also by the difference in surface area (Zhou et al. 2021).

Regarding *hydrophilic interactions*, these refer to those present in polar molecules or segments of them; that is, they are directly related to non-dispersive forces or specific interactions, which can be classified into three categories: dipole–dipole interactions, induced dipole–dipole interactions, and hydrogen bonding. Each type of interaction involves varying contributions from electrostatic forces and even chemical bonds, resulting in interaction energies ranging from weak to strong attractions (Li et al. 2025a). In the case of dipole interaction, this is generated between polar molecules that have a permanent dipole; this interaction happens when the positive pole of one dipole is aligned near the negative pole of another dipole (Rafa et al. 2023; Li et al. 2025a; Mohana et al. 2022; Zhang et al. 2024a). On the other hand, induced dipole–dipole interaction involves a molecule with a permanent dipole that can induce a temporary dipole in a neighboring molecule (Rafa et al. 2023; Li et al. 2025a; Mohana et al. 2022; Zhang et al. 2024a). Hydrogen bonding, on the other hand, is an attractive electrostatic force between an electronegative atom and a hydrogen atom that is bonded to another electronegative atom, such as oxygen or nitrogen (this also being an electron donor–acceptor process), such is the interaction that occurs between oxygen atoms of hydroxyl groups and the hydrogen of carboxylic acids

(Rafa et al. 2023; Li et al. 2025a; Mohana et al. 2022; Zhang et al. 2024a). Thus, hydrophilic interactions are related to the polar segments of the plastic material and the substance adsorbed on them; for instance, these interactions prevail between polymers such as bakelite, resulting from the condensation between formaldehyde and phenol, and emerging pharmaceutical contaminants such as ciprofloxacin, which is a fluoroquinolone with the presence of a carboxylic acid functional group (Rafa et al. 2023; Li et al. 2025a; Mohana et al. 2022; Zhang et al. 2024a). Comparable to π-π stacking interactions, hydrogen bonds can also condition the preferential adsorption of polar antibiotic molecules, including ciprofloxacin, along with additives like BPA, hormones such as 17 β-Estradiol, and agents like perfluoroalkylated and polyfluoroalkylated substances (PFAS) onto microplastics based on polyamide (PA) (as compared to PE, PS, PP, and PVC MPs) because the hydrogen of the amide group of the polymer can interact with heteroatoms such as oxygen and fluorine through these intermolecular forces (Mohana et al. 2022).

On the other hand, the *electron donor–acceptor interactions*, also known as Lewis acid–base interactions, can be understood from the perspective of chemical bond formation through Pearson's hard-soft theory. According to this theory, hard interactions are primarily electrostatic, while soft interactions are covalent, so acids and bases that mainly interact through electrostatic interactions are classified as hard, whereas soft acids and bases are more likely to form covalent bonds (Lee 1991). Regarding electrostatic interactions, this type of interaction occurs between atoms and molecules that carry electric charges; when these electric charges are analogous, the interaction is repulsive, while if they are opposite, the interaction is attractive. This interaction is essential in the adsorption process of metals on the surfaces of MNPs, particularly in the presence of ionizable functional groups, such as those found in sulfonic acid groups (common in polyelectrolyte-type polymers) because, in the environment, metals are usually found in their free ionic form as cations or within chemical complexes; where the speciation of these elements depends on various factors, including pH, temperature, redox potential, and the presence of ligands such as carbonates (Xie et al. 2025; Rafa et al. 2023; Wang et al. 2025). Furthermore, electrostatic interactions are not only responsible for adsorption processes but also for stability and interparticle interaction, which can be modeled through the classical Derjaguin-Landau-Verwey-Overbeek (DLVO) theory for colloids; this theory describes the interaction as a function of van der Waals forces and electrostatic repulsion, which can be extended to include Lewis acid–base interactions, making it applicable for evaluating colloidal aggregation rates (Permana et al. 2025; Zhang et al. 2024a). In this sense, the electrostatic strength of MNPs can be characterized through the Z potential, where at the isoelectric point, agglomeration phenomena tend to occur, and absolute values higher than 30 mV are indicative of electrostatically stable MNPs (Zhang et al. 2024a). Electrostatic interactions, meanwhile, occur when charged particles interact with polar molecules and during ion exchange processes, in which the displacement of the reaction depends on relative stability.

On the other hand, covalent interactions are those that are generated due to the sharing of electrons between atoms, whereas in the case of the adsorption of substances on MNPs, the most relevant covalent bonds are the coordinated covalent bonds, in which the electrons come from atoms within a single molecule, crucial for forming metal complexes. For example, surface groups containing nitrogen atoms, such as amines, and oxygen atoms, like carbonyls, can complex metal cations such as Hg^{2+} (Mohana et al. 2022; Zhang et al. 2024a; Wang et al. 2025). While interactions have been described separately, many can occur simultaneously; to illustrate this, consider the interaction between MNPs and PFAS, where hydrophobic interactions associated with the carbon chain and hydrophilic interactions derived from the fluorine atoms co-occur (Tanzil et al. 2024; Guo et al. 2023). It is also important to highlight that the strength of the hydrophilic interactions is greatest to that of the hydrophobic interactions, so it is to be expected that the adsorption capacity is congruent with this behavior; therefore, tetracycline molecules show a trend in their adsorption capacity as follows: PBAT (poly(butylene adipate-*co*-terephthalate) > PS > PP > PE, correspond with the strength of the predominant interactions, where PBAT presents hydrophilic interactions, PS of the π-π type, and PP and PE of the hydrophobic type (Guo et al. 2023). In summary, sorption processes between MNPs and substances depend on physisorption and chemisorption processes mediated by hydrophilic, hydrophobic, and electron donor–acceptor interactions. These, in turn, are modulated by factors related to the MNPs, chemicals, and environmental surroundings, such as surface aging, concentration, and ionic strength of the medium, respectively. Each of these parameters will be discussed in more detail below.

5.2.1 Influence of the Characteristics of Micro/Nanoplastics

One of the most influential factors in the sorption processes is the chemical composition of the MNPs, which exhibit significant variability, depending not only on the chemical nature of the plastic but also on several other factors, like the degree of polymerization, porosity, linear or branched structure, crystallinity, and the additives added to it according to its target application, such as pigments, plasticizers, flame retardants, and antioxidants. For example, PVC can be used both for the storage of blood products and the transport of water through pipes; in the first case, PVC must have an amorphous structure, be flexible and transparent, and have plasticizing and thermal stabilizing additives, while in the second case, it must be semi-crystalline and opaque, and it must have lubricating and thermal stabilizing additives (Palencia et al. 2016). However, it is common that aspects such as this one are overlooked in the classification and study of MNPs, emphasizing mainly the dimension and general composition since the area and surface chemistry of the material are the most determining aspects in the adsorption process (Sudarsan et al. 2024; Chellasamy et al. 2023). As for the surface area, a larger exposed surface area enhances the adsorption process, and smaller particle sizes lead to a relatively larger surface area, i.e., factors influencing this process include size, shape (such as fibers, ellipsoids, sheets, and spheres), porosity, and surface irregularities (Agboola and Benson 2021; Witczak and Pokorska-Niewiada 2023; Li et al. 2025a; Chen et al. 2023). Furthermore, although plastics are highly diverse and the particle distribution of them is even geographically dependent, the composition of MNPs derived from synthetic plastics is primarily dominated by:

- PE and PP MNPs, whose chemical composition is polyolefins, that is, saturated hydrocarbon chains, these materials exhibit hydrophobic characteristics, low density, and good chemical and mechanical resistance, as a result, they are commonly used in pipes and containers;
- MNPs based on PS, which is a polymer class that presents benzene rings in its structure favoring π-π stacking interactions, is characterized by its high resistance to biodegradability and its rigidity, being commonly employed as thermal insulator and disposable packaging material;
- Nylon MNPs, which are a type of PA, contain multiple amide groups (R-CONH-R) in their structure, these favor hydrophilic interactions, and the material exhibits high mechanical strength and elasticity, being commonly employed in textile fibers;
- MNPs consisting of polyethylene terephthalate (PET), a type of polyester that is, it has ester groups (R-COO-R) in its polymeric chain, which can interact with other molecules through dipole–dipole interactions and is characterized by its mechanical and thermal resistance, being useful in electronic components;
- PVC MNPs, which have chlorine groups bonded to the main carbon chain promoting dipole–dipole interactions, have high chemical and mechanical resistance and are widely used in coatings;
- Other relevant but less prevalent synthetic plastic polymers are polycarbonates, which feature carbonate groups in their main chain (R-O-COO-R) and are typically used in lenses, and polyurethanes (PUR), which contain urethane groups (R-NHCOO-R) and are often used in adhesives (Witczak and Pokorska-Niewiada 2023; Li et al. 2025a; Okunola et al. 2019; Zjačić et al. 2023). Although this chapter emphasizes these synthetic plastic materials considered non-biodegradable, it is important to mention that there are also MNPs of biodegradable polymers (such as polylactic acid) since these types of materials not only undergo biodegradation but also experience conventional degradation processes such as oxidation, photodegradation, erosion, and aging, which can generate microbioplastics and nanobioplastics (Zhou et al. 2021; Piyathilake et al. 2024). Early studies analyzing these substances indicate that they have a significant capacity to adsorb organic pollutants and may pose potential adverse effects in both exploratory and controlled environments (Zhou et al. 2021; Piyathilake et al. 2024).

The surface composition of pristine particles can be modified through aging effects derived from the interaction of PMs with biotic and abiotic components of the environment. These changes can influence various phenomena, including sorption and behavior, toxicity, surface reactivity, and overall fate of the particles in the natural environment and within living organisms (Permana et al. 2025; Rafa et al. 2023; Li et al. 2025a; Wang et al. 2025). Abiotic aging processes occur through various phenomena such as mechanical degradation (e.g., sand abrasion, tidal actions, and even soil tillage), photodegradation (caused by exposure to UV radiation from the sun, which is usually the primary cause of degradation), thermal degradation (resulting from cooling and heating processes), and oxidizing environments (which dependent on the redox potential of the environmental system) (Permana et al. 2025; Rafa et al. 2023; Li et al. 2025a; Wang et al. 2025). For example, the photoaging of PS involves UV-induced chain breakage, polymer chain reaction, auto-oxidation, and formation of inert end products. In air, this results in carbonyl bonds in the structure; conversely, in water, it forms phenolic hydroxyl groups, a behavior that can be attributed to the reduced availability of radiation in water (Li et al. 2025a; Mohana et al. 2022; Wang et al. 2025). In this sense, aging processes derive from physical and chemical changes; the physical surface alterations include surface cracking, fracture, erosion, roughness increases, deeper and irregular holes, wrinkles, and pores. In this way, the surface area is influenced by aging processes, where variations in adsorption between pristine and aged particles of the same material have been evidenced. For instance, the adsorption of tetracycline on pristine PP MPs is by the Langmuir model of monolayer adsorption with uniform distribution of active sites, and for PBAT, PE, and PS, it follows the Freundlich model, for nonlinear multilayer adsorption; which was modified after photodegradation, so that the Elovich model for non-uniform adsorption best described PP and for PBAT, PE, and PS by the Freundlich model (Guo et al. 2023).

Chemical modifications of the surface functional groups of MNPs are particularly favored when the polymer chains exhibit considerable mobility, influenced by the main chain structure, the presence of plasticizers, and the level of crystallinity (Permana et al. 2025; Rafa et al. 2023; Li et al. 2025a; Wang et al. 2025). The latter is a factor of particular interest since plastic polymers can be crystalline, semi-crystalline, and amorphous; the structure's degree of crystallinity directly affects the mobility of the polymer chains, and the higher mobility leads to more significant chemical modification and enhanced adsorption processes; therefore, more crystalline structures tend to undergo less surface chemical modification and adsorption compared to more amorphous structures (Agboola and Benson 2021; Mohana et al. 2022; Wang et al. 2025). Thus, it has been shown that a decrease in crystallinity from 35 to 30% is a factor that enhances the adsorption of hydrophobic organic pollutants, an effect that, together with consequences of the photodegradation of MPs such as the generation of oxygenated functional groups (which disturb the crystalline structure increasing amorphous segments), micropores, and roughness; then, it have been reported that the adsorption of chlorpyrifos increases by approximately 18% compared to pristine PE particles (Santana et al. 2024; Wang et al. 2025).

The chemical consequences of degradation processes are environment-dependent, so it has been reported that natural weathering of PVC MNPs results in the generation of carbonyl functional groups; photodegradation of PP and PE results in the presence of oxygen-containing functional groups such as aldehydes, hydroxyls, carboxylic acids, and carboxylates; the aging of PS with photo-Fenton-like reaction results in the presence of oxygen-containing functional groups; the aging of PA in the presence of sulfur results in the formation of carbonyl and thiol groups; and the alkaline aging of PET facilitates the cleavage of ester bonds resulting in the generation of hydroxyl and carbonyl groups (Permana et al. 2025; Santana et al. 2024; Xie et al. 2025; Zhao et al. 2023; Zhang et al. 2024a; Chen et al. 2023; Wang et al. 2025). Chemical modifications enhance heterogeneity, increase surface hydrophilic characteristics, and can confer to MNPs a negative charge, facilitating their interaction with cations and polar molecules. For instance, PS-aged particles exhibit a greater capacity for adsorbing antibiotics like tetracycline and oxytetracycline compared to pristine particles; this increased adsorption occurs through various interactions, including hydrogen bonds, electrostatic interactions, π-π stacking, and dipole–dipole interactions (Mohana et al. 2022; Zhang et al. 2024a; Chen et al. 2023).

Concerning the biotic aging processes of MNPs, this refers to modifications mediated by biological agents, including interaction with cellular secretions, enzymatic activity, and effects derived from the organisms' biological functions; thus, these processes depend on both the biological composition and the physicochemical conditions of the environment (Ruggieri et al. 2024; Wang et al. 2025; Verla et al. 2019). In particular, the formation of biofilms and the adsorption of biological substances from the natural environment on surfaces (such as exopolymers, lipids, nucleic acids, humified and fresh organic matter) result in a phenomenon known as the eco-crown, which is "hard" when the adhesion is strong, that depends on the medium and the properties of the MNPs, and "soft" when a dynamic exchange process occurs with the surrounding molecules (Permana et al. 2025; Costigan et al. 2022; Zhang et al. 2024a; Wang et al. 2025; Ge and Lu 2023; Ren et al. 2022). To illustrate this aspect, the adsorption of humic acids on PS NPs can be considered, where the aromatic rings present in the humic acids interact with the nanoparticles through hydrophobic interactions and π-π stacking, which influences the processes of both homo- and hetero-aggregation, resulting in modifications to the chemistry, area, and surface reactivity of the particles (Permana et al. 2025; Ge and Lu 2023). Consequently, the increased surface hydrophobicity of the nanoparticles enhances their affinity for more polar substances, promoting electrostatic interactions, hydrophilic interactions, and the formation of complexes, and this facilitates the complexation of metal ions with the functional groups present on the surface of the nanoparticles, such as hydroxyl groups (Permana et al. 2025; Ge and Lu 2023).

5.2.2 *Influence of Environmental Conditions*

Previous discussions have highlighted how the composition of the surrounding environment influences the aging processes of MNPs and how this affects adsorption. However, other environmental factors also have a notorious influence, the most relevant being pH, temperature, salinity, and ionic strength (Yu et al. 2021). As for pH, it refers to the concentration of hydronium cations in a solution; in this sense, pH is a crucial factor, both for MNPs and for the substances that can interact with them, since it can determine speciation, stability, solubility of adsorbates, and the electrical charge of ions and molecules with ionizable groups such as carboxylic acids (Rafa et al. 2023). Therefore, when considering antibiotic molecules such as amoxicillin as adsorbate, its adsorption to MNPs surfaces can be modified as a function of pH, where at pH below 2.7, the molecule carries a positive charge due to the protonation of the amine, giving it a cationic character; at pH between 2.7 and 7.4, the carboxylic acid group deprotonates, resulting in a net neutral charge and a zwitterionic character; and at pH above 7.4, the amine group has deprotonated, making the overall charge of the molecule anionic (Aranda and Rivas 2022; Gao et al. 2023). In turn, changes in pH affect the electrostatic interactions between the MNPs through variations in the Z potential so that, at a pH value higher than the zero charge point, the material will present a negative charge and, at lower values, a positive charge. Therefore, a pH above the zero charge point of PE MPs and below the pKa of a protonated amine, for example, sertraline, constitutes optimal conditions for high adsorption; in contrast, if the pH is higher than the pKa, electrostatic repulsion and competition for adsorption sites can lead to unfavorable conditions (Agboola and Benson 2021; Zhao et al. 2023; Chen et al. 2023; Wang et al. 2025; Gao et al. 2023). Additionally, in the case of metal ions, the pH can determine the solubility of the adsorbate, where, under basic conditions, it can precipitate in the medium as oxides and/or hydroxides, this being the case for Fe^{3+} (Agboola and Benson 2021; Zhao et al. 2023).

Conversely, the interactions between the adsorbate and the MNPs are influenced by salinity (the concentration of dissolved salts in the medium) and ionic strength (the measurement of the concentration of ions as a function of their ratio and charge) since an increase in ion concentration can influence colloidal stability, aggregation processes, and adsorption phenomena (Permana et al. 2025; Al Harraq and Bharti 2021; Tanzil et al. 2024). In this last aspect, salinity and ionic strength can have a dual impact; on the one hand, decreased adsorption can result from competition for adsorption sites; on the other hand, increased adsorption may occur due to the presence of ions that reduce repulsion between similarly charged particles; and even the salting-out effect can reduce the solubility of organic substances and favors hydrophobic interactions with MNPs that possess non-polar superficial characteristics (Agboola and Benson 2021; Zhou et al. 2021; Tanzil et al. 2024; Gao et al. 2023). For example, sodium cations decrease the adsorption of PFAS on MNPs with the presence of surface acidic groups due to their strong interaction with these groups. Additionally, high ionic strength derived from divalent cations such as Ca^{2+} and Mg^{2+} also decreases the adsorption of heavy metals, such as Cd^{2+}, Ni^{2+}, and Pb^{2+}, on aged MNPs due to competition for adsorption sites (Wang et al. 2025; Tanzil et al. 2024; Gao et al. 2023).

Referring to temperature, this is an influential factor in adsorption, as it can affect surface tension, the solubility of the adsorbate, and the thermodynamics of the adsorption process; where at low temperatures, physisorption is favored because molecules' kinetic energy decreases when they adsorb onto the surface (Mohana et al. 2022; Gao et al. 2023). In contrast, chemisorption is typically promoted at high temperatures, as this process is usually endothermic (Mohana et al. 2022; Gao et al. 2023). For instance, it has been evidenced that the adsorption of the pesticide α-endosulfan on PAN6 MPs presented a ratio of about 90% at 4 °C, decreasing to 50% at 20 °C (Gao et al. 2023). In addition, this physical parameter significantly affects desorption processes, such that it has been reported that desorption rates can increase by up to 30 times under conditions that mimic the physiological parameters of warm organisms (38 °C and pH 4), compared to seawater at lower pH levels and higher temperatures, leads to a greater availability of contaminants such as perfluorooctanoic acid and di-2-ethylhexylphthalate in PVC and PS MPs (Ruggieri et al. 2024). In this sense, isothermal sorption models are crucial in understanding how adsorbate molecules interact with adsorbent surfaces. These models determine the amount of adsorbate bound to the surface based on the apparent equilibrium concentration of the adsorbate at a constant temperature. They are essential for elucidating the interaction mechanisms and evaluating the dynamics of sorption and desorption of contaminants on MNPs. The most commonly used are the linear adsorption model (direct ratio between adsorption and adsorbate concentration), Freundlich model (adsorption on heterogeneous surfaces), Langmuir model (monolayer on the homogeneous surface), Temkin model (linear adsorption energy as a function of adsorption), Elovich model (decreasing adsorption rate), and Dubinine-Radushkevich model (free energy of adsorption in micropores); for the kinetic models, the most widespread are the first-order adsorption model, pseudo-second-order adsorption model, intraparticle diffusion model, and film diffusion model (Agboola and Benson 2021; Al Harraq and Bharti 2021).

5.3 Types of Pollutants Adsorbed in Micro/Nanoplastics

Environmental pollutants are "*any physical, chemical, biological, or radiological substance or matter that has an adverse effect on air, water, soil, or living organisms,*" including substances as varied as splinters, bones, radioactive materials, pharmaceuticals, viruses, and bacteria (Schultz 2014). In particular, this section focuses on chemical pollutants, which are of both natural and anthropogenic origin, where the processes of industrialization and human globalization have directly contributed to increased production and use of hazardous and potentially toxic substances. It is important to remember at this point the principle of Paracelsus: "*All things are poison, and nothing is without poison, only the dose permits something not to be poison*"—these substances often lack specific guidelines and regulations for optimal environmental control in production, usage, and final life (Tsatsakis et al. 2108). In this context, widespread activities such as the discharge of effluents and solid waste of industrial and domestic origin into soil and water represent a global public health and environmental problem, which is increasing over time, alongside the rising concentration and variety of chemical pollutants (Otalora et al. 2021). Briefly, these chemical pollutants can be classified according to their chemical nature as organic and inorganic contaminants, addressing substances as varied as contraceptive molecules, pesticides, hydrocarbons, and heavy metals. Substances that can be adsorbed on the surface of MNPs depend on factors such as environmental conditions and the surface composition of the particles (see Fig. 5.2), influencing their environmental mobility and acting as vectors in the transport of contaminants through the biotic and abiotic components of the environment (Xie et al. 2025; Tanzil et al. 2024; Otalora et al. 2021; Naqash et al. 2020).

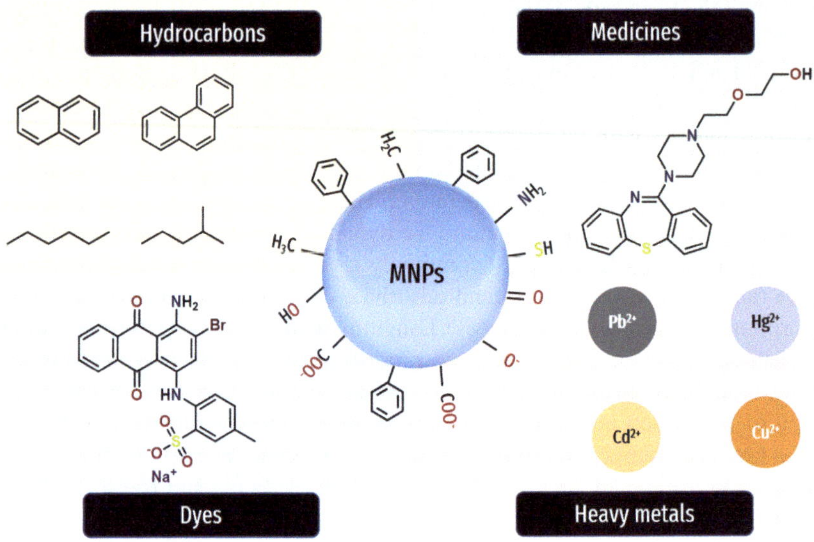

Fig. 5.2 Organic and inorganic contaminants commonly adsorbed on MNPs based on their surface composition

5.3.1 Inorganic Pollutants

Inorganic pollutants are substances such as salts and heavy metals, which are defined as metals and metalloids with densities greater than 4 g/cm^3; these materials can naturally occur in the Earth's crust and whose available concentration can increase due to environmental conditions, natural phenomena such as volcanic eruptions, as well as those of anthropogenic origin associated with improper waste disposal, extractive mining activities, and the use of agrochemicals (Xie et al. 2025; Naqash et al. 2020; Dagdag et al. 2023; Biedunkova and Kuznietsov 2024). Inorganic pollutants typically carry electrical charges that make them compatible with polar environments, presenting considerable environmental mobility through hydrological pathways and making them significant contributors to pollution in water bodies (Wang et al. 2022a; Zhou et al. 2022; Prasad et al. 2023). Despite the remarkable variety of inorganic salts found in the environment, heavy metals are the most relevant inorganic pollutants due to their prevalence, their persistence in the environment (as elemental entities, they cannot undergo degradation processes), and their high toxicity at concentrations in the order of ppm. The effects of heavy metals can range from inhibiting plant growth to causing serious health issues such as cardiac, neurological, and renal diseases (Wang et al. 2022a; Zhou et al. 2022; Prasad et al. 2023). These adverse effects derive from the high reactivity of heavy metals, which are usually found in their cationic form, allowing them to displace the active centers of biomolecules, generate reactive oxygen species, and interact through coordinated covalent bonds with functional groups such as thiols, amines, and carboxylates present in nucleic acids, proteins, metabolites, among others (Naqash et al. 2020; Dagdag et al. 2023; Biedunkova and Kuznietsov 2024; Wang et al. 2022a; Zhou et al. 2022; Prasad et al. 2023).

In the case of adsorption of heavy metal cations on the surface of MNPs, these are usually mediated by electrostatic interactions and the formation of coordination bonds; additionally, co-precipitation processes can also occur (Rafa et al. 2023; Li et al. 2025a; Wang et al. 2025). In this sense, MNPs derived from PA and with the presence of oxygenated functional groups are the adsorbents in which higher adsorption of heavy metals such as lead, cadmium, mercury, and arsenic have been reported; studies estimate that the concentration of these heavy metals adsorbed by MNPs can be approximately 800 times greater than their concentration in seawater (Ruggieri et al. 2024; Sun et al. 2022; Rafa et al. 2023; Li et al. 2025a). This statement implicitly establishes that aged MNPs, with oxygenated functional groups, possess higher heavy metal adsorption than their pristine counterparts (Permana et al. 2025). In this sense, research has shown that Ni^{2+} cations tend to adsorb more effectively on poly(lactic acid) MPs than on PS particles of similar size (Xie et al. 2025). However, when PS MPs undergo aging processes (such as freeze–thaw cycles, alkaline treatment, and dry–wet alternation), an increase in the specific surface area, crystallinity, pore size, and oxygenated functional groups is generated so that the adsorption of the metal increases notoriously (Xie et al. 2025). Furthermore, the metal can subsequently desorb from the surface of the MPs, with the amount and the desorption rate higher for the aged PS concerning the pristine one, leading to a higher overall migration capacity of Ni^{2+} (Xie et al. 2025). Another significant scenario of heavy metal adsorption on MNPs is the cotoxicity or synergy determined in some cases, which refers to the increased toxicity observed when combining individual components. This is the specific case of arsenic adsorbed on PS, PVC, and PE MNPs, in which the results showed a greater toxic effect of the composite pollutant on macrophytes, with observations showing an increase in arsenic adsorption by plants of up to 47% (Tang et al. 2022a). Table 5.1 provides several reported examples of the different inorganic contaminants that can adsorb on MNPs.

5.3.2 *Organic Pollutants*

Organic pollutants are substances with potentially harmful effects on public and environmental health. They are constituted primarily from carbon and hydrogen atoms organized in saturated, unsaturated, cyclic, or aromatic compounds; due to their polarity, these compounds present hydrophobic interactions that promote their affinity for pristine NMPs of PE, PP, and PS (Agboola and Benson 2021; Yu et al. 2021; Tanzil et al. 2024; Gao et al. 2023; Gaffney and Marley 2019). In turn, these chemical compounds can have a wide variety of functional groups with the presence of unsaturations, aromatic segments, and/or heteroatoms, which can introduce polar and hydrophilic characteristics in the structures, increasing their interaction with aged

Table 5.1 Examples of inorganic contaminants adsorbed on MNPs

Pollutant	Potential adverse effects	Type of MNPs	Remarks
Ni^{2+}	Lung cancer Dermatitis Inhibition of plant growth Alteration of the edaphic microbiota	PS PP	The effect of pH between 3 and 11 was evaluated, where values higher than 5 generate a competition between adsorption and precipitation of the metal as hydroxide. On the other hand, PS presented better adsorptive qualities than PP, possibly due to the more significant interaction of the cation with the electronic cloud of the aromatic rings in PS (Tenea et al. 2024)
Pb^{2+}	Anemia Neurological problems Reduced photosynthesis Decreased microbial activity in soil	PS PP PE PUR	The adsorption of the metal cation on the microplastic was evaluated on pristine and aged particles of the polymers, establishing a trend in the adsorption capacity of aged PS > aged PE > aged PS > aged PP > aged PP > PE > PP, which agrees with the trends in the hydrophilic and hydrophobic molecular interactions. In turn, the research showed that the process is modulated by pH, salinity, and the presence of fulvic acids, which affect the formation of surface complexes (Wang et al. 2022b) Higher adsorption of Pb^{2+} on PUR microplastics has been reported, in contrast to PP microplastics, due to the presence of carbonyl groups in PURs. Additionally, the adsorption process shows enhanced values after photoaging and biofilm formation (Tang et al. 2024)
Cd^{2+}	Malformations Osteomalacia Alteration in the ovaries Decrease in photosynthesis	PA PVC	The adsorption mechanism of Cd cations on the surface of PA6 MNPs is mediated by electrostatic interactions and complex formation, which are affected by pH, salinity, and the presence of humic acids (Tang et al. 2022b) For PVC, the interaction occurs mainly between the cadmium cation and the chlorine atoms attached to the polymer's carbon chain. An increase in pH and temperature leads to an increase and subsequent reduction in adsorption capacity (Bi et al. 2024)
Cr^{3+} and Cr^{6+}	Stomach ulcers DNA mutations Mortality in fish Alteration in microorganisms	PS PP PVC	The adsorption process on the MPs was controlled primarily by intraparticle diffusion and monolayer physical adsorption, which was favored at high temperatures and low salinity (Liu et al. 2022a)

(continued)

Table 5.1 (continued)

Pollutant	Potential adverse effects	Type of MNPs	Remarks
Cu^{2+}	Liver damage Behavioral disorders Alteration of algae metabolism Affection of enzymatic functions in fishes	PA PET PS PVC	The PMs were subjected to photodegradation by UVA irradiation in simulated environments of air, sand, soil, and seawater for three months, resulting in chemical (generation of oxygenated functional groups) and physical (increase in surface roughness) surface changes that favored the adsorption of cations on the particles' surfaces. The results in the seawater model showed that the adsorption trend was aged PVC > aged PS > aged PS > PS > PVC (Gao et al. 2021)
As^{3+} and As^{6+}	Arrhythmia Keratosis Reduction of soil fertility Inhibition of plant development	PS PE PVC	PVC and PS NPs presented a higher capacity to adsorb As cations over PE NPs due to interactions between the ions and the chlorine atoms of PVC and with the electronic cloud of the aromatic ring of PS. NPs present higher adsorption due to their large surface area with respect to MPs of the same composition (Tang et al. 2022a)

MNPs and with materials with heteroatoms as in the case of PAs and PURs. Such is the case of organic functions based on oxygen (such as carbonyls, carboxylic acids, aldehydes, ketones, esters, ethers, epoxides, and alcohols), nitrogen (such as amines, amides, azo compounds, nitriles, and nitro compounds), phosphorus (such as phosphates and phosphonates), sulfur (such as sulfones, thiols, and sulfonic acids), and halogens (such as chlorinated, fluorinated, iodinated, and brominated compounds) (Li et al. 2025a; Gao et al. 2023; Gaffney and Marley 2019). This structural diversity enables organic compounds to exhibit significant differences in their physicochemical properties, toxicity, and stability, resulting in a wide range of applications that facilitate the general classification of organic pollutants into:

- *Industrial precursors and fuels*: These are hydrocarbons and their aromatic derivatives that originate from petroleum and its by-products, such as those used to generate energy or during events like forest fires (Bartley and Fairbrother 2024; Patel et al. 2020; Yemele et al. 2024; Barathi et al. 2023). Due to their chemical composition, these substances are non-polar and, therefore, tend to be stored in lipidic tissues and undergo oxidative processes that lead to the formation of highly reactive functional groups, such as epoxides. These epoxides can interact with biomolecules and can cause liver, nervous system, and respiratory system damage (Bartley and Fairbrother 2024; Patel et al. 2020; Yemele et al. 2024; Barathi et al. 2023). Additionally, they can contaminate bodies of water and soil, resulting in bioaccumulation and negatively impacting environmental biodiversity. At the same time, the non-polar characteristics of hydrocarbons and their aromatic derivatives allow them to adsorb on NMPs with predominantly hydrophobic characteristics (Rafa et al. 2023). For example, PE MPs can act as adsorbates for polycyclic aromatic compounds and alkanes commonly found in crude oil, such as hydrocarbons from $C_{10}H_{22}$ to $C_{35}H_{72}$ and polyaromatics of 2 (naphthalene), 3 (phenanthrene and fluorene) and 4 rings (naphtobenzothiophene, pyrene, and chrysene), where the kinetics present rapid initial surface adsorption, followed by slower adsorption into the micropores of the MPs, and as the carbon chain length of the alkanes increases, the adsorption transitions from a monolayer to a multilayer process (Ahmad and Salam 2025). Also, as the molecular mass of polycyclic aromatic hydrocarbons increases, the adsorption on the surface of PE PMs increases (Ahmad and Salam 2025). Enzymes present in the simulated gastrointestinal fluid can desorb up to 28% of these compounds, which approximates the behavior and potential risk that represents the ingestion of microplastics with the presence of polyaromatic pollutants (Hou et al. 2023).

- *Pharmaceuticals and personal care products*: Pharmaceuticals refer to biologically active compounds that prevent and treat diseases, such as hormone regulators, antibiotics, and antipyretics (Adepipe et al. 2024; Malmqvist et al. 2023; Vatovec et al. 2021; Oliveira et al. 2021). In contrast, personal care products have an aesthetic and hygienic function that seeks to benefit the quality of life, such as deodorant, conditioner, and sunscreen. A concerning issue arises because many pharmaceuticals, metabolites, and personal care products are excreted, washed off, and ultimately discharged into wastewater and aquifer systems due to their intended functions (Adepipe et al. 2024; Malmqvist et al. 2023; Vatovec et al. 2021; Oliveira et al. 2021). This context has been recognized as an emerging contamination process since this concentration increases over time, exceeding the safety limits, which requires appropriate regulations to control the derived effects, among which are the disruption of the endocrine system, alterations in the phenotypic expression of marine organisms, and the generation of oxidative stress (Agboola and Benson 2021; Yang et al. 2025a). Moreover, pharmaceuticals and personal care products have a broad spectrum of properties depending on their applications; for instance, in this category are molecules such as the biguanide metformin used as an antidiabetic drug, which is highly hydrophilic, so its transport, absorption, and elimination require specific transporters; as well as lipophilic molecules such as sertraline, a structural analog of serotonin used for the treatment of depression, which can cross cell membranes through passive diffusion (Chhetri et al. 2024). In this sense, it has been reported that a wide diversity of pharmaceutical and personal care chemical compounds adsorbed on MNPs of diverse nature, depending on their affinity; for instance, it has been evidenced that PA particles allow the adsorption of antibiotic molecules such as ciprofloxacin, amoxicillin, and tetracycline in higher values than MNPs based on PE, PS, PP, and PVC due to the greater strength of the interactions between the adsorbate and the adsorbent, which include hydrogen bonding and dipole–dipole (Rafa et al. 2023; Martins et al. 2022). In turn, when MNPs undergo aging processes, their hydrophilic characteristics increase due to the generation of functional groups based on heteroatoms such as oxygen and sulfur (depending on the environmental conditions) and the increase of the amorphous phase. This results in higher adsorption of molecules with hydrophilic characteristics, such as ciprofloxacin and tetracycline, on aged PS and PVC (Li et al. 2025b).

- *Agrochemicals*: This category refers to a range of substances used in agri-food systems to promote and ensure favorable production conditions and animal and plant species protection. Thus, agrochemicals include substances as diverse as antibiotics, pesticides, growth regulators, herbicides, and hormones, which can undergo processes of leaching, runoff, volatilization, and/or metabolization, which can lead to issues such as eutrophication, selective pressures on microorganisms, bioaccumulation, and prevalence of neurological and immune diseases (Agathokleous et al. 2024; Devi et al. 2022; Saravanan et al. 2022; Mishra et al. 2022; Wang et al. 2020). Much like pharmaceuticals and personal care products, agrochemicals are a highly diverse group of substances found in MNPs from activities outside the agri-food system and derivatives of these. This is particularly the case of polyethylene films used in vegetable crops to improve the hydric and thermal conditions of the soil, which can undergo aging processes that lead to the formation of MNPs, whose co-exposure to pesticides such as diflubenzuron, difenoconazole, malathion, carbendazim, and dipterex, derives in the adsorption of these on the particle surface (Wang et al. 2020). This adsorption occurs through intermolecular interactions, mass transfer, and intraparticle diffusion, allowing the microplastics to act as carriers and transporters of potentially harmful substances, which could pose risks to public health and the environment (Wang et al. 2020).
- *Colorants*: They are substances with chromophore functional groups (unsaturated segments whose energy difference between two molecular orbitals is in the visible range) and auxochromes (substituents attached to a chromophore, which modifies its intensity and absorption wavelength) that can provide colors to cosmetic, food, and textile products (Dutta et al. 2024; Al-Tohamy et al. 2022; Lin et al. 2023; Garg and Chopra 2022; Tubić et al. 2022; Wang et al. 2023). It is in the latter industry, where these molecules are primarily used in dyeing processes, with high concentrations of these discharged into the environment, leading to a decline in water quality and safety, a high chemical and biological demand for oxygen, and the presence of highly toxic derivatives such as naphthoquinones and carcinogenic aromatic amines (Dutta et al. 2024; Al-Tohamy et al. 2022; Lin et al. 2023; Garg and Chopra 2022; Tubić et al. 2022; Wang et al. 2023; Bahrami et al. 2024). A scenario that has shown effects to the detriment of the photosynthetic process of terrestrial and aquatic plants has been linked to increased kidney and liver health issues (Dutta et al. 2024; Al-Tohamy et al. 2022; Lin et al. 2023; Garg and Chopra 2022; Tubić et al. 2022; Wang et al. 2023; Bahrami et al. 2024). One of these textile dyes is Congo Red, which is phytotoxic, mutagenic, and cytotoxic, and it can be adsorbed in MPs of high-density PE, PVC, low-density PE, PP, and PET, in that specific order of capacity (Nadarajah et al. 2025). The adsorption process is predominantly influenced by temperature, pH, intraparticle diffusion, and film diffusion. Additionally, it was found that the desorption capacity of the dye was 5.7 mg/L with an adsorbent dosage of 2 g/L, indicating the MPs' ability to transport dyes and potentially generate diffuse contamination (Nadarajah et al. 2025). On the other hand, as anticipated, MPs that have undergone aging

processes, with the consequent surface increase of active sites for molecules with polar functional groups, have improved their adsorption capacity for dyes such as Malachite Green; this is the case of PP and PET-based particles aged in a simulated seawater environment (Bai et al. 2024).

- *Chemical additives*: They are substances intentionally included in products during their manufacture to give them specific properties, such as improving their durability, appearance, or functionality, including antioxidants, plasticizers, preservatives, stabilizers, foaming agents, and humectants, used in industries as diverse as materials production, construction, polymers, and agri-food systems (Agboola and Benson 2021; Cortés-Arriagada et al. 2023). These include a variety of substances, some of which are harmless, such as Vitamin C, which acts as an antioxidant; natural carotenes, which serve as colorants; pectin, used as a thickener; and citric acid, which functions as an acidulant. Despite this panorama, there are also chemical additives that have adverse effects on human health; in the case of the polymer industry, one of the chemical additives of most significant concern are plasticizers such as BPA and the wide range of phthalate esters (such as di-ethyl phthalate, dimethylglycol phthalate, and butylbenzyl phthalate), which have been recognized as endocrine disruptors that can interfere in the metabolism, development, and reproduction of organisms, as well as adsorb on the surface of MNPs, where phthalate esters and BPA tend to adsorb mostly on PS-based MPs, concerning those of PE (Agboola and Benson 2021; Cortés-Arriagada et al. 2023). This difference is primarily due to the stronger stacking interactions of the aromatic rings in PS compared to the dispersive interactions in PE (Agboola and Benson 2021; Cortés-Arriagada et al. 2023). Table 5.2 provides a few reported examples of the different organic contaminants that can adsorb on MNPs.

Although the chemical substances have been mentioned individually, it is important to emphasize that the MNPs are not selective to only one type of entity, but rather, depending on the composition of the environment, they can co-absorb multiple substances on their surface, presenting mixed contamination of heavy metals, dyes, additives, and pharmaceutical compounds, resulting in the generation of increasingly complex contamination systems (Zhao et al. 2023). For instance, the adsorption of binary mixtures of sulfonamide antibiotics on PS PMs has been reported, where these mixtures were constituted from the combination of sulfapyridine, sulfamethazine, sulfamethoxypyridazine, and sulfamethoxazole; it was observed that, in general, the PMs adsorbed with the binary mixtures presented a higher toxicity than the pristine MNPs and a lower toxicity than the individual antibiotics against *Scenedesmus obliquus* (Yang et al. 2025b). On the other hand, research on the adsorption of pollutants on MNPs is primarily conducted at the laboratory level and often focuses on fundamental concepts. Most studies simulate aqueous systems with a single contaminant and MNPs with low polydispersity and a uniform composition. However, this approach overlooks that MNPs contamination is often heterogeneous, and these particles can either heteroaggregate or homoaggregate depending on environmental factors and aging. For instance, a reduction in the Zeta potential has been observed, indicating decreased stability and an increased likelihood of aggregation. While

Table 5.2 Examples of organic contaminants adsorbed on MNPs

Pollutant	Potential adverse effects	Type of MNPs	Remarks
Chlorpyrifos	Organophosphate insecticide that generates the accumulation of acetylcholine in the nervous system, resulting in neurological, digestive, and respiratory problems	PS	PS NPs did not present acute toxicity in *Daphnia Magna*. However, adsorption of insecticide resulted in the generation of an IC_{50} of 400 μg/L, an increase concerning the value of the insecticide (50.8 μg/L), indicating an antagonistic effect of the NPs on the toxicity of CPF (Zhang et al. 2024b)
Triphenyl phosphate	Flame retardant and plasticizer, which is an endocrine disruptor that also affects neural development and kidney function	PS	The presence of MPs and PS NPs with triphenyl phosphate adsorption, higher in NPs than in MPs, led to increased cell apoptosis and oxidative stress in HepG2 cells, where NPs exhibited more significant effects due to their internalization capacity (Huang et al. 2024)
Ciprofloxacin	Antibiotic of the fluoroquinolone family that can cause adverse effects such as hallucinations, convulsions, cardiac risks, and liver damage	PS-COOH	Ciprofloxacin's adsorption mechanism was mediated by hydrogen bonds and hydrophobic interactions with PS NPs containing carboxylic acid functional groups, particles that were more toxic to *Caenorhabditis elegans* than their individual components (Yilimulati et al. 2020)
Triclosan	An antimicrobial agent that affects development, reproduction, and the immune system; its prolonged use has been shown to reduce the effectiveness of medical treatments	PE PS	The plastics, in their macroscopic size, underwent an aging process through in-vitro digestion that simulates the digestive process of the human body, which resulted in the formation of MPs and NPs with an increased adsorption and transport capacity of triclosan (Krasucka et al. 2022)

(continued)

Table 5.2 (continued)

Pollutant	Potential adverse effects	Type of MNPs	Remarks
Reactive Orange Dye	Synthetic dye that has adverse effects on the respiratory system and dermis and are potential carcinogens and mutagens	PA	The adsorption process of the Orange Dye Reagent onto the PA MPs occurred from diffusion processes with a subsequent dynamic equilibrium, and the formation of this composite system resulted in an increase in removal efficiency by coagulation of up to 4.7% (Duan et al. 2024)
N-desmethylclarithromycin	Bioactive metabolite of clarithromycin that can alter microbial communities and promote the emergence of resistant bacteria	PE PET	The adsorption of the metabolite is higher in PET MPs than in PE MPs, values that decreased as the composition of the medium increased in complexity so that it presented the following order of adsorption in PET: tap water > surface water > effluent wastewater > influent wastewater (Mejías et al. 2024)

evaluations based on pristine or aged particles under a single parameter can provide valuable insights, they may not accurately reflect potential environmental impacts (Agboola and Benson 2021; Permana et al. 2025; Wang et al. 2025).

5.4 Impact of Pollutant-Micro/Nanoplastics Interactions

In the environment, plastic materials can be found as megaplastics (> 100 mm), macroplastics (20–100 mm), mesoplastics (5–20 mm), microplastics 0.1–5000 μm), and nanoplastics (1–100 nm) (Sheriff et al. 2023; Upadhyay et al. 2024). Each of these dimensions represents different environmental challenges and risks; nonetheless, the presence of MNPs in the environment signifies a higher risk to public health associated with a potential inherent toxicity linked to its dimension (Agboola and Benson 2021; Hajji et al. 2024; Yu et al. 2021; Sun et al. 2022; Hale 2017; Wagner et al. 2014). Smaller particles have greater mobility, allowing them to be widely distributed across different environmental compartments. This mobility has led to detecting MNPs in soil, bodies of water, air, microorganisms, and macroorganisms. The presence of MNPs generates a variety of effects, both positive and negative; these effects primarily depend on the nature of the MNPs, their concentration, size, exposure time, any substances or organisms adsorbed on their surface, and the physicochemical and biological characteristics of the surrounding environment (Li et al. 2025a). For example, it is possible to consider the case of poly(acrylamide), a low-toxicity material typically used as a soil conditioner and flocculant for water treatment, where the effects of degradation derive in the decrease of the size of the polymeric chains with the potential release of the monomer acrylamide, which presents neurotoxic and carcinogenic activity (Adegoke et al. 2023). Furthermore, additives added to synthetic plastics are usually found physically interacting with the material, which facilitates their release into the environment, even more so when the surface-to-volume ratio increases (as in the case of MNPs). Thus, the probability of releasing additives such as antioxidants, pigments, flame retardants, and plasticizers increases, allowing these to be available in the environment (Agboola and Benson 2021; Yu et al. 2021; Li et al. 2025a). This is a concern because some of these substances can harm public health such as endocrine disruptors, such as nonylphenol plasticizers and brominated flame retardants (Agboola and Benson 2021; Adegoke et al. 2023; Li et al. 2025a). Regarding the surface adsorption of pollutants in MNPs, a higher surface concentration of particles has been evidenced of several orders of magnitude concerning the environment, which favors the transport and mobilization of biological, chemical, and physical pollutants, thus being a vector that has facilitated the spread of diseases such as COVID-19 (Komorowska-Kaufman and Marciniak 2024; Al Harraq and Bharti 2021; Adegoke et al. 2023; Zhi et al. 2025; Naqash et al. 2020; Massahi et al. 2024).

This section will briefly address the effects of MNPs and co-MNP-pollutant interactions across different environmental compartments to provide an overview of the effects and impact of chemical pollutants-MNPs interactions in the environment. Readers are encouraged to consult specialized literature on the subject for a more detailed understanding of a specific natural system. The discussion will be organized into three simplified categories: air, soil, and water bodies.

In the case of *air*, this is the gas that constitutes the Earth's atmosphere, being composed mainly of diatomic nitrogen molecules, followed by oxygen and with the presence of components in smaller proportions, but with an impact on the temperature conditions of the planet, such as carbon dioxide and water vapor; where MNPs are airborne pollutants that pose risks to respiratory health and environmental quality (Amato-Lourenço et al. 2020; Torres-Agullo et al. 2021). On the other hand, one of the most relevant effects in the air is the wind, which has a speed and direction and includes phenomena like turbulence, convection lifting, updrafts, and downdrafts (Bank and Hansson 2019; Liu et al. 2019; Zhang et al. 2019). These processes are important vectors that alter the flow mechanisms and source-sink dynamics of plastic contamination in terrestrial and marine ecosystems by influencing the movement of MNPs (Bank and Hansson 2019; Liu et al. 2019; Zhang et al. 2019). The size, shape, and density of MNPs vary significantly in the environment, and these physical characteristics can influence their aerodynamics and atmospheric deposition. For instance, heavy MPs may not be carried very far from their origins because they drop from the air quickly; in contrast, MPs with greater surface-to-volume ratios or higher masses (such as lines and fragments) tend to settle more slowly and travel farther into the atmosphere; studies have shown that the MPs found in onshore wind samples are probably of marine origin, offering preliminary proof of MNPs interaction between the atmosphere and oceans (Brahney et al. 2021; Allen et al. 2020). On the other hand, MPs in the atmosphere have the potential to impact the formation of clouds and precipitation because they generate the development of cloud condensation nuclei, which are microscopic aerosols or particles in the atmosphere that serve as surfaces for the condensation of water vapor, resulting in the formation of cloud droplets (Brahney et al. 2021; Allen et al. 2020; Aeschlimann et al. 2022). MPs may also absorb and reflect sunlight, affecting cloud formation and heat dynamics (Aeschlimann et al. 2022). In addition, it has been shown that atmospheric pollutants can be adsorbed on the surface of MNPs; for example, PET MNPs, which are almost ubiquitous in the atmosphere, can spontaneously adsorb gases such as carbon dioxide, ammonia, nitrogen, and sulfur oxides, acting as transport vectors for these substances, whose adsorption capacity increases due to the effects of ozone on surface oxidation and the breakdown of the material's structure (Ortega and Cortés-Arriagada 2022). This is relevant not only when considering the co-transport of the composite pollutant through the atmosphere, but MNPs also make their way back to the Earth's surface through precipitation, where they are mixed into snowflakes or raindrops, contaminating water and soil systems; thus, the interaction between MNPs and pollutants results in the diffusion and dispersion of various forms and pollutants through the different environmental compartments (Liu et al. 2019; Zhang et al. 2019; Österlund et al. 2023; Allen et al. 2019; Klein and Fischer 2019).

In the case of *soil*, this is one of the major natural systems with the presence of MNPs, which come mostly from direct emissions, degradation of macroplastics, disintegration of textile fibers, and even through irrigation water and tire wear (Parsaeimehr et al. 2023; Zhai et al. 2024; Zhang and Liu 2018). In this natural system, particles can interact not only with the contaminants in the environment but also with their biotic and abiotic components, such as plants and humified organic acids. Additionally, these particles can transfer to groundwater and move to different bodies of water, resulting in a wide range of affectations (Komorowska-Kaufman and Marciniak 2024; Mesquita et al. 2023; Parsaeimehr et al. 2023; Zhai et al. 2024). In general, MPs can generate alterations in soil texture, derived from the formation of aggregates with organic matter and modifications in the distribution of soil particles; the more prominent MPs can act in a similar way to sand particles, favoring the enlargement of pores and increasing water infiltration, which can promote water erosion, and the smaller MPs can be located in the interstitial spaces and pores, decreasing soil aeration, hydraulic conductivity and permeability, presenting a direct effect on evapotranspiration and microbiological activity (Zhai et al. 2024; Ren et al. 2021; Liu et al. 2022b; Wang et al. 2022c). Also, the presence of MPs can lead to variations in pH, which can modify the stability of soil aggregates, favor erosive processes, and affect the bioavailability of nutrients to plants, thus decreasing agricultural yields, which are also derived from the adsorption of nutrients on the surface of the MPs, decreasing their availability to plant roots (Komorowska-Kaufman and Marciniak 2024; Talukdar et al. 2024; Massahi et al. 2024; Mesquita et al. 2023; Zhai et al. 2024; Souza Machado et al. 2018; Li et al. 2021; Das 2023). Nonetheless, plants can also adsorb, translocate, and accumulate MPs, reporting beneficial effects, such as increased root biomass, and harmful effects, such as decreased photosynthetic efficiency, depending on the characteristics of the PMs, the environmental conditions and the plant organism under analysis, a context that requires further research on the mode of action and mechanisms of phytotoxicity of MPs (Komorowska-Kaufman and Marciniak 2024; Talukdar et al. 2024; Massahi et al. 2024; Mesquita et al. 2023; Zhai et al. 2024; Souza Machado et al. 2018; Li et al. 2021; Das 2023).

This also applies to soil microorganisms, where both positive and negative alterations of microbial communities have been reported. For example, in clay soil, at 5% w/w, PE MPs (150 μm) reduced the variety and richness of bacterial and fungal communities, whereas smaller MPs (13 μm) increased the populations of both bacteria and fungi (Ren et al. 2022; Li et al. 2025b). In addition to these effects, it should be considered that if MNPs have organic and inorganic contaminants on their surface, these can be transported, desorbed, and/or internalized by organisms. For instance, it has been reported that MNPs with surface antibiotics have reduced the soil microbial community and increased the diversity of antibiotic-resistance genes; these genes can adsorb onto the surfaces of MNPs, serving as a vector for their transport (Zhou et al. 2021). This situation poses a significant threat because it facilitates the easy transfer of these genes between bacteria, resulting in more resistant infections to pharmacological treatments (Zhou et al. 2021). In the case of plants, an antagonistic effect between PE PMs and the fungicide carbendazim has been evidenced, so carbendazim's ability to control *Verticillium* wilt of cotton was reduced (Mateos-Cárdenas et al. 2021). Additionally, the presence of these contaminants has been found to have inhibitory effects on chlorophyll content and biomass in cotton. It also affects soil enzymatic activity by altering the structural composition of the bacterial community present in the soil (Mateos-Cárdenas et al. 2021). In turn, in higher organisms such as earthworms, co-MNPs-pollutant increases the accumulation and toxicity of heavy metals such as cadmium and the industrial surfactant perfluorooctanoic acid (Sun et al. 2022).

Furthermore, one of the most studied environmental settings in this field is *water* bodies, which are heavily affected by plastic waste. Plastic garbage disintegrates into particles in various aquatic environments and inside aquatic organisms, resulting in the presence of MNPs that have been identified in surface water, groundwater, wastewater, seabed, and along coastlines (Agboola and Benson 2021; Talukdar et al. 2024; Witczak and Pokorska-Niewiada 2023; Xie et al. 2025). In the case of wastewater, plastic micro/nanoparticles enter municipal and industrial wastewater systems through various sources, including stormwater runoff, industrial discharges, and residential effluents, as wastewater treatment facilities cannot wholly remove MPs and fail to remove NPs (Komorowska-Kaufman and Marciniak 2024; Xie et al. 2025; Liu et al. 2022b; Hu et al. 2024). In addition, studies have shown that many personal care products (such as facial powders, vanishing creams, toothpaste, detergents, and shaving foams) containing plastic particles smaller than 0.5 mm in size are routinely discharged into waterways via municipal wastewater. These systems have become the main entry points of plastic nanoparticles into aquatic ecosystems, where the main types of plastic nanoparticles identified in wastewater correspond, in descending order, to PS, PE, and PP (Komorowska-Kaufman and Marciniak 2024; Xie et al. 2025; Liu et al. 2022b; Hu et al. 2024). As for marine and coastal environments, numerous sources contribute to the presence of microplastics, which come from land-based sources (it is estimated that about 13 million tons of plastic waste are released into the aquatic environment), sewage spills, wastewater discharges, and marine operations such as accidental or illegal dumping during offshore drilling or fishing, which can be transported through the ocean, transferred to the food and trophic chain, and spread contamination composed of MNPs and organic and inorganic pollutants (see Fig. 5.3) (Komorowska-Kaufman and Marciniak 2024; Sudarsan et al. 2024; Talukdar et al. 2024; Monira et al. 2023; Acarer 2023; Murray and Cowie 2011; Costa et al. 2016; Sheavly and Register 2007). In these ecosystems, MPs can spread both vertically and horizontally across large distances in seas and oceans due to water, wind, and human activity. They are widely dispersed horizontally throughout aquatic environments and often gather in gyres and coastal areas (Murray and Cowie 2011; Falco et al. 2019; Auta et al. 2017). Where MPs can affect surface and benthic ecosystems as they move vertically through the water column; this movement occurs due to factors like turbulence, the sinking of organic matter, or biofouling. Also, species such as filter feeders, suspension feeders, and surface-feeding birds in the upper water column have access to these low-density MPs (Murray and Cowie 2011; Cole et al. 2011). On the other hand, high-density MPs that sink in the water column are consumed by deposit feeders, benthic suspension, and detritivores, where the biofouling boosts the nutritional value of microplastic particles, which promotes ingestion and grazing (Murray and Cowie 2011; Falco et al. 2019; Acampora et al. 2016; Wright et al. 2013; Vroom et al. 2017). Additionally, since zooplankton uses chemoreceptors to identify their algal food, the foulants cause them to react, which results in a greater rate of ingestion of fouled MPs than pure MPs (Wright et al. 2013). For example, blue mussels absorbed 7–30 μm PS beads more efficiently after they were incorporated into marine snow, going from 340 particles per mussel as freely floating beads to 1.6×10^5 particles per mussel (Courtene-Jones et al. 2017).

On the other hand, it has been found that microplastics can absorb organic contaminants and heavy metals with concentrations ranging from 10^{-1} to 10^4 μg/g in marine and coastal ecosystems, depending on their hydrophobicity (Porter et al. 2018), in which not only the particles but also the presence of these is identified with the adsorption of chemical substances such as persistent organic pollutants, dyes, and heavy metals, exhibiting an effect superior to that of their constituents. For instance, the case of mRNA expression levels of MMP2, MMP9, and IL-8 (associated with hepatocellular carcinoma and colonic adenocarcinoma) in the liver and intestinal tissues of grass carp against PMs with the presence of tetracycline; as well as in oxidative stress (increased superoxide dismutase, catalase and ascorbate peroxidase) in *Salvinia cucullata* exposed to Ps PMs with glyphosate adsorption and in *Danio rerio* exposed to PS MPS with triclosan (Hu et al. 2024; Guo and Wang 2019; Parenti et al. 2020). Also, a decrease in toxicity of individual compounds has been reported; for instance, *Danio rerio* embryos exposed to MPs with metformin and guanylate exhibited reduced embryotoxicity by blocking drug entry into embryos, as well as the

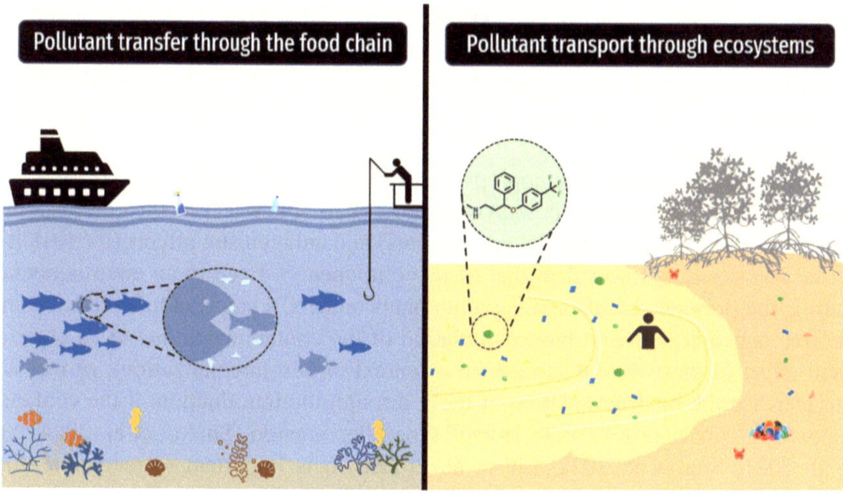

Fig. 5.3 Primary effects of MNPs adsorbed with pollutants in marine and coastal ecosystems

combined exposure of NPs and PCBs reduced the acute toxicity of the contaminant at a low concentration of NPs (Sun et al. 2022; Parenti et al. 2020).

Due to the context described above, it is well-known that living beings are constantly exposed to contact with MNPs through ingestion, inhalation, and direct interaction (e.g., through the skin), generating a wide variety of toxicological effects, which have been mainly associated with their size and morphology (where sharply edges derive in inflammatory processes), and described at the cellular and human tissue level, as well as in microbiological species, and more complex entities such as annelids, fish, birds and mammals; some reported effects include necrosis, oxidative stress responses, digestive tract issues, alterations in lipid metabolism, and changes in protein folding (Komorowska-Kaufman and Marciniak 2024; Talukdar et al. 2024; Hajji et al. 2024; Sheriff et al. 2023; Jiang et al. 2022; Fan et al. 2023; Zhi et al. 2025; Zhou et al. 2021; Hu et al. 2024; Wu et al. 2024; Prata et al. 2020). MNPs can generate adverse effects through direct mechanisms derived from their presence, interaction with biological entities, direct ingestion (for example, the average density of PS MPs resembles that of microplankton algae being easily confused by aquatic species), or through transfer in the trophic chain (lower, middle, upper trophic level, and top predators such as humans who ingest contaminated food) and bioaccumulation; or indirect, related to the release of chemical and/or biological substances present in their structure and/or adsorbed on the surface, this being one of the major concerns of the interaction between MNPs with potentially harmful substances, where positive correlations have been established between the ingestion of MPs with the potential accumulation of contaminants in related tissues (Talukdar et al. 2024; Yu et al. 2021; Zhi et al. 2025; Rafa et al. 2023; Li et al. 2025b; Acampora et al. 2016). This is associated with the fact that the morphology and nature of the particle can determine its organ of accumulation, and the process of desorption of a specific contaminant derives more significant effects due to the high localized contamination; for example, granular MPs are commonly located in abdominal lymph nodes where they can desorb emerging contaminants such as endocrine disruptors, generating inflammatory processes and defects in the functioning of the lymphatic system (Yu et al. 2021; Li et al. 2021).

Despite this scenario, it is appropriate to consider that the presence of surface-adsorbed MNPs can lead to both synergistic effects (the effect of the composite is higher than that of the individual components) and antagonistic effects (the toxicity of the composite is lower than that of its components). Even in an environmental setting, the presence of adsorbed contaminants on MNPs can lead to a reduction in the free concentration and bioaccumulation of the contaminants due to an enrichment effect of these on the surface (in a context of non-internalization), as well as can increase the bioaccumulation of these through the introduction of the contaminants as a carrier in a "Trojan Horse" type phenomenon (Shukla et al. 2024; Yu et al. 2021; Sun et al. 2022; Liu et al. 2022a). It is important to recognize that many of the toxicological effects identified in the studies arise from laboratory-scale tests conducted under simplified conditions using pristine or artificially aged MNPs. These tests often overlook significant factors influencing how these materials interact with living organisms and non-living components in various environmental settings. Therefore, claims regarding the toxicity of MNPs should be analyzed with caution, considering the specific conditions and scope of the research (Sudarsan et al. 2024; Chellasamy et al. 2023). In particular, many of the tests tend to use concentrations of the MNPs (the average concentration in the ocean is 0.01–10/m^3 and in soil < 1000/kg) and of the pollutant that may be significantly higher than those predicted under real conditions, which may lead to conclusions that are not fully extrapolated to natural ecosystems but are fully valuable for establishing acute rather than chronic toxic effects (Li et al. 2025a, 2021). In this sense, research on the effect of MNPs and co-MNPs-pollutants is an area in continuous development that seeks a comprehensive and long-term understanding of the dynamic effects of the interaction between pollutants, for which it has begun to evaluate more complex environments that simulate environmental conditions in terms of biotic and abiotic substances present in it, as well as in terms of environmental conditions. It is important to mention that evaluating the transfer in the trophic chain, transgenerational effects, and effects on higher macroscopic organisms are mainly conditioned by ethical limitations derived from the study in large organisms (Li et al. 2025a, 2021; Flores-Díaz et al. 2025).

5.5 Conclusions and Final Remarks

Synthetic plastics are highly relevant materials in today's societies since they manufacture manifold items and instruments for medical, agri-food, educational, and research purposes. This widespread use has led to an increase in their production, application, and final disposal, where this last aspect is particularly problematic in their life cycle, as the deficiencies in their management have resulted in their accumulation in various environmental compartments, notably in water bodies like marine and coastal ecosystems, which facilitate the mobility of these particles. This has increased the presence of primary and secondary MNPs, which interact with the different biotic and abiotic components of the environment, with potentially harmful effects, including inflammatory processes and decreased bioavailability of nutrients, undergoing phenomena such as surface and internal aging, which lead to changes in their physical and chemical properties, notably in terms of size, crystallinity, and surface composition.

In turn, MNPs can act as adsorbents of biological substances in the formation of eco-crown, and notably, they can present sorption processes of a wide variety of inorganic and organic chemical pollutants, due to their assorted composition, as well as their large surface area, and the modifications in the properties derived from environmental interaction and aging. The generation of these co-contaminants is of immense concern since the pollutants can adsorb, concentrate, transport, transfer up the trophic chain, and desorb from the surface, potentially increasing the harmful effects on public and environmental health.

Under this framework, substantial investigations have been conducted to understand the nature and factors that influence the sorption of chemical contaminants on the surface of MNPs, establishing the relevance of hydrophilic and hydrophobic interactions and chemical bonds, as well as pH, temperature, ionic force, and salinity in the modulation of this phenomenon. Studies have also shown that the interaction between MNPs and chemical contaminants on the environment and model biological organisms can present synergistic or antagonistic effects, depending on factors such as concentration, contact time, exposure conditions, and model entity. However, it is essential to analyze these results with wariness and understand the scope and limitations of the research because most studies are oriented to provide and build fundamental knowledge about the interaction and effect of chemical contaminants in MNPs. As a result, they typically involve simplified model systems with high concentrations of a single contaminant, which do not accurately reflect real-world conditions. These studies also often utilize MNPs with low polydispersity and single composition, overlooking important parameters such as additives, crystallinity, structure, aging processes, and mixed contamination. Therefore, current research aims to continue building fundamental knowledge and progressively include a wider range of variables, parameters, and conditions to understand the phenomenon in increasingly realistic and long-term environments.

Acknowledgements The authors acknowledge Mindtech s.a.s., Universidad del Valle, Analytical Science and Analytical Technology Institute Golden-Hammer, and the National Planning Department of Colombia, specifically to the General Royalty System (Sistema General de Regalías, SGR) for project BPIN2020000100261.

References

Acampora H, Lyashevska O, Van Franeker JA, O'Connor I (2016) The use of beached bird surveys for marine plastic litter monitoring in Ireland. Mar Environm Res 120:122–129. https://doi.org/10.1016/j.marenvres.2016.08.002

Acarer S (2023) Microplastics in wastewater treatment plants: sources, properties, removal efficiency, removal mechanisms, and interactions with pollutants. Water Sci Technol 87(3):685–710. https://doi.org/10.2166/wst.2023.022

Adegoke KA, Adu FA, Oyebamiji AK, Bamisaye A, Adigun RA et al (2023) Microplastics toxicity, detection, and removal from water/wastewater. Mar Pollut Bull 187:114546. https://doi.org/10.1016/j.marpolbul.2022.114546

Adepipe D, Chen C, Lai R, Xu S, Luo Q et al (2024) Occurrence and potential risks of pharmaceutical contamination in global estuaries: a critical review and analysis. Environ Int 192:109031. https://doi.org/10.1016/j.envint.2024.1090310

Aeschlimann M, Li G, Kanji ZA, Mitrano DM (2022) Potential impacts of atmospheric microplastics and nanoplastics on cloud formation processes. Nat Geosci 15(12):967–975. https://doi.org/10.1038/s41561-022-01051-9

Agathokleous E, Guedes R, Calabrese E (2024) Reimagining agrochemical pollution mitigation: leveraging hormesis for sustainable environmental solutions. Sci Total Environ 949:175165. https://doi.org/10.1016/j.scitotenv.2024.175165

Agboola OD, Benson NU (2021) Physisorption and chemisorption mechanisms influencing micro (Nano) plastics-organic chemical contaminants interactions: a review. Front Environ Sci 9. https://doi.org/10.3389/fenvs.2021.678574

Ahmad A, Salam D (2025) Adsorption behavior of crude oil hydrocarbons on polyethylene microplastics in batch experiments. Mar Pollut Bull 215:117832. https://doi.org/10.1016/j.marpolbul.2025.117832

Ajithkumar V, Philomina A, Meena K, Pothiaraj G, Dey D et al (2024) Insights of recent developments in microplastics pollution and its degradation in wastewater and other environment. J Taiwan Inst Chem Eng 105504. https://doi.org/10.1016/j.jtice.2024.105504

Al Harraq A, Bharti B (2021) Microplastics through the lens of colloid science. ACS Environ Au 2(1):3–10. https://doi.org/10.1021/acsenvironau.1c00016

Allen S, Allen D, Phoenix VR, Le Roux G, Durántez Jiménez P, Simonneau A, Binet S et al (2019) Atmospheric transport and deposition of microplastics in a remote mountain catchment. Nat Geosci 12(5):339–344. https://doi.org/10.1038/s41561-019-0335-5

Allen S, Allen D, Moss K, Le Roux G, Phoenix VR et al (2020) Examination of the ocean as a source for atmospheric microplastics. PLoS ONE 15(5):e0232746. https://doi.org/10.1371/journal.pone.0232746

Al-Tohamy R, Ali S, Li F, Okasha K, Mahmoud Y, Elsamahy T et al (2022) A critical review on the treatment of dye-containing wastewater: ecotoxicological and health concerns of textile dyes and possible remediation approaches for environmental safety. Ecotoxicol Environ Saf 231:113160. https://doi.org/10.1016/j.ecoenv.2021.113160

Amato-Lourenço LF, dos Santos Galvão L, de Weger LA, Hiemstra PS, Vijver MG, Mauad T (2020) An emerging class of air pollutants: potential effects of microplastics to respiratory human health? Sci Total Environ 749:141676. https://doi.org/10.1016/j.scitotenv.2020.141676

Aranda F, Rivas B (2022) Removal of amoxicillin through different methods, emphasizing removal by biopolymers and its derivatives. J Chil Chem Soc 67:5643–5655. https://doi.org/10.4067/s0717-97072022000305643

Arregocés-Garcés R, Garcés-Ordóñez O, Vivas-Aguas LJ, Canals M (2024) Microplastics transfer from a malfunctioning municipal wastewater oxidation pond into a marine protected area in the Colombian Caribbean. Regional Stud Mar Sci 69:103361. https://doi.org/10.1016/j.rsma.2023.103361

Auta HS, Emenike CU, Fauziah SH (2017) Distribution and importance of microplastics in the marine environment: a review of the sources, fate, effects, and potential solutions. Environ Int 102:165–176. https://doi.org/10.1016/j.envint.2017.02.013

Bahrami M, Amiri M, Rajabi S, Mahmoudi M (2024) The removal of methylene blue from aqueous solutions by polyethylene microplastics: modeling batch adsorption using random forest regression. Alexandria Eng J 95:101–113. https://doi.org/10.1016/j.aej.2024.03.100

Bai H, Liu B, Jiang Y, Zhang J, Zhang M et al (2024) Adsorption-desorption behavior of malachite green on aged microplastics in seawater environment. Sep Purif Technol 354:128991. https://doi.org/10.1016/j.seppur.2024.128991

Bank MS, Hansson SV (2019) The plastic cycle: a novel and holistic paradigm for the anthropocene. Environ Sci Technol 53(13):7177–7179. https://doi.org/10.1021/acs.est.9b02942

Barathi S, Gitanjali J, Rathinasamy G, Sabapathi N, Aruljothi K et al (2023) Recent trends in polycyclic aromatic hydrocarbons pollution distribution and counteracting bio-remediation strategies. Chemosphere 337:139396. https://doi.org/10.1016/j.chemosphere.2023.139396

Bartley T, Fairbrother M (2024) Tackling toxins: case studies of industrial pollutants and implications for climate policy. Regul Gov. https://doi.org/10.1111/rego.12626

Beggel S, Kalis EJJ, Geist J (2024) Towards harmonized ecotoxicological effect assessment of micro- and nanoplastics in aquatic systems. Environ Pollut 366:125504. https://doi.org/10.1016/j.envpol.2024.125504

Bi S, Liu S, Liu E, Xiong J, Xu Y et al (2024) Adsorption behavior and mechanism of heavy metals onto microplastics: a meta-analysis assisted by machine learning. Environ Poll 360:124634. https://doi.org/10.1016/j.envpol.2024.124634

Biedunkova O, Kuznietsov P (2024) Dataset on heavy metal pollution assessment in freshwater ecosystems. Sci Data 11. https://doi.org/10.1038/s41597-024-04116-z

Bodus B, O'Malley K, Dieter G, Gunawardana C, McDonald W (2024) Review of emerging contaminants in green stormwater infrastructure: antibiotic resistance genes, microplastics, tire wear particles, PFAS, and temperature. Sci Total Environ 906:167195. https://doi.org/10.1016/j.scitotenv.2023.167195

Brahney J, Mahowald N, Prank M, Cornwell G, Klimont Z et al (2021) Constraining the atmospheric limb of the plastic cycle. Proc Natl Acad Sci 118(16). https://doi.org/10.1073/pnas.2020719118

Chellasamy G, Ramasundaram S, Veerapandian M, Chandran M, Dhanasekaran B, et al (2023) Systematic review on fate and behavior of microplastics towards the environment. TrAC, Trends Anal Chem 169:117390. https://doi.org/10.1016/j.trac.2023.117390

Chen Z, Yang J, Huang D, Wang S, Jiang K et al (2023) Adsorption behavior of aniline pollutant on polystyrene microplastics. Chemosphere 323:138187. https://doi.org/10.1016/j.chemosphere.2023.138187

Chhetri H, Thapa P, Van Schepdael A (2024) Simple HPLC-UV method for the quantification of metformin in human plasma with one step protein precipitation. Saudi Pharm J 22:483–487. https://doi.org/10.1016/j.jsps.2013.12.011

Cole M, Lindeque P, Halsband C, Galloway TS (2011) Microplastics as contaminants in the marine environment: a review. Mar Poll Bullet 62(12):2588–2597. https://doi.org/10.1016/j.marpolbul.2011.09.025

Cortés-Arriagada D, Ortega D, Miranda-Rojas S (2023) Mechanistic insights into the adsorption of endocrine disruptors onto polystyrene microplastics in water. Environ Poll 319:121017. https://doi.org/10.1016/j.envpol.2023.121017

Costigan E, Collins A, Hatinoglu M, Bhagat K, MacRae J et al (2022) Adsorption of organic pollutants by microplastics: overview of a dissonant literature. J Hazard Mater Adv 6:10009. https://doi.org/10.1016/j.hazadv.2022.100091

Courtene-Jones W, Quinn B, Gary SF, Mogg AOM, Narayanaswamy BE (2017) Microplastic pollution identified in deep-sea water and ingested by benthic invertebrates in the rockall trough, North Atlantic Ocean. Environ Poll 231:271–280. https://doi.org/10.1016/j.envpol.2017.08.026

da Costa JP, Santos PSM, Duarte AC, Rocha-Santos T (2016) (Nano)plastics in the environment—sources, fates and effects. Sci Total Environ 566–567:15–26. https://doi.org/10.1016/j.scitotenv.2016.05.041

Dagdag O, Quadri T, Haldhar R, Kim S, Daoudi W et al (2023) An overview of heavy metal pollution and control. ACS Symp Ser 3–24. https://doi.org/10.1021/bk-2023-1456.ch001

Das A (2023) The emerging role of microplastics in systemic toxicity: involvement of reactive oxygen species (ROS). Sci Total Environ 895:165076. https://doi.org/10.1016/j.scitotenv.2023.165076

De Falco F, Di Pace E, Cocca M, Avella M (2019) The contribution of washing processes of synthetic clothes to microplastic pollution. Sci Rep 9(1):6633. https://doi.org/10.1038/s41598-019-43023-x

de Oliveira H, dos Santos R, Rabelo G, dos Santos M (2021) Pharmaceutical pollution and sustainable development goals: going the right way? Sustain Chem Pharm 21:100428. https://doi.org/10.1016/j.scp.2021.100428

de Souza Machado AA, Lau CW, Till J, Kloas W, Lehmann A et al (2018) Impacts of microplastics on the soil biophysical environment. Environ Sci Technol 52(17):9656–9665. https://doi.org/10.1021/acs.est.8b02212

Desidery L, Lanotte M (2022) Polymers and plastics: types, properties, and manufacturing. In: Giustozzi F, Nizamuddin S (eds) Plastic waste for sustainable asphalt roads. Woodhead Publishing Series in Civil and Structural Engineering, pp 3–28. https://doi.org/10.1016/b978-0-323-85789-5.00001-0

Devi P, Manjula M, Bhavani R (2022) Agrochemicals, environment, and human health. Annu Rev Environ Resour 47:399–421. https://doi.org/10.1146/annurev-environ-120920-111015

Duan Y, Sun S, Zhao J, Yuan H (2024) Microplastics affect the removal of dye in textile wastewater: adsorption capacity and its effect on coagulation behavior. Sep Purif Technol 130505. https://doi.org/10.1016/j.seppur.2024.130505

Dutta S, Adhikary S, Bhattacharya S, Roy D, Chatterjee S et al (2024) Contamination of textile dyes in aquatic environment: adverse impacts on aquatic ecosystem and human health, and its management using bioremediation. J Environ Manage 353:120103. https://doi.org/10.1016/j.jenvman.2024.120103

Fan L, Mohseni A, Schmidt J, Evans B, Murdoch B, Gao L (2023) Efficiency of lagoon-based municipal wastewater treatment in removing microplastics. Sci Total Environ 876:162714. https://doi.org/10.1016/j.scitotenv.2023.162714

Flores-Díaz A, Arango J, Calvo D, Rangel-Mendez R et al (2025) Physicochemical and microbial treatments for plastics, microplastics, and nanoplastics. Environ Dev 101181. https://doi.org/10.1016/j.envdev.2025.101181

Forero D, Fabiani M, Lassalle VL, Spetter CV, Fernandez D (2022) Critical review of the characteristics, interactions, and toxicity of micro/nanomaterials pollutants in aquatic environments. Mar Pollut Bull 174:113276. https://doi.org/10.1016/j.marpolbul.2021.113276

Franco AA, Arellano JM, Albendín G, Rodríguez-Barroso R, Zahedi S, Quiroga JM, Coello MD (2020) Mapping microplastics in Cadiz (Spain): occurrence of microplastics in municipal and industrial wastewaters. J Water Process Eng 38:101596. https://doi.org/10.1016/j.jwpe.2020.101596

Gaffney J, Marley N (2019) The chemistry of carbon. In: Gaffney JS, Marley AA (eds) General chemistry for engineers. Elsevier, pp 419–458. https://doi.org/10.1016/b978-0-12-810425-5.00013-8

Gao L, Fu D, Zhao J, Wu W, Wang Z et al (2021) Microplastics aged in various environmental media exhibited strong sorption to heavy metals in seawater. Mar Pollut Bull 169:112480. https://doi.org/10.1016/j.marpolbul.2021.112480

Gao D, Junaid M, Chen X, Liao H, Chen G et al (2023) Interaction of micro(nano)plastics and bisphenols in the environment: a recent perspective on adsorption mechanisms, influencing factors and ecotoxic impacts. TrAC, Trends Anal Chem 165:117132. https://doi.org/10.1016/j.trac.2023.117132

Garg A, Chopra L (2022) Dye waste: a significant environmental hazard. Mater Today: Proc 48:1310–1315. https://doi.org/10.1016/j.matpr.2021.09.003

Ge Z, Lu X (2023) Impacts of extracellular polymeric substances on the behaviors of micro/nanoplastics in the water environment. Environmental Poll 338:122691. https://doi.org/10.1016/j.envpol.2023.122691

Guo X, Wang J (2019) The chemical behaviors of microplastics in marine environment: a review. Marine Poll Bull 142:1–14. https://doi.org/10.1016/j.marpolbul.2019.03.019

Guo C, Wang L, Lang D, Qian Q, Wang W (2023) UV and chemical aging alter the adsorption behavior of microplastics for tetracycline. Environ Pollut 318:120859. https://doi.org/10.1016/j.envpol.2022.120859

Hajji S, Ben-Haddad M, Abelouah MR, Rangel-Buitrago N, Alla AA (2024) Microplastic characterization and assessment of removal efficiency in an urban and industrial wastewater treatment plant with submarine emission discharge. Sci Total Environ 945:174115. https://doi.org/10.1016/j.scitotenv.2024.174115

Hale RC (2017) Analytical challenges associated with the determination of microplastics in the environment. Anal Methods 9(9):1326–1327. https://doi.org/10.1039/C7AY90015E

Hou G, Zhao X, Zhao T, Wu X, Pu S et al (2023) The adsorption of PAHs on microplastics and desorption in the simulated human digestive system. Chem Eng J 473:145157. https://doi.org/10.1016/j.cej.2023.145157

Hu X, Hua Z, Ding Z, Sun J, Wang T et al (2024) Carbendazim adsorption on polyethylene microplastics and the toxicity mechanisms on cotton plants, soil enzyme activity and rhizosphere bacterial community under combined stress conditions. J Environ Chem Eng 114219. https://doi.org/10.1016/j.jece.2024.114219

Huang J, Wang L, Liu J, Qian X, Wu Y (2023) Abundance, characteristics, and removal of microplastics in different wastewater treatment plants in a Yangtze River Delta City of China. J Water Process Eng 54:103987. https://doi.org/10.1016/j.jwpe.2023.103987

Huang W, Yang Y, Tang S, Yin H, Yu X et al (2024) The combined toxicity of polystyrene nano/micro-plastics and triphenyl phosphate (TPHP) on HepG2 cells. Ecotoxicol Environ Saf 279:116489. https://doi.org/10.1016/j.ecoenv.2024.116489

Jiang L, Chen M, Huang Y, Peng J, Zhao J, Chan F, Yu X (2022) Effects of different treatment processes in four municipal wastewater treatment plants on the transport and fate of microplastics. Sci Total Environ 831:154946. https://doi.org/10.1016/j.scitotenv.2022.154946

Kim SW, Kim YT, Tsang YF, Lee J (2023) Sustainable ethylene production: recovery from plastic waste via thermochemical processes. Sci Total Environ 903:166789. https://doi.org/10.1016/j.scitotenv.2023.166789

Klein M, Fischer EK (2019) Microplastic abundance in atmospheric deposition within the Metropolitan area of Hamburg, Germany. Sci Total Environ 685:96–103. https://doi.org/10.1016/j.scitotenv.2019.05.405

Komorowska-Kaufman M, Marciniak W (2024) Removal of microplastic particles during municipal wastewater treatment: a current review. Desalin Water Treat 317:100006. https://doi.org/10.1016/j.dwt.2024.100006

Krasucka P, Bogusz A, Baranowska-Wójcik E, Czech B, Szwajgier D et al (2022) Digestion of plastics using in vitro human gastrointestinal tract and their potential to adsorb emerging organic pollutants. Sci Total Environ 843:157108. https://doi.org/10.1016/j.scitotenv.2022.157108

Lee L (1991) Hard—Soft Acid—Base (HSAB) principle for solid adhesion and surface interactions. Springer eBooks, pp 349–362. https://doi.org/10.1007/978-1-4899-2073-7_12

Li S, Wang T, Guo J, Dong Y, Wang Z, Gong L, Li X (2021) Polystyrene microplastics disturb the redox homeostasis, carbohydrate metabolism and phytohormone regulatory network in barley. J Hazard Mat 415:125614. https://doi.org/10.1016/j.jhazmat.2021.125614

Li M, Ma W, Kar J, Mo J, Li L et al (2025) A review on the combined toxicological effects of microplastics and their attached pollutants. Emerging Contam 100486. https://doi.org/10.1016/j.emcon.2025.100486

Li H, Bai L, Liang S, Chen X, Gu X et al (2025) The wheel of time: the environmental dance of aged micro- and nanoplastics and their biological resonance. Eco-Environment Health 100138. https://doi.org/10.1016/j.eehl.2025.100138

Liang D, Shan R, Gu J, Cheng P, Wang S et al (2024) Analysis, risk assessment and treatment of aquatic micro/nanoplastics: a critical review. Sep Purif Technol 354:129418. https://doi.org/10.1016/j.seppur.2024.129418

Lin J, Ye W, Xie M, Seo D, Luo J et al (2023) Environmental impacts and remediation of dye-containing wastewater. Nat Rev Earth Environ 4:785–803. https://doi.org/10.1038/s43017-023-00489-8

Liu K, Wu T, Wang X, Song Z, Zong C et al (2019) Consistent transport of terrestrial microplastics to the ocean through atmosphere. Environ Sci Technol 53(18):10612–10619. https://doi.org/10.1021/acs.est.9b03427

Liu W, Zhang J, Liu H, Guo X, Zhang X et al (2021) A review of the removal of microplastics in global wastewater treatment plants: characteristics and mechanisms. Environ Int 146:106277. https://doi.org/10.1016/j.envint.2020.106277

Liu Q, Wu H, Chen J, Guo B, Zhao X et al (2022a) Adsorption mechanism of trace heavy metals on microplastics and simulating their effect on Microalgae in River. Environ Res 214:113777. https://doi.org/10.1016/j.envres.2022.113777

Liu Y, Wang B, Pileggi V, Chang S (2022b) Methods to recover and characterize microplastics in wastewater treatment plants. Case Stud Chem Environ Eng 5:100183. https://doi.org/10.1016/j.cscee.2022.100183

Malmqvist E, Fumagalli D, Munthe C, Larsson J (2023) Pharmaceutical pollution from human use and the polluter pays principle. Public Health Ethics 16:152–164. https://doi.org/10.1093/phe/phad012

Martins A, Santos A, Araújo D, Rodrigues R, Rocha T (2022) Micro(Nano)Plastics as a vector of pharmaceuticals in aquatic ecosystem: historical review and future trends. J Hazard Mater Adv 6:100068. https://doi.org/10.1016/j.hazadv.2022.100068

Massahi T, Amin A, Meshabaz A, Pirsaheb M, Terry L et al (2024) Microplastic occurrence and its potential role as a carrier for SARS-CoV-2 in health center wastewater treatment plant and surface water. J Sea Res 198:102477. https://doi.org/10.1016/j.seares.2024.102477

Mateos-Cárdenas A, van Pelt F, O'Halloran J, Jansen M (2021) Adsorption, uptake and toxicity of micro- and nanoplastics: effects on terrestrial plants and aquatic macrophytes. Environ Pollut 284:117183. https://doi.org/10.1016/j.envpol.2021.117183

Mejías C, Martín J, Martín-Pozo L, Santos J, Aparicio I, Alonso E (2024) Adsorption of macrolide antibiotics and a metabolite onto polyethylene terephthalate and polyethylene microplastics in aquatic environments. Antibiotics 13:408. https://doi.org/10.3390/antibiotics13050408

Mesquita D, Quintelas C, Ferreira E (2023) Fate and occurrence of microplastics in wastewater treatment plants. Environ Sci: Adv 2:1616–1628. https://doi.org/10.1039/d3va00167a

Mishra C, Samal S, Samal R, Behera B, Pallavini P et al (2022) Polyvinylchloride and polypropylene as adsorbents of the pesticide monocrotophos enhance oxidative stress in *Eudrillus Eugeniae* (Kinberg). Chemosphere 295:133837. https://doi.org/10.1016/j.chemosphere.2022.133837

Mohana A, Rahman M, Sarker S, Haque N, Gao L et al (2022) Nano/microplastics: fragmentation, interaction with co-existing pollutants and their removal from wastewater using membrane processes. Chemosphere 309:136682. https://doi.org/10.1016/j.chemosphere.2022.136682

Monira S, Roychand R, Hai FI, Bhuiyan M, Dha BR, Pramanik BK (2023) Nano and microplastics occurrence in wastewater treatment plants: a comprehensive understanding of microplastics fragmentation and their removal. Chemosphere 334:139011. https://doi.org/10.1016/j.chemosphere.2023.139011

Murray F, Cowie PR (2011) Plastic contamination in the decapod crustacean *Nephrops norvegicus* (Linnaeus, 1758). Mar Poll Bull 62(6):1207–1217. https://doi.org/10.1016/j.marpolbul.2011.03.032

Nadarajah K, Jeganathan Y, Ramakrishnan S, Bandala E (2025) Understanding the interaction between selected microplastics and the toxic dye "congo red" in water. Chemosphere 372:144097. https://doi.org/10.1016/j.chemosphere.2025.144097

Naqash N, Prakash S, Kapoor D, Singh R (2020) Interaction of freshwater microplastics with biota and heavy metals: a review. Environ Chem Lett 18:1813–1824. https://doi.org/10.1007/s10311-020-01044-3

Okunola A, Kehinde I, Oluwaseun A, Olufiropo E (2019) Public and environmental health effects of plastic wastes disposal: a review. J Toxicol Risk Assess 5. https://doi.org/10.23937/2572-4061.1510021

Ortega D, Cortés-Arriagada D (2022) Atmospheric microplastics and nanoplastics as vectors of primary air pollutants—a theoretical study on the polyethylene terephthalate (PET) case. Environ Poll 318:120860. https://doi.org/10.1016/j.envpol.2022.120860

Österlund H, Blecken G, Lange K, Marsalek J, Gopinath K et al (2023) Microplastics in urban catchments: review of sources, pathways, and entry into stormwater. Sci Total Environ 858:159781. https://doi.org/10.1016/j.scitotenv.2022.159781

Otalora A, Lerma T, Arrieta-Urango Y, Palencia M (2021) Emerging organic pollutants in aqueous environments: detection, monitoring, and removal techniques. J Sci Tech Appl 10:92–153. https://doi.org/10.34294/j.jsta.21.10.68

Palencia M, Berrio M, Melendrez M (2016) Nanostructured polymer composites with potential applications into the storage of blood and hemoderivates. J Sci Technol Appl 1:4–14. https://doi.org/10.34294/j.jsta.16.1.1

Palencia M, Lerma TA, Garcés V, Mora MA, Martínez JM, Palencia SL (2021) Eco-friendly functional polymers: an approach from application-targeted green chemistry. Elsevier, p 448. https://doi.org/10.1016/C2019-0-01709-8

Parenti C, Magni S, Ghilardi A, Caorsi G, Torre C et al (202) Does triclosan adsorption on polystyrene nanoplastics modify the toxicity of single contaminants? Environ Sci Nano 8:282–296. https://doi.org/10.1039/d0en00961j

Parsaeimehr A, Miller CM, Ozbay G (2023) Microplastics and their interactions with microbiota. Helyon 9(4):e15104. https://doi.org/10.1016/j.heliyon.2023.e15104

Patel A, Shaikh S, Jain K, Desai C, Madamwar D (2020) Polycyclic aromatic hydrocarbons: sources, toxicity, and remediation approaches. Front Microbiol 11. https://doi.org/10.3389/fmicb.2020.562813

Permana R, Chakraborty S, Valsami-Jones E (2025) Nanoplastics in aquatic environments: the hidden impact of aging on fate and toxicity. Environ Chem Ecotoxicol 7:429–444. https://doi.org/10.1016/j.enceco.2025.02.007

Piyathilake U, Lin C, Bolan N, Bundschuh J, Rinklebe J et al (2024) Exploring the hidden environmental pollution of microplastics derived from bioplastics: a review. Chemosphere 355:141773. https://doi.org/10.1016/j.chemosphere.2024.141773

Porta R (2021) Anthropocene, the plastic age and future perspectives. FEBS Open Biol 11(4):948–953. https://doi.org/10.1002/2211-5463.13122

Porter A, Lyons BP, Galloway TS, Lewis C (2018) Role of marine snows in microplastic fate and bioavailability. Environ Sci Technol 52(12):7111–7119. https://doi.org/10.1021/acs.est.8b01000

Prasad B, Das P, Srivastava V, Kumar M (2023) Perspectives of heavy metal pollution indices for soil, sediment, and water pollution evaluation: an insight. Total Environ Res Themes 6:100039. https://doi.org/10.1016/j.totert.2023.100039

Prata JC, da Costa JP, Lopes I, Duarte AC, Rocha-Santos T (2020) Environmental exposure to microplastics: an overview on possible human health effects. Sci Total Environ 702:134455. https://doi.org/10.1016/j.scitotenv.2019.134455

Rafa N, Ahmed B, Zohora F, Bakya J, Ahmed S et al (2023) Microplastics as carriers of toxic pollutants: source, transport, and toxicological effects. Enviro Poll 343:123190. https://doi.org/10.1016/j.envpol.2023.123190

Ren Z, Gui X, Xu X, Zhao X, Qiu H et al (2021) Microplastics in the soil-groundwater environment: aging, migration, and co-transport of contaminants—a critical review. J Hazard Mater 419:126455. https://doi.org/10.1016/j.jhazmat.2021.126455
Ren J, Andrikopoulos N, Velonia K, Tang H, Cai R, Ding F, Ke PC, Chen C (2022) Chemical and biophysical signatures of the protein corona in nanomedicine. J Am Chem Soc 144(21):9184–9205. https://doi.org/10.1021/jacs.2c02277
Ruggieri F, Battistini B, Sorbo A, Senofonte M, Leso V et al (2024) From food-to-human microplastics and nanoplastics exposure and health effects: a review on food, animal and human monitoring data. Food Chem Toxicol 115209. https://doi.org/10.1016/j.fct.2024.115209
Santana GB, Conceição KGA, Silva EM, Diaz GZ, Oliveira JT et al (2024) Photoaging effects on polyethylene microplastics: structural changes and chlorpyrifos adsorption. Mar Environ Res 203:106844. https://doi.org/10.1016/j.marenvres.2024.106844
Saravanan A, Kumar P, Jeevanantham S, Anubha M, Jayashree S (2022) Degradation of toxic agrochemicals and pharmaceutical pollutants: effective and alternative approaches toward photocatalysis. Environ Poll 298:118844. https://doi.org/10.1016/j.envpol.2022.118844
Schultz M (2014) Chemicals of environmental concern. In: Wexler P, Encyclopedia of toxicology, pp 805–809. https://doi.org/10.1016/b978-0-12-386454-3.01004-6
Sheavly SB, Register KM (2007) Marine debris and plastics: environmental concerns, sources, impacts and solutions. J Polym Environ 15(4):301–305. https://doi.org/10.1007/s10924-007-0074-3
Sheriff I, Yusoff MS, Halim HB (2023) Microplastics in wastewater treatment plants: a review of the occurrence, removal, impact on ecosystem, and abatement measures. J Water Process Eng 54:104039. https://doi.org/10.1016/j.jwpe.2023.104039
Shukla S, Khanna S, Khanna K (2024) Unveiling the toxicity of micro-nanoplastics: a systematic exploration of understanding environmental and health implications. Toxicol Rep 14:101844. https://doi.org/10.1016/j.toxrep.2024.101844
Sudarsan JS, Dogra K, Kumar R, Raval NP, Leifels M et al (2024) Tricks and tracks of prevalence, occurrences, treatment technologies, and challenges of mixtures of emerging contaminants in the environment: with special emphasis on microplastic. J Contam Hydrol 265:104389. https://doi.org/10.1016/j.jconhyd.2024.104389
Sun N, Shi H, Li X, Gao C, Liu R (2022) Combined toxicity of micro/nanoplastics loaded with environmental pollutants to organisms and cells: role, effects, and mechanism. Environ Int 171:107711. https://doi.org/10.1016/j.envint.2022.107711
Talukdar A, Kundu P, Bhattacharya S, Dutta N (2024) Microplastic contamination in wastewater: sources, distribution, detection and remediation through physical and chemical-biological methods. Sci Total Environ 916:170254. https://doi.org/10.1016/j.scitotenv.2024.170254
Tang N, Li X, Gao X, Liu X, Xing W (2022a) The adsorption of arsenic on micro- and nano-plastics intensifies the toxic effect on submerged macrophytes. Environ Poll 311:119896. https://doi.org/10.1016/j.envpol.2022.119896
Tang S, Yang X, Zhang T, Qin Y, Cao C et al (2022b) Adsorption mechanisms of metal ions (Pb, Cd, Cu) onto polyamide 6 microplastics: new insight into environmental risks in comparison with natural media in different water matrices. Gondwana Res 110:214–225. https://doi.org/10.1016/j.gr.2022.06.017
Tang S, Ma S, Lin L, Ding Y, Zhang X et al (2024) Carrier effects of face mask-derived microplastics on metal ions: enhanced adsorption by photoaging combined with biofilms, exemplified with Pb(II). J Hazard Mater 477:135311. https://doi.org/10.1016/j.jhazmat.2024.135311
Tanzil K, Ealias A, George G, Panigrahi S (2024) Microplastics: a hidden carrier of per- and polyfluoroalkyl substances and their effect on soil properties. J Environ Chem Eng 114974. https://doi.org/10.1016/j.jece.2024.114974
Tenea A, Dinu C, Rus P, Ionescu I, Gheorghe S et al (2024) Exploring adsorption dynamics of heavy metals onto varied commercial microplastic substrates: isothermal models and kinetics analysis. Heliyon 10(15):e35364. https://doi.org/10.1016/j.heliyon.2024.e35364

Thomas KV (2022) Understanding the plastics cycle to minimize exposure. Nat Sustain 5:282–284. https://doi.org/10.1038/s41893-021-00814-3

Thompson RC, Swan SH, Moore CJ, von Saal FS (2009) Our plastic age. Phylosophical Trans R Soc B 364:1973–1976. https://doi.org/10.1098/rstb.2009.0054

Torres-Agullo A, Karanasiou A, Moreno T, Lacorte S (2021) Overview on the occurrence of microplastics in air and implications from the use of face masks during the COVID-19 pandemic. Sci Total Environ 800:149555. https://doi.org/10.1016/j.scitotenv.2021.149555

Tsatsakis A, Vassilopoulou L, Kovatsi L, Tsitsimpikou C, Karamanou M et al (2018) The dose response principle from philosophy to modern toxicology: the impact of ancient philosophy and medicine in modern toxicology science. Toxicol Rep 5:1107–1113. https://doi.org/10.1016/j.toxrep.2018.10.001

Tubić A, Vujić M, Gvoić V, Agbaba J, Vasiljević S et al (2022) Sorption potential of microplastics for azo- and phthalocyanine printing dyes. Dyes Pigm 209:110884. https://doi.org/10.1016/j.dyepig.2022.110884

Um M, Weerackody D, Gao L, Mohseni A, Evans B, Murdoch B, Schmidt J, Fan L (2023) Investigating the fate and transport of microplastics in a lagoon wastewater treatment system using a multimedia model approach. J Hazard Mater 446:130694. https://doi.org/10.1016/j.jhazmat.2022.130694

Upadhyay S, Sharma PK, Dogra K, Bhattacharya P, Kumar M et al (2024) Microplastics in freshwater: unveiling sources, fate, and removal strategies. Groundw Sustain Dev 26:101185. https://doi.org/10.1016/j.gsd.2024.101185

Vatovec C, Kolodinsky J, Callas P, Hart C, Gallagher K (2021) Pharmaceutical pollution sources and solutions: survey of human and veterinary medication purchasing, use, and disposal. J Environ Manage 285:112106. https://doi.org/10.1016/j.jenvman.2021.112106

Verla A, Enyoh C, Verla E, Nwarnorh K (2019) Microplastic-toxic chemical interaction: a review study on quantified levels, mechanism and implication. SN Appl Sci 1:1400. https://doi.org/10.1007/s42452-019-1352-0

Vroom RJE, Koelmans AA, Besseling E, Halsband C (2017) Aging of microplastics promotes their ingestion by marine zooplankton. Environ Pollut 231:987–996. https://doi.org/10.1016/j.envpol.2017.08.088

Wagner M, Scherer C, Alvarez-Muñoz D, Brennholt N, Bourrain X et al (2014) Microplastics in freshwater ecosystems: what we know and what we need to know. Environ Sci Eur 26(1):12. https://doi.org/10.1186/s12302-014-0012-7

Wang T, Yu C, Chu Q, Wang F, Lan T et al (2020) Adsorption behavior and mechanism of five pesticides on microplastics from agricultural polyethylene films. Chemosphere 244:125491. https://doi.org/10.1016/j.chemosphere.2019.125491

Wang Z, Chen Y, Wang S, Yu Y, Huang W et al (2022a) Pollution risk assessment and sources analysis of heavy metal in soil from bamboo shoots. Int J Environ Res Public Health 19:14806. https://doi.org/10.3390/ijerph192214806

Wang L, Guo C, Qian Q, Lang D, Wu R et al (2022b) Adsorption behavior of UV aged microplastics on the heavy metals Pb(II) and Cu(II) in aqueous solutions. Chemosphere 313:137439. https://doi.org/10.1016/j.chemosphere.2022.137439

Wang F, Wang Q, Adams CA, Sun Y, Zhang S (2022c) Effects of microplastics on soil properties: current knowledge and future perspectives. J Hazard Mater 424:127531. https://doi.org/10.1016/j.jhazmat.2021.127531

Wang K, Kou Y, Wang K, Liang S, Guo C et al (2023) Comparing the adsorption of methyl orange and malachite green on similar yet distinct polyamide microplastics: uncovering hydrogen bond interactions. Chemosphere 340:139806. https://doi.org/10.1016/j.chemosphere.2023.139806

Wang Q, Gao Y, Tang X, Yang Z et al (2025) How aging microplastics influence heavy metal environmental fate and bioavailability: a systematic review. Environm Res 121128. https://doi.org/10.1016/j.envres.2025.121128

Witczak A, Pokorska-Niewiada K (2023) Food safety. CRC Press eBooks, pp 419–432. https://doi.org/10.1201/9781003265955-15

Wright SL, Thompson RC, Galloway TS (2013) The physical impacts of microplastics on marine organisms: a review. Environ Pollut 178:483–492. https://doi.org/10.1016/j.envpol.2013.02.031

Wu W, Du R, Chen Z, Li W, Huang X et al (2024) Unlocking the combined impact of microplastics and emerging contaminants on fish: a review and meta-analysis. Aq Toxicol 278:107176. https://doi.org/10.1016/j.aquatox.2024.107176

Xie A, Chen S, Liang X, Li L, Song Y, Lv M et al (2025) Influence of microplastic aging on the adsorption and desorption behavior of Ni(II) under various aging conditions. Environ Geochem Health 47(3). https://doi.org/10.1007/s10653-025-02403-1

Yang J, Yang Z, Wang J, Liang Y, Zeng H et al (2025a) Toxic effects and mechanisms of nanoplastics and sulfonamide antibiotics on scenedesmus obliquus. Ecotoxicol Environ Saf 289:117685. https://doi.org/10.1016/j.ecoenv.2025.117685

Yang J, Yang Z, Wang J, Liang Y, Zeng H, Qin L et al (2025b) Toxic effects and mechanisms of nanoplastics and sulfonamide antibiotics on *Scenedesmus Obliquus*. Ecotoxicol Environ Saf 289:117685. https://doi.org/10.1016/j.ecoenv.2025.117685

Yemele O, Zhao Z, Nkoh J, Ymele E, Usman M (2024) A systematic review of polycyclic aromatic hydrocarbon pollution: a combined bibliometric and mechanistic analysis of research trend toward an environmentally friendly solution. Sci Total Environ 926:171577. https://doi.org/10.1016/j.scitotenv.2024.171577

Yilimulati M, Wang L, Ma X, Yang C, Habibul N (2020) Adsorption of ciprofloxacin to functionalized nano-sized polystyrene plastic: kinetics, thermochemistry and toxicity. Sci Total Environ 750:142370. https://doi.org/10.1016/j.scitotenv.2020.142370

Yu Y, Mo WY, Luukkonen T (2021) Adsorption behaviour and interaction of organic micropollutants with nano and microplastics—a review. Sci Total Environ 797:149140. https://doi.org/10.1016/j.scitotenv.2021.149140

Zhai Y, Bai J, Chang P, Liu Z, Wang Y et al (2024) Microplastics in terrestrial ecosystem: Exploring the menace to the soil-plant-microbe interactions. TrAC, Trends Anal Chem 174:117667. https://doi.org/10.1016/j.trac.2024.117667

Zhang GS, Liu YF (2018) The distribution of microplastics in soil aggregate fractions in southwestern China. Sci Total Environ 642:12–20. https://doi.org/10.1016/j.scitotenv.2018.06.004

Zhang Y, Gao T, Kang S, Sillanpää M (2019) Importance of atmospheric transport for microplastics deposited in remote areas. Environ Pollut 254:112953. https://doi.org/10.1016/j.envpol.2019.07.121

Zhang D, Chen Q, Xu T, Yin D (2024a) Current research status on the distribution and transport of micro(nano)plastics in hyporheic zones and groundwater. J Environ Sci 151:387–409. https://doi.org/10.1016/j.jes.2024.03.042

Zhang Y, Gao R, Zhu F, Fu L, Zhang J (2024b) Combined effect of polystyrene nanoparticles and chlorpyrifos to Daphnia Magna. Chemosphere 369:143765. https://doi.org/10.1016/j.chemosphere.2024.143765

Zhao Y, Tao S, Liu S, Hu T, Zheng K et al (2023) Research advances on impacts micro/nanoplastics and their carried pollutants on algae in aquatic ecosystems: a review. Aquat Toxicol 264:106725. https://doi.org/10.1016/j.aquatox.2023.106725

Zhi L, Li Z, Li R, Su Z, Chen F et al (2025) Micro/nanoplastics in aquaponic systems: sources, features and and their effects on the immune system of fish. TrAC Trends Anal Chem 184:118158. https://doi.org/10.1016/j.trac.2025.118158

Zhou D, Cai Y, Yang Z (2021) Key factors controlling transport of micro- and nanoplastic in porous media and its effect on coexisting pollutants. Environ Pollut 293:118503. https://doi.org/10.1016/j.envpol.2021.118503

Zhou Q, Yang N, Li Y, Ren B, Ding X et al (2022) Total concentrations and sources of heavy metal pollution in global river and lake water bodies from 1972 to 2017. Glob Ecol Conserv 22:e00925. https://doi.org/10.1016/j.gecco.2020.e00925

Zhu J, Wang C (2020) Biodegradable plastics: green hope or greenwashing? Mar Pollut Bull 161:111774. https://doi.org/10.1016/j.marpolbul.2020.111774

Zjačić J, Tonković S, Pulitika A, Katančić Z, Kovačić M et al (2023) Effect of aging on physicochemical properties and size distribution of PET microplastic: influence on adsorption of diclofenac and toxicity assessment. Toxics 11:615. https://doi.org/10.3390/toxics11070615

Chapter 6
Advanced Functional Nanomaterials for Marine Anti-fouling Technology

Yuhan Guo and Qinghua Zhang

Abstract Marine biofouling poses a significant threat to maritime industries. This chapter reviews the development of nanocomposite antifouling coatings as environmentally responsible alternatives to traditional toxic solutions. It examines various mechanisms, including low surface energy, superhydrophobicity, controlled biocide release, and photocatalysis. The analysis highlights that incorporating nanomaterials enhances mechanical properties and fouling resistance. A key finding is the superior performance of multi-factor synergistic coatings, which combine several antifouling strategies for broad-spectrum, long-lasting efficacy. However, challenges such as nanoparticle dispersion stability and long-term durability in marine environments must be resolved to enable widespread commercial application. Nanocomposite coatings represent a promising, green direction for future antifouling technology.

Keywords Nanomaterials · Antifouling mechanisms · Marine biofouling · Synergistic effect

6.1 Introduction

Marine biofouling and corrosion are two major problems faced by marine engineering equipment. Due to the diversity and complexity of marine organisms, marine biological fouling is more difficult to prevent and control than corrosion. To prevent marine biofouling, anti-fouling agents such as organotin, cuprous thiocyanate, and zinc oxide have been widely used. However, the toxicity of these anti-fouling agents poses a threat of irreversible damage to the entire ecosystem.

Since the advent of nanomaterials, and with advances in understanding their characteristics and manufacturing technologies, many countries and regions have investigated their feasibility for anti-fouling applications in marine structures. The results

Y. Guo · Q. Zhang (✉)
Multifunctional Nanocomposite Materials, College of Chemical and Biological Engineering, Zhejiang University (Zijingang Campus), Hangzhou, China
e-mail: qhzhang@zju.edu.cn

S. K. Panda and V. P.M. (eds.), *Bionanocomposites from Marine Origin*, Advanced Structured Materials 247, https://doi.org/10.1007/978-3-032-10749-7_6

have shown that the nano-anti-fouling materials have excellent anti-fouling performance and are environmentally friendly. These nanomaterials can circumvent the environmental pollution issues associated with organotin-based antifouling coatings and provide a new approach to antifouling for ships and marine engineering structures. This chapter reviews the serious threat of biofouling to the marine industry, nanocomposite antifouling technology, new approaches in antifouling strategies, and future prospects and challenges of antifouling nanomaterials.

6.2 Biofouling: A Serious Threat to Marine Industry

The fouling of marine organisms originated from the adhesion of plankton to the marine engineering equipment. The plankton in the sea is dominated by tiny protozoa and algae. Among them, marine algae is the main member of fouling organisms. Algae are the fastest growing plants in the world. They grow explosively on a large scale and can clog the equipment system. It also provides a food source for other fouling organisms, such as mussels, oysters, and barnacles. Mussels belong to the phylum *Mollusca*, class *Bivalvia*, and family *Mytilidae*. There are many species of mussels, and the shell length of different species varies from a few millimeters to tens of millimeters. Oysters belong to the mollusk phylum, bivalve class, pearl shellfish order. The bivalves of the oysters are unequal, the left shell or lower shell is large and concave, and the left shell is fixed on a solid surface to survive. After death, the left shell of the oyster can still be firmly attached to the solid surface. Barnacle belongs to the arthropod, Crustacea, barnacle suborder. Barnacles are widely distributed in different sea areas in the world and have strong adaptability to different sea water quality. The surface of hard objects can be attached by it. They cluster together in groups and cling to the solid surface. These organisms adhere firmly to the surface of marine engineering equipment, which brings great harm to the marine industry. First of all, biological fouling reduces the flow cross-sectional area of seawater transport pipelines, resulting in reduced seawater flow and water transport efficiency, and can completely block water transport pipelines in serious cases, affecting the normal operation of water facilities. Secondly, biological fouling increases resistance, improves energy consumption of water transport equipment, reduces the efficiency of the cooler, increases equipment maintenance costs, and reduces the economy of production and operation. In addition, biological fouling causes corrosion of equipment gaps and accelerates the corrosion rate of equipment, or causes abnormalities or damage to valves, instruments, and other equipment in the key parts.

6.3 Nanocomposite-Based Anti-fouling Technology

Among the many anti-fouling technologies, the application of coatings (paints) is the most cost-effective and widely applicable method, accounting for about 76% of the market share. However, traditional antifouling coatings release harmful antifouling agents (such as organotin) that are difficult to decompose, polluting the ecological environment. Moreover, these substances can enter the human food chain through fish, shellfish, and other organisms, posing a serious threat to human health. Currently, most countries have restricted the use of traditional antifouling agents. The development of green anti-fouling coatings has become an important direction in marine antifouling.

Due to the extremely small size of nanomaterials, they possess unique physical and chemical properties, and the resulting effects are commonly referred to as the nano-effect. By combining nanomaterials with polymers, nanocomposite materials can be produced. Nanocomposite coatings refer to the addition of nanomaterials (with at least one dimension in the three-dimensional space within the size range of 1–100 nm) to coatings (Nguyen-Tri et al. 2018), enhancing the physical and chemical properties of the coatings through their small size effects, surface effects, quantum size effects, and dielectric confinement effects. Currently, nanomaterials can be applied in fields such as anti-microbial coatings, antifouling coatings, waterproof coatings, weather-resistant coatings, stealth coatings, wear-resistant coatings, and anti-corrosion coatings (Necolau and Pandele 2020). For antifouling coatings, nanoscale particles form strong hydrogen bonding with the coating, improving the compactness and resistance to ion penetration of the coating. Nanoparticles can not only enhance the rheological properties of the coating, hardness, and aging resistance but also modify its surface morphology. For instance, they can be used to reduce surface roughness or, conversely, to create a micro/nanocomposite structure, thereby improving the sample's hydrophobicity and achieving a self-cleaning effect. Additionally, some nanomaterials have catalytic functions and can induce the apoptosis of microorganisms or catalytically decompose adhered substances through the generation of reactive groups, effectively preventing biological fouling. The reactive groups produced during use can be rapidly degraded, ensuring good safety and not causing environmental pollution. In response to the basic requirements of environmental friendliness and stability of nano-antifouling coatings, the development of organic hybrid coatings containing nanofillers is an environmentally friendly alternative to traditional antifouling methods. By embedding nanoparticles into the polymer matrix, the mechanical properties, surface morphology, and active sites of the coating are fundamentally improved, with the aim of achieving better performance. At the same time, the doping of nanoparticles supports the preparation of ternary or quaternary nanocomposite materials, realizing the synergistic antifouling effect. This chapter focuses on the research progress of low surface energy nanocomposite antifouling coatings, superhydrophobic nanocomposite antifouling coatings, release-type nanocomposite anti-fouling coatings, catalytic nano-anti-fouling coatings, and multi-factor synergistic nano-antifouling coatings according to their different anti-fouling mechanisms.

6.3.1 Low Surface Energy Nano-anti-fouling Coatings

In recent years, with the rapid development of science and technology, the demand for high-performance materials has increased year by year. The performance of ordinary low surface energy resin coatings cannot meet the expected application requirements in some fields. Therefore, researchers are actively seeking for methods to improve their performance. Nanotechnology, as a rapidly developing emerging technology, has been applied in many research fields due to its special physical and chemical properties. Researchers have combined nanotechnology with low surface energy coatings to prepare nanocomposite high-performance materials with low surface energy, making them widely applicable in special situations.

Fouling organisms mainly adhere firmly to the target surface or the already formed fouling film through one or more interaction forces such as chemical bonding, electrostatic interaction, physical adsorption, mechanical interlocking, and diffusion. This process is influenced by the surface energy of the target surface (Lejars et al. 2012). It is generally believed that low surface energy coatings can utilize their low surface free energy to make it difficult for fouling substances to adhere to the material surface; even if fouling organisms attach to the surface, their adhesion is very weak and can be removed under low water flow shear force, achieving self-cleaning of the ship's surface and anti-fouling effects. Low surface energy coatings achieve anti-fouling based on the physical properties of the material surface, without the issue of toxic substance release and depletion, and have gained attention for their environmental friendliness and good anti-fouling properties. Low surface energy anti-fouling materials are classified into three categories according to the matrix resin: silicone series, fluorine series, and silicone-fluorine synergistic modified low surface energy anti-fouling materials. Among them, silicone elastomers, such as polydimethylsiloxane (PDMS) with a surface energy of 20 mJ/m^2 and polymethylphenylsiloxane (PMPS) with a surface energy of 26 mJ/m^2, are close to the surface energy range (22–24 mJ/m^2) corresponding to the minimum amount of fouling adhesion (Fig. 6.1) described in Baier's curve, thus gaining widespread application (Nurioglu et al. 2015).

Although silicone resins have many advantages, their mechanical properties, adhesion to substrates, and re-coatability are poor, and they do not work well when used alone as film-forming materials. Therefore, chemical methods, surface modification techniques, plasma technology, and nanotechnology are often used to modify the low surface energy coatings of silicone. Doping with inorganic nanomaterials is one of the most promising solutions to improve the mechanical properties and anti-fouling performance of silicone low surface energy coatings. In recent years, various nanomaterials such as Al_2O_3 nanoparticles (Zhao et al. 2019), carbon nanotubes (Cavas et al. 2018), and ZnO nanoparticles (Qiu et al. 2018) have been introduced

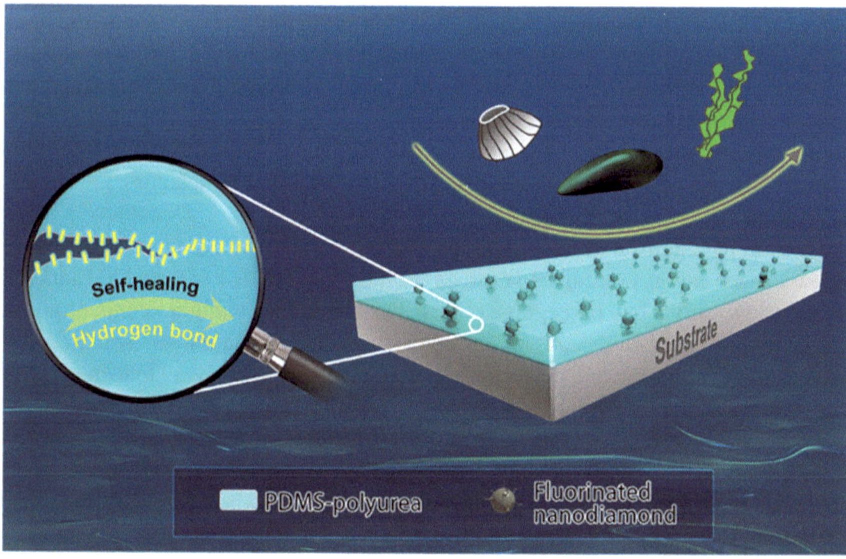

Fig. 6.1 Anti-fouling mechanism of the low surface energy nano-anti-fouling coatings (Xie et al. 2020)

into low surface energy coatings to enhance the physical, chemical, mechanical, and anti-fouling properties of the coatings.

Xie et al. (2020) added fluorinated nanodiamonds to polydimethylsiloxane-based polyurea (PDMS-PUa) to improve its mechanical properties. The introduced fluorinated nanodiamonds increased the tensile strength of PDMS-PUa from 1.2 MPa to 1.6 MPa and the bonding strength from 1.5 MPa to 2.1 MPa, significantly improving the mechanical properties. Due to the self-enrichment of fluorine-based and polysiloxane on the coating surface, the contact angle of the PDMS-PUa coating increased from 110° to 124°, and the surface energy was reduced to 18 mJ/m^2. The lower surface energy allowed the coating to maintain good fouling release ability after 4 months of exposure in the real sea (Fig. 6.1).

Chen et al. (2020) used dodecyl fluoroheptyl methacrylate (DFMA), polyethylene glycol methyl ether methacrylate (PEGMA), triethoxyoctylsilane (KH832), and tetraethyl orthosilicate (TEOS) as raw materials to prepare a non-elastic organic–inorganic hybrid coating by the sol–gel method. The prepared coating had high adhesion to the substrate (1.3 MPa) and high hardness (57 MPa), effectively preventing mechanical damage. The surface energy of the coating was 20 mJ/m^2, the elastic modulus was 420 MPa, and the surface roughness was 3 nm. The suitable low surface energy, higher elastic modulus, and smooth surface made it very easy for marine bacteria, biofilms, and diatoms to detach from the non-elastic coating surface.

Ba et al. (2019) added multiwalled carbon nanotubes (MWCNTs), nano-sized Fe_2O_3, and silicone oil to PDMS to construct an anti-fouling coating. The addition of Fe_2O_3 was to prevent shrinkage of PDMS during the curing process, while

MWCNTs could form diffusion channels within the coating, transforming the transport of silicone oil from molecular network diffusion to channel diffusion, thereby improving the efficiency of silicone oil leaching. Although the surface roughness increased with the concentration of MWCNTs, the channel diffusion enhanced the stable leaching efficiency of the silicone oil, leading to more effective biofouling release. The rate of silicone oil exudation can be controlled by selecting appropriate nanofillers.

Overall, the addition of nanomaterials can improve the anti-fouling performance of low surface energy coatings. In most cases, the concentration of the added nanomaterials is low, only a few percentage points. This is because at higher concentrations, nanoparticle aggregation leads to an increase in surface roughness, which in turn reduces the fouling release performance. It is evident that only a small number of nanomaterials needs to be added to have a positive impact on the durability and fouling release performance of the coating. This inorganic–organic hybrid method is beneficial to the development of low surface energy anti-fouling technology. However, it still cannot fully prevent surface fouling when low-surface-energy antifouling materials remain in a static state for extended periods.

6.3.2 Superhydrophobic Nano-anti-fouling Coatings

In the past decade or so, attention has increasingly focused on the unique structure and properties of superhydrophobic materials and their related application fields, including corrosion resistance, self-cleaning, and anti-fouling. Superhydrophobic coatings have almost no wetting properties, and water droplets easily roll off their surfaces. In recent years, nanoparticles have often been used to adjust surface roughness to obtain superhydrophobic surfaces. The preparation of micro- and nanoscale superhydrophobic surfaces has attracted growing interest from researchers. Generally, nanoparticles are used to produce sub-micron-scale rough structures, followed by chemical treatment or modification with low surface energy substances to increase the hydrophobic effect; or nanoparticles are dispersed in a polymer solution and sprayed onto the target surface to obtain a hydrophobic coating (Fig. 6.2) (He et al. 2019a). Currently, a series of superhydrophobic anti-fouling nano-coatings have been prepared based on inorganic nanoparticles such as SiO_2, ZnO, TiO_2, and Fe_3O_4.

Wang et al. (2020) used TiO_2 nanoparticles with sizes ranging from 25 to 300 nm to prepare nano-coatings with different particle sizes through the spray method, followed by fluorination treatment of the coatings. Using Streptococcus mutans as a typical fouling organism, they studied the synergistic effects of nanoparticles of different sizes and surface fluorination treatment on the wettability and bacterial adhesion behavior of the TiO_2 nano-coatings. It is generally believed that the proteins on the surface of the Streptococcus mutans cell membrane have many hydrophobic functional groups and tend to adhere to hydrophobic surfaces (Qiu et al. 2018). The 25 nm TiO_2 nanoparticle-formed nano-coating exhibited superhydrophobic properties after fluorination treatment and could effectively inhibit the adhesion of Streptococcus

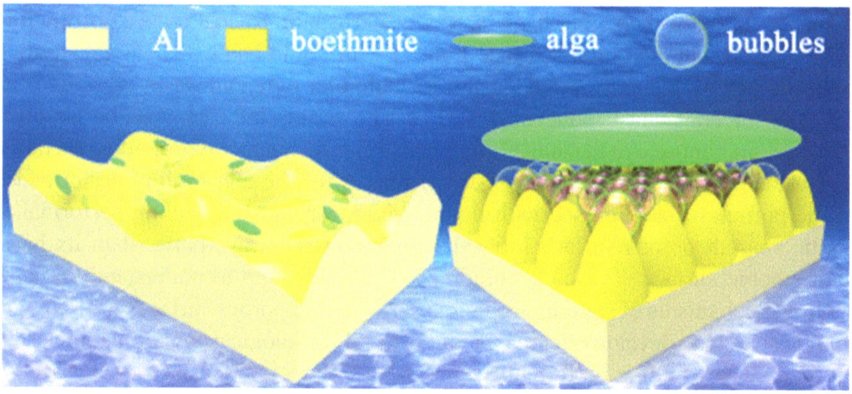

Fig. 6.2 Anti-fouling mechanism of the superhydrophobic nano-anti-fouling coatings (He et al. 2019a)

mutans in phosphate buffer solution; while the 300 nm TiO_2 nanoparticle-formed nano-coating also exhibited superhydrophobic properties after fluorination treatment, but the anti-fouling efficiency was reduced. This result indicates that smaller nanoparticles are more likely to form superhydrophobic nano-anti-fouling coatings.

Wang et al. (2016a) successfully prepared stable superhydrophobic PSU/PVDF-based composite coatings by designing micro/nanostructures and using chemical modification methods. By adding PSU/PVDF (6:4) and 12 wt% MMT-PDMS to the prepared coating, excellent superhydrophobicity (WCA = 159°, SA = 3.5°) and adhesion (grade 1) could be achieved. The cross-linking ability of PVDF and the special multilayer structure of MMT effectively improved the ductility and mechanical properties of the PSU composite coating. The resulting composite coating exhibited excellent self-cleaning, anti-fouling, and durability.

He et al. (2019b) subjected the sandblasted aluminum surface to high-pressure sterilization treatment at 120 °C and grew nano-sized alumina in situ on its surface, successfully preparing a micro/nano-binary structure superhydrophobic surface on the aluminum substrate. The prepared surface could repel water, egg white, and yolk. This further demonstrates that the micro/nano-binary structure plays an important role in achieving superhydrophobicity. A water contact angle of 160° and a contact angle hysteresis of 1.8° were observed on the fluorinated aluminum surface with a hierarchical structure. At the same time, the surface exhibited excellent anti-fouling properties against algae through the trapped air layer within the micro/nanostructure. This result will provide clues for the in situ construction of micro/nano-hierarchical topological structures and the modification of low surface energy polymers on metal structures to enhance anti-fouling performance.

He et al. (2019c) successfully constructed disordered layers on stainless steel substrates using a layer-by-layer assembly method through electrostatic attraction between polycations and negatively charged SiO_2 nanoparticles. Subsequently, based on the disordered layer, superhydrophobic surfaces and liquid-infused surfaces were prepared to study their dominant factors in anti-fouling performance. Biofouling tests showed that the porous surface infused with liquid had excellent anti-fouling performance by reducing the adhesion of Chlorella pyrenoidosa by 93% and the adhesion of Navicula tripunctata by 71%. The results indicate that the air layer plays a dominant role in the anti-fouling performance of superhydrophobic surfaces. The presence of a lubricating layer is the reason for the anti-fouling performance of the porous surface infused with liquid. Moreover, the anti-fouling performance of the smooth porous surface infused with liquid is always better than its basic hydrophobic surface. These encouraging results may help to understand the anti-fouling mechanisms of superhydrophobic surfaces and porous surfaces infused with liquid and help to develop effective methods to solve biological contamination issues to improve anti-fouling performance.

Li et al. (2020) fabricated a superhydrophobic stainless steel meshes with antibacterial silver nanoparticles by combining dual-potential deposition and surface modification. According to the inhibition zone method, the growth time had a significant impact on the antibacterial performance, with the inhibition zone diameter for Escherichia coli being 11.28 ± 0.88 mm. The stainless steel mesh exhibited a superhydrophobic state, with a contact angle (CA) of 160.6° and a sliding angle (SA) of 3°. Meanwhile, a comparative analysis under static and flowing conditions (50 rpm) at 37 °C was conducted to explore the bacterial anti-adhesion properties of the prepared stainless steel surface. Under static incubation, the superhydrophobic surface reduced bacterial adhesion by nearly five orders of magnitude. This demonstrates that the superhydrophobic specimen can effectively repel the adhesion of Staphylococcus aureus under static culture conditions, with the effect being more pronounced than under dynamic conditions. In addition, through long-term exposure to air and the sandpaper abrasion process, the superhydrophobic stainless steel mesh demonstrated good long-term stability and mechanical stability.

6.3.3 Release-Type Nano-anti-fouling Coatings

Fouling-release antifouling coatings are non-biocides that possess non-stick and fouling-release properties. These materials can minimize the adhesion of fouling organisms to the surface or reduce the strength of adhesion, making it easy to desorb and remove under the action of flowing seawater or other external forces, and are widely used in marine facilities, especially on ship surfaces. Such materials typically have hydrophobicity and low surface energy, and their anti-fouling capabilities are also related to factors such as surface wettability and coating thickness. Among them, hydrophobicity can significantly reduce the resistance during ship navigation, thereby reducing fuel consumption, which is also of great importance for greenhouse gas emissions. Currently, fluorinated materials and silicone materials are the main types.

Perfluoropolyether polymers (PFPE) are formed by the polymerization of multiple perfluoropolyether monomers and have higher flexibility and friction coefficients, making them smoother while also effectively reducing surface energy. Caliskan et al. (2022) deposited poly(glycidyl methacrylate) (PGMA) on the silicon surface as an anchoring layer before attachment and then grafted perfluoropolyether-based poly-acetic acid (PFPE-COOH) onto the PGMA surface to prepare a nano-coating based on perfluoropolyether.

Silicone rubber and ketones (R_2CO) have similar structures. PDMS is the most common low surface energy silicone anti-fouling coating. PDMS has good flexibility and low surface energy, which can reduce the attachment of fouling organisms, but its adhesion strength to the substrate and mechanical properties are poor, making it easily damaged by external forces and shortening its service life. Therefore, researchers have improved silicone materials by introducing nanoparticles and other resins. Acrylic resins are chemically excellent in solvent resistance and are widely used in the coating field.

The addition of nanoparticles can improve the mechanical properties of silicone materials. Currently, commonly used nanoparticles include graphene nanofillers, carbon nanotubes, cellulose, nano-titanium dioxide, and nano-silica. Selim et al. (2018) prepared a ternary composite silicone coating with high flexibility and high mechanical strength by adding graphene oxide sheets, aluminum oxide nanorods, and titanium dioxide nanospheres to the silicone elastomer.

Nanoparticles of silver have broad-spectrum bactericidal properties and are widely used in various fields, making them one of the oldest anti-fouling strategies (Zhang et al. 2018). As a release-type anti-fouling agent, the direct release of Ag^+ or its formation from the oxidation of metallic silver can effectively kill microorganisms at concentrations as low as 1 μg/mL, including drug-resistant bacteria. It is generally believed that Ag^+ can attach to the cell membrane to disrupt cell structure, bind to the sulfhydryl groups of intracellular enzymes to interfere with metabolism, bind to DNA to prevent normal DNA replication, or form ROS and other methods to kill microorganisms.

Additionally, Vasilev et al. (2010) have shown that mammalian cells have significantly higher tolerance to silver than bacteria, making the use of silver nanoparticles at low concentrations safe.

Zhang et al. (2020) used natural phenol as a reducing agent, dispersant, and surfactant to synthesize silver nanoparticles (AgNPs) in situ. Meanwhile, silver nitrate catalyzed the polymerization of phenol to form polyphenol (PUL). This in situ reduction method allowed AgNPs to be uniformly distributed in the polymer matrix, and the binding between AgNPs and PUL ensured the stable release of Ag^+, providing long-lasting anti-fouling properties. After 120 days of marine panel testing, the blank substrate surface was covered with a large amount of fouling organisms such as algae and barnacles, while the polymer coating containing 0.3% AgNPs showed no attachment of any fouling organisms. This demonstrates that the sample effectively inhibited marine biological fouling.

Tian et al. (2019) prepared a nanocomposite hydrogel coating by blending silane elastomer, hydrophilic polymer, and silver trifluoromethanesulfonate. The Ag nanoparticles in the coating served as a cross-linking agent, distributing the hydrophilic polymer uniformly in the PDMS matrix and enhancing the interfacial

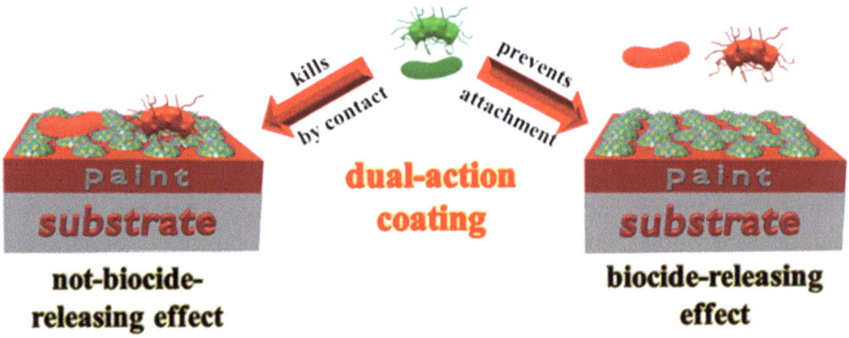

Fig. 6.3 Anti-fouling mechanism of the release-type nano-anti-fouling coatings (Michailidis et al. 2020)

compatibility of the two polymers. The coating achieved a near 100% kill rate against *E. coli* and reduced the attachment of Chattonella marina, Navicula, and Chlorella vulgaris by 37%, 37%, and 52%, respectively. Marine panel tests in the South China Sea also showed that the nanocomposite hydrogel coating had good anti-fouling performance. In addition to silver nanoparticles, copper and copper-based coatings also have bactericidal and anti-fouling effects and are the most widely used broad-spectrum anti-fouling coatings (Vucko et al. 2012, 2014; Piola et al. 2009).

Nanomaterials can also be used as carriers for anti-fouling agents added to the polymer matrix to regulate the release of anti-fouling agents (Fu et al. 2019). Michailidis et al. (2020) loaded 2-n-octyl-4-isothiazolin-3-one (DCOIT), dimethyl octadecyl [3-(trimethoxysilyl)propyl] ammonium chloride, and dimethyl tetradecyl [3-(triethoxysilyl)propyl] ammonium chloride (quaternary ammonium salts) onto spherical mesoporous SiO_2 nanoparticles and then blended these nano-capsules with Jotaguard paint to prepare coatings with 2 and 5 wt% nano-capsule content (Fig. 6.3). Six months of real-sea panel testing showed that the fouling coverage rate on the surface of the undoped coating was 49%, while the fouling coverage rate on the surface of the 5 wt% nano-capsule doped coating was 6.9%, greatly inhibiting the attachment of marine organisms. Moreover, the coating not only relied on the release of DCOIT to inhibit the attachment of fouling organisms but also reduced the attachment of fouling organisms through the contact bactericidal action of quaternary ammonium salts.

6.3.4 Catalytic Nano-anti-fouling Coatings

When exposed to light, certain nanomaterials can exhibit strong oxidation and reduction characteristics on their surfaces, providing a theoretical basis for using photocatalytic materials to prevent biofouling. In recent years, the strategy of applying photocatalytic materials to the field of marine anti-fouling has attracted the attention of many researchers (Sun et al. 2020). When the energy of light is greater than or equal to the band gap width, semiconductor materials will be excited, and electrons on their valence band will absorb energy and transition to the conduction band. At the same time, an equal number of holes will be formed on the valence band, which have strong oxidation capabilities. These holes can not only directly participate in oxidation reactions but also adsorb OH^- or H_2O on the material's surface, generating strong oxidizing free radicals OH. Electrons on the conduction band have strong reduction capabilities and can capture O_2 to generate O_2^-. When these free radicals with strong oxidizing activity come into contact with fouling microorganisms, they can destroy their cell membranes and cell walls in a short time, leading to the efflux of cytoplasm and ultimately causing bacterial death. Additionally, these free radicals can also enter the cell through the cell membrane, destroy intracellular macromolecules and organelles, and oxidize and decompose endotoxins and intracellular substances. Although the photocatalytic bactericidal ability of semiconductor materials has been proven, research on the mechanism of photocatalytic bactericidal action is still controversial. The three widely recognized mechanisms mainly involve the oxidation of coenzyme A, the destruction of cell walls and membranes, and the destruction of genetic material.

Titanium dioxide (TiO_2) is a typical photocatalytic material that, under light conditions, produces reactive oxygen species (ROS) such as H_2O_2 and OH, which can kill bacteria and other microorganisms, thereby achieving anti-fouling effects [55]. Due to its high photocatalytic activity, TiO_2 is one of the most studied photoactive anti-fouling materials.

Selim and colleagues (Selim et al. 2016) prepared a PDMS/TiO_2 hybrid nanocomposite material. When light shines on TiO_2, electrons in TiO_2 move from the valence band to the conduction band, while the holes remain in the valence band. The electrons that have moved to the conduction band reduce oxygen to superoxide anion radicals, and the more oxidizing holes in the valence band can oxidize water to hydroxyl radicals. The more oxidizing hydroxyl radicals attack the unsaturated bonds in organic matter or extract hydrogen from organic matter, generating new free radicals. The variety of free radicals formed by a series of chain reactions initiated under light exposure can inhibit bacterial growth and reproduction, achieving antibacterial and anti-fouling effects.

Saravanan et al. (2016) used 3-aminopropyltriethoxysilane (APTES) as a coupling agent to surface treat TiO_2 nanoparticles and combined them with epoxy diglycidyl ester resin to prepare a surface-functionalized nano-epoxy hybrid coating. The inorganic nanoparticles improved the coating's corrosion resistance and anti-fouling properties. The 3 wt% APTES-TiO_2 nano-hybrid coating had anti-fouling effects lasting up to 6 months.

Unlike TiO_2 and ZnO, which utilize their photocatalytic properties to achieve anti-fouling performance, some nanoparticles can use their own enzyme-like activity for anti-fouling (Chen et al. 2018). Nanoparticles such as V_2O_5 (Andre et al. 2011) and CeO_2 (Hu et al. 2018; He et al. 2020) have catalytic activities similar to haloperoxidases. In the presence of H_2O_2, they catalyze the reaction $H_2O_2 + X^- + H^+$

= HOX + H_2O, generating the corresponding hypohalous acids (HOCl or HOBr) (Crans et al. 2004; Almeida et al. 2001) from halogen elements (with ~ 1 mM Br^- and 500 mM Cl^- present in seawater (Natalio et al. 2012)). Hypohalous acids can directly kill microorganisms or affect the quorum sensing signals of fouling organisms, thereby inhibiting the formation of initial biofilms and preventing biological fouling (Almeida et al. 2001).

Natalio et al. (2012) synthesized V_2O_5 nanorods, which, in the presence of H_2O_2, catalyze the oxidation of bromide ions to HOBr and the production of singlet molecular oxygen ($^1O_2$). These reactive oxygen species have strong antibacterial activity and can prevent marine biological pollution. The acute toxicity value of V_2O_5 nanowires is 1.2 mg/mL, which is much lower than the acute toxicity values of common anti-fouling agents in the marine anti-fouling field, such as zinc pyrithione (0.085 mg/mL) and copper pyrithione (0.0012 mg/mL). The toxicity of V_2O_5 nanorods is 14 times and 1000 times smaller than that of zinc pyrithione and copper pyrithione, respectively, and can replace commercial anti-fouling agents. Natalio added V_2O_5 nanorods to commercial coatings to prepare a V_2O_5 nano-coating. After 60 days of real-sea panel testing, the V_2O_5 nano-coating effectively inhibited biological fouling (Fig. 6.4).

Jotun has launched the SeaLion Repulse type of fouling release anti-fouling coating, which is based on the nanorepellent TM patented technology to construct a

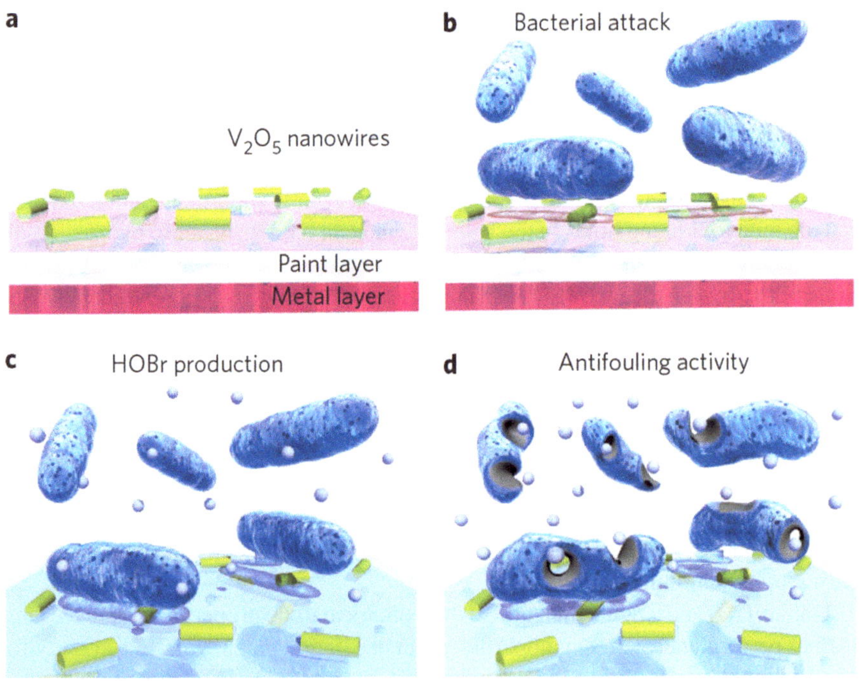

Fig. 6.4 Anti-fouling mechanism of the catalytic nano-anti-fouling coatings (Natalio et al. 2012)

nanostructure on the surface of silicone materials, forming a dynamic elastic layer upon contact with water to resist the adhesion of biological mucus and fouling organisms. SeaLion Repulse coating can still exhibit superior anti-fouling performance under static conditions, with the longest protection period of up to 10 years (recoating every 5 years) (Jotun 2023).

6.4 New Approaches in Anti-fouling Technology

With the advancement of technology and the continuous deepening of scientific research, different anti-fouling technologies are showing a trend of specialization, non-toxicity, and synergy.

(1) Specialization. The species, quantity, and growth of fouling organisms vary with different sea areas and seasons, presenting species diversity. Therefore, it is necessary to further improve the relevant marine biology and environmental condition databases and establish specialized anti-fouling technologies for specific sea areas or facilities (Wang et al. 2016b). This targeted anti-fouling technology can effectively enhance anti-fouling capabilities.
(2) Non-Toxicity. Toxic chemicals generally have good anti-fouling effects on fouling organisms but can also poison other marine life, thereby damaging the marine ecological environment. Today, as humans enter an era of large-scale exploitation and utilization of the oceans, we are also committed to balancing development and protection, pollution control, and ecological restoration. The use of toxic chemicals is inevitably restricted (Wang et al. 2016b). Therefore, developing green, long-lasting, and non-toxic anti-fouling technologies is the focus and mainstream research direction for future anti-fouling efforts.
(3) Synergy. Every anti-fouling technology has its strengths and weaknesses, so using a single anti-fouling technology often has very limited effects. Therefore, combining multiple anti-fouling principles to leverage synergistic effects and complementary advantages can achieve superior outcomes. For example, combining biomimetic anti-fouling technology with low surface energy anti-fouling technology to construct specific microstructures on the surface of low surface energy materials can significantly enhance the material's anti-fouling capabilities.

All of the above coatings can significantly inhibit biological attachment, but in the long run, biofouling is inevitable. Low surface energy coatings do not have anti-fouling effects at low speeds; superhydrophobic anti-fouling coatings are unstable, and the air layer is gradually replaced by seawater; release-type anti-fouling coatings will slowly lose their anti-fouling activity due to the depletion of anti-fouling agents; catalytic anti-fouling coatings mainly rely on ROS to kill microorganisms and reduce adhesion, but their active sites may be covered by adhered biological macromolecules, thus losing anti-fouling activity. Therefore, it is necessary to develop synergistic anti-fouling coatings. Currently, some research has been conducted on multi-factor nanocomposite anti-fouling coatings. Soleimani et al. (2021) used Avicennia marina and $AgNO_3$ to reduce graphene oxide, obtaining Ag-loaded reduced graphene oxide (GOH@Ag). By doping 0.5 wt% GOH@Ag particles into PDMS, a nanocomposite coating was prepared. The coating's surface energy was 16 mJ/m^2, and the surface adhesion was 0.16 MPa; the attached fouling organisms could also be easily detached from the coating surface. In addition, the coating used released Ag^+ to kill some fouling organisms, reducing adhesion and achieving static anti-fouling effects, overcoming the defects of low surface energy anti-fouling coatings. The coating achieved synergistic anti-fouling effects through low surface energy and the release of Ag^+.

Tan et al. (2020) synthesized (N-methoxyethyl)-3-aminopropyltriethoxysilane (MAPS) and bis(N-methoxyethyl)-3-aminopropyltriethoxysilane (BMAPS) through an aza-Michael addition reaction and grafted these seawater-responsive silanes onto SiO_2 nanoparticles through a silylation reaction, obtaining zwitterionic-modified SiO_2 nanoparticles (M-SiO_2 and BM-SiO_2). Subsequently, the modified SiO_2 was embedded in PDMS to construct a biomimetic heterogeneous structure coating. The coating could rapidly switch from superhydrophobic in air to superoleophobic under water, and after mechanical wear, outdoor exposure tests, or artificial immersion for 5 months, the coating still maintained its superoleophobic property under water. Laboratory fouling tests and real-sea panel tests showed that coatings doped with M-SiO_2 and BM-SiO_2 could effectively inhibit the attachment of fouling organisms. This is mainly due to the synergistic effect of the coating's surface hydration barrier layer and the zwitterionic antibacterial properties, which effectively enhance the coating's anti-fouling performance.

Bharathidasan et al. (2015) dispersed ZnO nanoparticles in a porous boron carbonitride (BCN) framework to form a new type of functional hybrid. The loading of BCN improved the photocatalytic performance of ZnO in the visible light region. Subsequently, ZnO-BCN two-dimensional materials were dispersed in PDMS, forming a stable coating on aluminum alloy. When the content of ZnO-BCN was 66.6 wt%, the coating's surface contact angle was 157.6°, and the rolling angle was less than 6°, making it a superhydrophobic anti-fouling coating. In addition, the photocatalytic performance of ZnO could also kill adhered bacteria. The synergistic anti-fouling effect of superhydrophobicity and photocatalysis makes this coating promising in the field of anti-fouling.

WANG et al. (2022) modified $Ti_3C_2T_x$ with 3-sulfopropyl methacrylate potassium (PSPMA) through surface-initiated atom transfer radical polymerization and then blended it with epoxy resin to form an epoxy resin coating. The coating could reduce bacterial adhesion by 55% and microalgal adhesion by 71% through the hydration layer formed by PSPMA and the contact bactericidal effect of $Ti_3C_2T_x$.

It is evident that the doping of nanoparticles can achieve the preparation of anti-fouling coatings with multiple anti-fouling strategies working in synergy. Nanoparticles can utilize their inherent antibacterial and anti-fouling mechanisms or serve as carriers for other anti-fouling agents, combined with the impact of nanoparticles on the surface energy or hydrophobic properties of the coating, to achieve synergistic anti-fouling effects. Coatings with multiple anti-fouling strategies working in synergy are beneficial for realizing the long-term and broad-spectrum effectiveness of anti-fouling.

6.5 Future Outlook and Challenges

Currently, with the development of nanotechnology, the use of nanomaterials to enhance the mechanical properties and anti-fouling performance of marine anti-fouling coatings has received significant attention. By embedding nanoparticles into the polymer matrix, the mechanical properties, surface morphology, and active sites of the coating are fundamentally improved, resulting in superior performance. A comprehensive summary of the application of inorganic nanomaterial-modified polymer coatings in the field of anti-fouling, both domestically and internationally, has been conducted, and the main conclusions are as follows:

(1) For low surface energy nano-anti-fouling coatings, the addition of a few percentage points of nanoparticles can enhance the durability and fouling release properties of the coating.
(2) For superhydrophobic nanocomposite anti-fouling coatings, the addition of nanoparticles can effectively improve the hydrophobicity of the coating, thereby enhancing its anti-fouling performance. The content of nanoparticles involved in the superhydrophobic nano-coatings studied to date ranges from a few tenths of a percent to several tens of percent, showing a wide variation.
(3) For release-type nanocomposite anti-fouling coatings, nanoparticles act as anti-fouling agents to endow the coating with efficient anti-fouling performance. However, release-type nano-anti-fouling coatings have a fatal flaw: The slow release of the anti-fouling agent leads to the gradual loss of anti-fouling activity in the coating.
(4) For catalytic-type nano-anti-fouling coatings, nanoparticles effectively inhibit the attachment of fouling organisms through the generation of reactive groups.
(5) For multi-factor synergistic nano-anti-fouling coatings, the doping of nanoparticles supports the preparation of ternary or quaternary nanocomposite materials, utilizing their inherent antibacterial and anti-fouling mechanisms or as carriers for other anti-fouling agents, combined with the impact of nanoparticles on the surface properties of the coating, to achieve synergistic anti-fouling effects.

Environmental friendliness and stability are fundamental requirements for nano-antifouling coatings. They are a green and efficient anti-fouling technology and represent the future development direction of anti-fouling materials. However, to achieve large-scale application, the following issues need to be addressed:

(1) Ensuring the stable dispersion of inorganic nanoparticles in polymer matrices. Nanomaterials have a high surface energy and are highly unstable, prone to aggregation. Interparticle forces such as Van der Waals forces, electrostatic attractions, liquid bridges formed by ion absorption bridges, and hydrogen

bonds can increase adhesion between particles, leading to aggregation. Nanoparticles are difficult to wet and disperse in organic media, which limits the manifestation of their unique effects. Therefore, how to improve the dispersibility and stability of nanoparticles in the medium is crucial.

(2) Optimizing inorganic nanocomposite coatings based on the synergistic action of multiple anti-fouling mechanisms. The types of fouling involved in biological fouling are numerous, and the fouling conditions are complex. Nanocomposite anti-fouling coatings based on a single anti-fouling mechanism struggle to achieve broad-spectrum and long-lasting anti-fouling effects. Therefore, using synergistic anti-fouling mechanisms and combining two or more anti-fouling strategies to construct multifunctional nano-anti-fouling coatings is a key development focus.

(3) Ensuring the long-term anti-fouling performance of inorganic nanocomposite coatings. Currently, the testing periods for the anti-fouling performance of nanocomposite coatings are not long, and the fouling targets mainly focus on bacteria, algae, and other fouling organisms. However, in the early stages of biological fouling, substances such as proteins and polysaccharides in the environment can adsorb onto the coating surface, forming a conditioning film, which alters the physical and chemical properties of the coating surface, such as surface energy and hydrophobicity/hydrophilicity, or can combine with released ions, weakening its anti-fouling performance. Therefore, how to restrain the adsorption of biological macromolecules such as proteins and polysaccharides through appropriate material design is crucial for ensuring the anti-fouling performance of nano-coatings.

References

Almeida M, Filipe S, Humanes M et al (2001) Vanadium haloperoxidases from brown algae of the Laminariaceae family. Phytochemistry 57(5):633–642

Andre R, Natalio F, Humanes M et al (2011) V_2O_5 nanowires with an intrinsic peroxidase-like activity. Adv Func Mater 21(3):501–509

Ba M, Zhang Z, Qi Y (2019) The influence of MWCNTs-OH on the properties of the fouling release coatings based on polydimethylsiloxane with the incorporation of phenylmethylsilicone oil. Prog Org Coat 130:132–143

Bharathidasan T, Mandalam A, Balasubramanian M et al (2015) Zinc oxide-containing porous boron-carbon-nitrogen sheets from glycine-nitrate combustion: synthesis, self-cleaning, and sunlight-driven photocatalytic activity. ACS Appl Mater Interfaces 7(33):18450–18459

Caliskantd, Hukum KO, Caykara T et al (2022) Toward the replacement of long-chain perfluoroalkyl compounds: perfluoropolyether-based low surface energy grafted nanocoatings. ACS Appl Polym Mater 4(2):980–986

Cavas L, Yildiz PG, Mimigianni P et al (2018) Reinforcement effects of multiwall carbon nanotubes and graphene oxide on PDMS marine coatings. J Coat Technol Res 15(1):105–120

Chen Z, Wang Z, Ren J et al (2018) Enzyme mimicry for combating bacteria and biofilms. Acc Chem Res 51(3):789–799

Chen RZ, Xie QY, Zeng HH et al (2020) Non-elastic glassy coating with fouling release and resistance abilities. J Mater Chem A 8(1):380–387

Crans DC, Smee JJ, Gaidamauskas E et al (2004) The chemistry and biochemistry of vanadium and the biological activities exerted by vanadium compounds. Chem Rev 104(2):849–902

Fu Y, Wang WC, Zhang LQ et al (2019) Development of marine antifouling epoxy coating enhanced with clay nanotubes. Materials 12(24):4195

He X, Cao P, Tian F et al (2019a) Autoclaving-induced in-situ grown hierarchical structures for construction of superhydrophobic surfaces: a new route to fabricate antifouling coatings. Surf Coat Technol 357:180–188

He XY, Cao P, Tian F et al (2019b) Autoclavinginduced in-situ grown hierarchical structures for construction of superhydrophobic surfaces: a new route to fabricate anti- fouling coatings. Surf Coat Technol 357:180–188

He X, Tian F, Bai X et al (2019c) Role of trapped air and lubricant in the interactions between fouling and SiO_2 nanoparticle surfaces. Colloids SurfS B: Bio-Interfaces 184:110502

He X, Tian F, Chang J et al (2020) Haloperoxidase mimicry by CeO2–x nanorods of different aspect ratios for antibacterial performance. ACS Sustain Chem Eng 8(17):6744–6752

Hu M, Korschelt K, Viel M et al (2018) Nanozymes in nanofibrous mats with haloperoxidase-like activity to combat biofouling. ACS Appl Mater Interfaces 10:44722–44730

Jotun (2023) Sealion repulse nanorepellent technology [EB/OL]. https://pdf.nauticexpo.com/pdf/jotun/sealionrepulse-brochure/31197-58657.html

Lejars M, Margaillan A, Bressy C (2012) Fouling release coatings: a nontoxic alternative to biocidal antifouling coatings. Chem Rev 112(8):4347–4390

Li S, Liu Y, Tian Z et al (2020) Biomimetic superhy-drophobic and antibacterial stainless-steel mesh via doublepotentiostatic electrodeposition and modification. Surf Coat Technol 403:126355

Michailidis M, Gutner-Hoch E, Wengier R et al (2020) Highly effective functionalized coatings with antibacterial and antifouling properties. ACS Sustain Chem Eng 8(24):8928–8937

Natalio F, Andre R, Hartog AF et al (2012) Vanadium pentoxide nanoparticles mimic vanadium haloperoxidases and thwart biofilm formation. Nat Nanotechnol 7(8):530–535

Necolau MI, Pandele AM (2020) Recent advances in graphene oxide-based anticorrosive coatings: an overview. Coatings 10(12):1149

Nguyen-Tri P, Nguyen TA, Carriere P et al (2018) Nanocomposite coatings: preparation, characterization, properties, and applications. Int J Corros 2018:1–19

Nurioglu AG, Esteves ACDC, With GD (2015) Non-toxic, non-biocide-release antifouling coatings based on molecular structure design for marine applications. J Mater Chem B 3(32):6547–6570

Piola RF, Dafforn KA, Johnston EL (2009) The influence of antifouling practices on marine invasions. Biofouling 25(7):633–644

Qiu H, Hoelken I, Gapeeva A et al (2018) Development and characterization of mechanically durable siliconepolythiourethane composites modified with tetrapodal shaped ZnO particles for the potential application as fouling-release coating in the marine sector. Materials 11(12):2413

Saravanan P, Jayamoorthy K, Ananda KS (2016) Design and characterization of non-toxic nano-hybrid coatings for corrosion and fouling resistance. J Sci: Adv Mater Devices 1(3):367–378

Selim MS, El-Safty SA, El-Sockary MA et al (2016) Smart photo-induced silicone/TiO_2 nanocomposites with dominant [110] exposed surfaces for self-cleaning foul-release coatings of ship hulls. 101:218–225

Selim MS, El-Saftysa, Fatthallah NA et al (2018) Silicone/graphene oxide sheet-alumina nanorod ternary composite for superhydrophobic antifouling coating. Prog Org Coat 121:160–172

Soleimani S, Jannesari A, Yousefzadi M et al (2021) Eco-friendly foul release coatings based on a novel reduced graphene oxide/Ag nanocomposite prepared by a green synthesis approach. Prog Org Coat 151:106107

Sun Z, Li Z, Li W et al (2020) Mesoporous cellulose/TiO_2/SiO_2/TiN-based nanocomposite hydrogels for efficient solar steam evaporation: low thermal conductivity and high light-heat conversion. Cellulose 27(1):481–491

Tan J, Xu J, Wang D et al (2020) Seawater-responsive SiO_2 nanoparticles for in situ generation of zwitterionic polydimethylsiloxane antifouling coatings with underwater superoleophobicity. J Mater Chem A 8(45):24086–24097

Tian S, Jiang D, Pu J et al (2019) A new hybrid silicone-based antifouling coating with nanocomposite hydrogel for durable antifouling properties. Chem Eng J 370:1–9

Vasilev K, Sah V, Anselme K et al (2010) Tunable antibacterial coatings that support mammalian cell growth. Nano Lett 10(1):202–207

Vucko MJ, King PC, Poole AJ et al (2012) Cold spray metal embedment: an innovative antifouling technology. Biofouling 28(3):239–248

Vucko MJ, King PC, Poole AJ et al (2014) Assessing the antifouling properties of cold-spray metal embedment using loading density gradients of metal particles. Biofouling 30(6):651–666

Wang HY, Zhang XG, Liu ZJ et al (2016a) A super- robust superhydrophobic PSU composite coating with selfcleaning properties, wear resistance and corrosion resistance. RSC Adv 6(13):10930–10937

Wang B, Ju P, Zhang D et al (2016b) Colorimetric detection of H_2O_2 using flower-like $Fe_2(MoO_4)_3$ microparticles as a peroxidase mimic. Microchim Acta 183(11):3025–3033

Wang G, Weng D, Chen C et al (2020) Influence of TiO_2 nanostructure size and surface modification on surface wettability and bacterial adhesion. Colloid Interface Sci Commun 34:100220

Wang P, He B, Du Y et al (2022) Functionalized Ti_3C_2Tx-based nanocomposite coatings for anticorrosion and antifouling applications. Chem Eng J 448:137668

Xie Q, Liu C, Lin X et al (2020) Nanodiamond reinforced poly(dimethylsiloxane)-based polyurea with self-healing ability for fouling release coating. ACS Appl Polym Mater 2(8):3181–3188

Zhang DY, Hao Q, Liu J et al (2018) Antifouling polyimide membrane with grafted silver nanoparticles and zwitterion. Sep Purif Technol 192:230–239

Zhao W, Zhu R, Jiang J et al (2019) Environmentallyfriendly superhydrophobic surface based on Al_2O_3@ KH560@SiO_2 electrokinetic nanoparticle for long-term anti-corrosion in sea water. Appl Surf Sci 484:307–316

Zheng L, Lin YC, Wang DH et al (2020) Facile one-pot synthesis of silver nanoparticles encapsulated in natural polymeric urushiol for marine antifouling. RSC Adv 10(24):13936–13943

Chapter 7
Enhancing Bioremediation of Oil Spills Using Nanotechnology

Abhay B. Fulke, Siddant Ratanpal, Sanjana Varma, and Bhushan P. Chaudhari

Abstract Oil spills pose a major environmental and economic challenge, affecting ecosystems and communities worldwide. This chapter offers a thorough review of both traditional and emerging techniques for oil spill remediation, examining the effectiveness of physical, thermal, chemical, and biological methods. While conventional approaches have had some success, they often struggle under varying environmental conditions. Bioremediation, which utilizes microbes to break down hydrocarbons, is a promising solution, particularly when enhanced with modern strategies like bioaugmentation and bio stimulation. Recent breakthroughs in nanotechnology have introduced innovative solutions, including magnetic nanomaterials, nano-based emulsifiers, and specialized nutrient delivery systems, all of which improve oil absorption and stimulate microbial activity. The combination of nanotechnology and bioremediation shows promise for accelerating the breakdown of hydrocarbons while minimizing ecological harm. Future efforts will focus on developing biodegradable, eco-friendly nanomaterials, refining nanostructure designs for better oil capture, and extending their use in remote areas. This chapter advocates for a combined approach that integrates traditional remediation methods with advanced nanotechnology to offer effective, sustainable solutions for oil spill cleanup.

Keywords Oil spills · Nanotechnology · Bioremediation · Nano-bioremediation · Hydrocarbons

A. B. Fulke (✉) · S. Ratanpal
Microbiology Division, CSIR-National Institute of Oceanography (CSIR-NIO), Mumbai, Maharashtra, India
e-mail: afulke.nio@csir.res.in

A. B. Fulke · S. Varma · B. P. Chaudhari
Academy of Scientific and Innovative Research (AcSIR), Ghaziabad, India

S. Varma · B. P. Chaudhari
Biochemical Sciences Division, CSIR-National Chemical Laboratory, Pune, India

S. K. Panda and V. P.M. (eds.), *Bionanocomposites from Marine Origin*, Advanced Structured Materials 247, https://doi.org/10.1007/978-3-032-10749-7_7

7.1 Introduction

Marine pollution caused by oil spills and seepage poses significant threats to ecosystems and the economies dependent on them. Transporting oil across global waterways inherently risks spills, with each incident having the potential for catastrophic impacts. The long-term damage from an oil spill often depends more on the speed of cleanup efforts than on the volume of oil spilled. Once oil reaches the water's surface, its composition and properties begin to change immediately, undergoing processes such as evaporation, oxidation, emulsification, sedimentation, biodegradation, and dispersion. Although oil spills have been a recurring issue for decades, cleanup techniques have largely remained stagnant. Current post-spill response methods include mechanical (booms, sorbents, and skimmers), thermal (in situ burning), chemical (dispersants), and biological (bioremediation) approaches. The effectiveness of these techniques varies based on external factors like oil type, viscosity, volume, location, and weather conditions.

Crude oil, composed primarily of aliphatic and aromatic hydrocarbons, is a natural product derived from aquatic algae deposited between 180 and 85 million years ago (Dell'Anno et al 2021). These hydrocarbons, often released into the environment through spills such as the Exxon Valdez and BP Deepwater Horizon disasters, also seep naturally from underground reservoirs. Since petroleum hydrocarbons occur naturally in marine environments, microorganisms capable of degrading these substances have evolved over time. These oil-degrading microorganisms—bacteria, archaea, and fungi—are ubiquitous but often represent only a small fraction of the pre-spill microbial community. Oil spills affect both marine and terrestrial environments, endangering ecosystems and human health. They can contaminate drinking water, pose fire and explosion risks, degrade water and air quality, destroy recreational areas, and lead to substantial economic losses. The environmental consequences of spills are far-reaching and long-lasting, affecting natural resources, wildlife, tourism, and fisheries. For example, oil on the sea's surface blocks sunlight and oxygen from penetrating deeper layers, threatening marine flora and fauna. Shorelines can also be contaminated by tidal and wave actions, as seen in spills like the 1978 Amoco Cadiz disaster, which polluted 320 km of shoreline in Brittany, France (Djønne 2023).

Despite various regulations and response guidelines, accidental spills remain inevitable. Governments must be prepared with robust contingency plans, such as the U.S. National Contingency Plan (NCP), which outlines strategies for oil spill and hazardous substance response. Prevention and preparedness are key priorities for organizations like the United States Environmental Protection Agency (EPA). Effective remediation methods include physical, chemical, and biological approaches, with bioremediation being one of the most promising due to its eco-friendliness and ability to harness natural processes.

Historically, oil pollution has been a persistent issue. The first recorded major spill occurred in 1907, releasing 7400 tons of paraffin oil into UK waters (Shyam et al. 2022). Since then, over 140 significant spills have occurred, releasing an estimated seven million tons of oil into the environment. Not all oil pollution is accidental;

over 90% originates from natural seepage, land-based runoff, or ship operations like tank washing. Major spills, such as the Gulf War oil spill, Deepwater Horizon, and Ixtoc I blowouts, have caused widespread ecological damage. For instance, the 1988 Ashland oil spill resulted in diesel oil contaminating the Monongahela River after a storage tank collapsed. (Joeckel 2023).

Nanotechnology offers new solutions for oil spill remediation. This emerging field focuses on designing and fabricating materials at the nanoscale (1–100 nm). Magnetic nanomaterials (MNMs) have garnered significant attention for their unique properties, such as superparamagnetic, low coercivity, and high magnetic susceptibility, distinguishing them from their bulk counterparts. These properties arise from their nanoscale size, which alters their internal structure, allowing magnetic moments to fluctuate under thermal energy. Such innovations hold great promise for improving oil spill cleanup techniques, offering faster, more efficient, and environmentally friendly solutions.

7.2 Conventional Oil Spill Response Techniques

a. **Physical Methods**

Physical remediation techniques are widely regarded as effective solutions for managing marine oil spills. These approaches rely on mechanical tools and specialized materials to physically contain, recover, and remove oil from affected areas. They are particularly valued for their capacity to recover substantial amounts of oil quickly while minimizing immediate environmental harm. However, their success is often influenced by factors such as environmental conditions and the characteristics of the spilled oil.

Booms: Containment Barriers.

Booms are essential for controlling the spread of oil slicks on the water's surface. Acting as floating barriers, they prevent oil from dispersing, enabling recovery efforts to proceed effectively. Modern designs feature enhanced buoyancy and stability to withstand harsh conditions like strong winds, fast currents, and rough waves. Advanced materials, such as double-skirted configurations and durable polymers, have improved boom performance under adverse weather conditions. However, extreme scenarios, such as storms or strong wave action, can lead to oil breaching the booms through "over wash" or "underflow," reducing their effectiveness (Dhaka and Chattopadhyay 2021).

Skimmers: Oil Recovery Devices

Skimmers are employed to recover oil contained by booms or accumulated in specific areas. These devices operate by skimming oil from the water's surface and transferring it into storage tanks for disposal or recycling. Their efficiency depends significantly on the oil's viscosity. Low-viscosity oils are easier to collect, while thicker oils can clog skimmers, reducing efficiency.

Innovations like positive displacement pumps have enhanced skimmers' ability to manage viscous oils by maintaining a consistent flow rate during operation. Self-propelled skimmers, equipped with independent propulsion systems, have been developed for use in calm, confined environments like harbors and marinas, where precise maneuverability is crucial. Despite these advancements, their effectiveness in open water remains limited due to environmental factors such as wave height and oil thickness (Sharma et al. 2021).

Sorbent Materials: Absorption and Adsorption

Sorbents are vital for removing smaller oil patches or cleaning up in areas inaccessible to skimmers and booms. These materials work by absorbing or adsorbing oil from the water.

- Traditional Sorbents: Affordable and easy to deploy, options like polypropylene pads, sawdust, and peat moss are commonly used.
- Natural Sorbents: Materials such as sugarcane bagasse and human hair have shown excellent oil absorption capabilities. These biodegradable and abundant options are particularly suitable for large-scale cleanup in ecologically sensitive areas.
- Advanced Sorbents: Recent innovations in materials science have produced high-efficiency sorbents such as aerogels, zeolites, and polymer-based products. Aerogels, for instance, are lightweight and highly porous, enabling them to absorb significant amounts of oil while being reusable after cleaning. Hydrophobic and oleophilic polymers or nanoparticles are also effective, selectively absorbing oil and minimizing water uptake (Zamparas et al. 2020).

b. **Thermal Methods**

Thermal remediation, such as in situ burning, is a rapid and equipment-light approach to oil removal. It involves igniting the oil within fire-resistant booms using specialized ignitors, which can be deployed from vessels, land, or helicopters. These ignitors often utilize fuels like kerosene, propane, or butane, or oil-soaked sorbents as combustion aids.

In situ burning is most effective in calm waters and before the oil emulsifies. For ignition to occur, the oil slick must be at least ~ 3 mm thick, and its water content should not exceed 25%. Under ideal conditions, removal efficiency can reach up to 95%. However, this method has limitations, including a narrow operational time frame before natural emulsification reduces its feasibility.

While efficient, in situ burning has significant drawbacks. These include severe air pollution, the sinking of burnt residue, and the potential for secondary fires or explosions. These risks necessitate careful consideration when deploying this technique (Dhaka and Chattopadhyay 2021).

c. **Chemical Methods**: Chemical Methods for Oil Spill Remediation Chemical techniques play a significant role in managing oil spills by using chemical agents to reduce their impact and prevent the oil from spreading. These methods are particularly valuable in challenging conditions, such as rough or icy waters, where mechanical approaches may be less effective.

Emulsifiers and Dispersants

Emulsifiers and dispersants are commonly used to break down oil slicks into smaller droplets. These surface-active agents, or surfactants, reduce the oil–water interfacial tension, stabilizing the resulting droplets. This process does not remove the oil but helps to disperse it into the water column, which can increase its bioavailability for microbial degradation.

A notable example is Corexit, a dispersant extensively used during the Deepwater Horizon spill cleanup. Studies on Corexit's efficacy and effects have produced mixed results. While it has been shown to effectively disperse oil, concerns have been raised about its potential to inhibit bioremediation processes. Additionally, adverse effects have been observed in human health studies, including reduced cell viability and DNA damage in cultured human bronchial cells after Corexit exposure. However, some findings have been challenged, citing inconsistencies in laboratory methods, high-exposure thresholds, and reliance on self-reported data for human health impacts. Despite these concerns, dispersants remain a key tool for addressing large-scale oil spills, as they can enhance biodegradation by increasing oil droplet surface area (Dhaka and Chattopadhyay 2021).

Advantages of Chemical Methods

Chemical methods such as dispersants and oil herders are versatile and can be deployed in various environmental conditions, including those where traditional mechanical methods fail. Their ability to break down or contain oil slicks plays a critical role in minimizing environmental damage and facilitating subsequent cleanup efforts. However, concerns about long-term environmental persistence and potential health impacts highlight the importance of careful application and ongoing research (Fig. 7.1).

7.3 Bioremediation of Oil Spills

Bioremediation, the use of microorganisms to break down environmental contaminants, was first utilized six decades ago by George Robinson, who deployed microbes to address oil spills along the Santa Barbara coast in California (Upadhyay 2024). This process involves microorganisms and their metabolic byproducts to eliminate pollutants from aquatic systems and sediments. Indigenous microorganisms in sediments play a critical role in this process, as they can degrade organic compounds into simpler inorganic molecules or their elemental forms through a process known as mineralization.

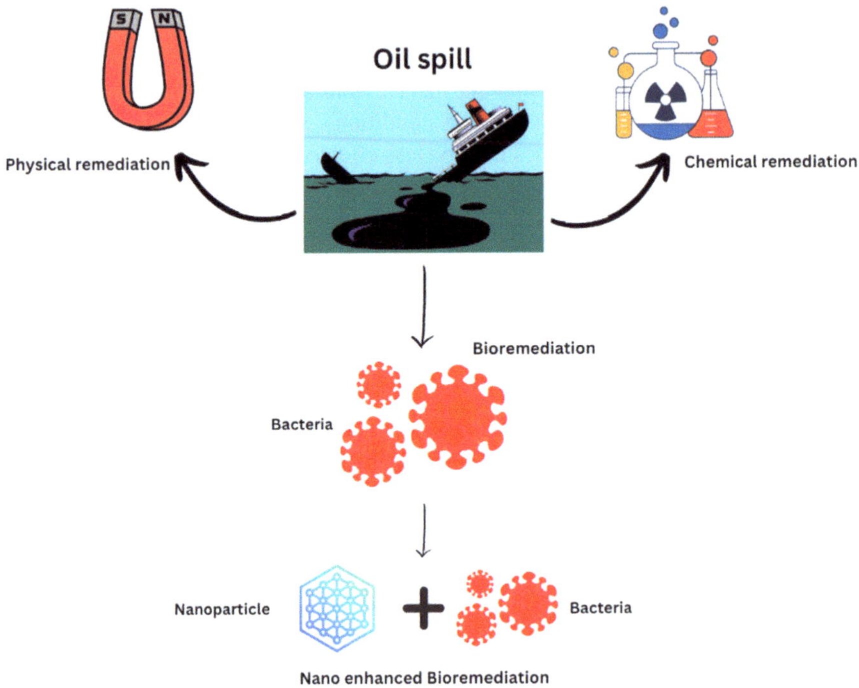

Fig. 7.1 Different oil spill remediation methods

Microorganisms in bioremediation can neutralize, reduce, and transform oil contaminants into naturally occurring compounds in air, sediment, water, sludge, or soil. This process often occurs alongside detoxification, where microorganisms immobilize or remove waste materials through transformations and mineralization. For microbial cells to reproduce and remain viable, they require energy, typically derived from redox reactions such as respiration. Essential elements for this energy system include nutrients and electron donors/acceptors, such as oxygen, carbon dioxide, iron (III), and sulfate.

The success of hydrocarbon (HC) bioremediation hinges on finding suitable microbial strains or consortia with the metabolic capacity to degrade hydrocarbons effectively. Microbial communities rapidly adapt to HC exposure, undergoing genetic changes and selective enrichment within hours, which enhances biodegradation rates compared to HC-free environments.

Bioremediation often requires a consortium of microbial strains because oil spills consist of a complex mixture of hydrocarbons, and individual strains can metabolize only specific substrates. Effective bioremediation also depends on optimal environmental conditions, including nutrient availability, pH, water content, and temperature. High contaminant concentrations may inhibit microbial growth due to toxicity, while very low concentrations may fail to activate the necessary bacterial enzymes.

Temperature plays a pivotal role, with biodegradation rates typically increasing with higher temperatures and declining in colder conditions.

Various microorganisms, including bacteria, fungi, green algae, and cyanobacteria, are capable of degrading hydrocarbons under different environmental conditions (e.g., aerobic, anaerobic, and varying salinities). Hydrocarbon degradation occurs progressively as the compounds are metabolized in sequence. The genetic information for hydrocarbon degradation is located on plasmid or chromosomal DNA.

Hydrocarbons such as aliphatic and aromatic compounds can be degraded under both aerobic and anaerobic conditions. In anaerobic settings, microorganisms such as sulfate-reducing bacteria utilize various terminal electron acceptors, including nitrate, iron (III), and sulfate, to catalyze degradation. Aerobic degradation, on the other hand, involves oxygenase enzymes, which incorporate oxygen atoms into hydrocarbons. Dioxygenases introduce two oxygen atoms, while monooxygenases introduce one. Since oxygen is a more efficient electron acceptor, aerobic degradation is generally faster than anaerobic processes.

Bioaugmentation Since the 1970s, bioaugmentation has been employed as a strategy to enhance the degradation of hydrocarbons in marine environments by introducing hydrocarbon-degrading microorganisms. This approach is particularly useful in areas where native microbial communities are either unable to degrade all hydrocarbons present or exhibit prolonged lag phases following oil exposure. The introduced organisms may be isolated from other contaminated sites or genetically engineered to improve their biodegradation efficiency.

Despite its potential, bioaugmentation has certain limitations. The introduced microorganisms often compete with native species for critical nutrients, which can disrupt the local ecosystem. For successful implementation, it is essential to have a detailed understanding of the oil's composition, the capabilities of indigenous microbial communities, and the compatibility of the introduced species with the local ecosystem to ensure synergistic interactions.

Efforts to optimize bioaugmentation have shown promise, particularly through the modification or immobilization of microbial cells in laboratory and mesocosm studies. Immobilized cells offer several advantages, including higher cell concentrations, faster growth rates, better protection against environmental stressors, and enhanced biofilm production. These benefits collectively improve the efficiency of bioremediation, making immobilized cells a valuable tool in tackling oil spills more effectively (Adetunji and Anani 2021).

Bio Stimulation Bio stimulation is a technique used to improve the bioremediation of oil spills by supplementing growth-limiting nutrients like nitrogen and phosphorus, often in the form of fertilizers. This method promotes the proliferation of native oil-degrading microbial communities, accelerating the breakdown of oil contaminants. For example, following the Exxon Valdez oil spill, shorelines treated with oleophilic fertilizers containing nitrogen and phosphorus achieved 44% degradation of total polynuclear aromatic hydrocarbons and 63% degradation of total detectable hydrocarbons within 109 days. In comparison, untreated control sites showed no significant degradation even after 70 days.

However, the effectiveness of bio stimulation can be hindered by the dilution of added nutrients in water. To address this, water-insoluble sources like uric acid have been suggested as alternatives, as they can serve as a nitrogen source near the oil–water interface and are less susceptible to dilution. Despite its benefits, bio stimulation may sometimes encourage the growth of non-target microbial species, leading to competition and potentially reducing the efficiency of oil-degrading microbes. Targeted nutrient delivery to oil droplets has been shown to mitigate this issue, enhancing the effectiveness of oil-degrading communities.

Bio stimulation remains one of the most effective strategies for oil spill remediation. Recent advancements, including the development of water-insoluble nutrient sources, have further improved its efficiency, making it a valuable tool in environmental recovery efforts. (Pete et al. 2021).

7.4 Challenges in Bioremediation of Oil Spills

The study of hydrocarbon-degrading bacteria dates back nearly seventy years. In 1946, Zobell published the seminal work "Action of Microorganisms on Hydrocarbons," which explored their occurrence, characteristics, biochemical activity, and potential economic benefits. Despite this long history, marine bioremediation remains an underutilized commercial opportunity. While environmental agencies express interest in innovative biotechnologies, they must adhere to officially approved guidelines.

One major challenge in advancing bioremediation within the blue growth strategy is aligning regulations and legal frameworks with intervention strategies that are efficient, cost-effective, and environmentally friendly. For instance, many countries allow the use of native bacterial consortia for soil treatment but prohibit their application in marine environments. The lack of field testing significantly limits robust data needed to evaluate the effectiveness and environmental impact of such treatments. Furthermore, regulations in bioremediation often focus on genetically modified organisms (GMOs).

Traditional soil bioremediation methods are increasingly being replaced by advanced techniques involving GMOs. Genetic engineering can produce highly efficient microorganisms capable of degrading pollutants more rapidly. GMOs can also be designed to metabolize toxic intermediates of bioremediation. By manipulating genetic pathways, these organisms can enhance enzyme specificity and optimize metabolic processes (Basarkar et al. 2022). Given the complex nature of petroleum, which consists of a variety of hydrocarbons, developing engineered bacteria capable of degrading multiple petroleum hydrocarbons represents a promising direction for advancing oil spill bioremediation.

Genetically engineered strains for petroleum biodegradation have been developed. However, the use of GMOs in environmental applications remains a contentious issue, as it poses risks similar to those associated with the introduction of non-native species. To date, there are no documented cases of GMO application in marine environments. To make this technology more acceptable and feasible for regulatory authorities, end users, and the public, it is crucial to provide robust data, clear descriptions of its practical applications, and realistic expectations of its outcomes.

Currently, further research is needed to better understand the hydrocarbon-degrading abilities of marine microbes. Their degradation efficiency must be demonstrated at larger scales and validated by policymakers for real-world applications. Advances in molecular tools and high-throughput screening have uncovered remarkable biodiversity and functional potential within marine microbial communities in oil-polluted areas (Acosta-González and Marqués 2016). Additionally, significant data is now available on microbial biocatalysis, biosurfactant production, substrate affinities, and other characteristics. This enables predictions of the extent, time, and rate of biodegradation based on in vitro or micro/mesoscale simulation studies (Nikolaivits et al. 2017).

Nevertheless, while microorganisms can theoretically degrade almost any organic pollutant, certain hydrocarbons, such as polycyclic aromatic hydrocarbons (PAHs), are particularly resistant to biodegradation, especially in natural environments like marine sediments. Additionally, since bioremediation relies on natural processes, the time required for effective cleanup can often exceed practical or customer needs.

Some key challenges faced by conventional methods of bioremediation of oil spills are listed below.

- Poor point of contact between oil and microbial surfaces: Hydrocarbons are highly hydrophobic and poorly soluble in water, making direct contact between microbial cells and hydrocarbons essential for degradation. The bioavailability of hydrocarbons is a critical factor in initiating microbial degradation, as it enables the activation of molecular mechanisms and enzyme activity required for the process. Consequently, low bioavailability is a key limiting factor for efficient hydrocarbon biodegradation (Gao and Gu 2021)
- Environmental factors: Environmental factors like temperature, salinity, nutrients, electron acceptors, and metabolic substrates significantly impact bioremediation. While some microbes show promising lab-scale results, these factors can limit their effectiveness in field conditions. Temperature, crucial for bacterial growth and pollutant behavior, is typically optimal between 30–40 °C. Microorganisms also require essential nutrients like nitrogen, phosphorus, oxygen, sulfur, and trace elements for growth and metabolism, highlighting the importance of these factors for effective bioremediation (Saha et al. 2019)
- Rate of bioremediation: Microorganisms typically do not use hydrocarbons as a primary carbon source, turning to hydrocarbons only when other carbon sources are unavailable. This process relies on hydrocarbonoclastic enzymes, whose expression depends on the bacteria's physiology. Microbes require time to acclimate to contaminated environments and produce these enzymes, which catalyze

hydrocarbon degradation. However, bacteria that perform well in laboratory conditions often show reduced efficiency in field conditions due to environmental complexities (Patowary et al. 2023)

- Hinderance of metabolic pathways: The concentration and composition of hydrocarbons significantly impact the bioremediation process. High levels of crude oil can inhibit microbial growth and metabolism due to toxicity, emphasizing the need for ecotoxicological studies. Simpler hydrocarbons, such as linear alkanes, are more easily degraded by microbes, whereas aromatic hydrocarbons and complex polycyclic aromatic hydrocarbons (PAHs) are more resistant to degradation. The physicochemical properties and bioavailability of hydrocarbons, influenced by their size and structure, play a critical role in determining their biodegradability (Patowary et al. 2023).

7.5 Nanomaterials for Oil Spill Remediation

Over the past five years, there has been increasing global interest in exploring nanomaterials as effective solutions for cleaning up oil spills. Among these, nanoporous sorbents such as thermally exfoliated graphite oxide (TEGO), CF3-functionalized aerogels, and carbon nanotube (CNT) sponges have demonstrated exceptional oil absorption capacities, surpassing many traditional adsorbents. Examples of nanomaterials utilized for oil spill remediation include aerogels, magnetic nanocomposites, carbon-based nanostructures like CNTs and graphene tubes, specialized membranes, and nano-dispersants.

A critical requirement for sorbent, membrane, or filter nanomaterials in oil spill applications is their selectivity for organic solvents and oils to effectively separate oil from water. Traditional sorbents made from polypropylene, silicon-coated glass fibers, or raw cotton tend to absorb both water and organic solvents, limiting their efficiency. In contrast, nanomaterials like CNT sponges, nanowire membranes, recams, and hydrophobic core–shell magnetic nanoparticles (e.g., Fe_2O_3@C) selectively absorb oil due to their unique combination of super hydrophobicity and superoleophobicity. (Rehman et al. 2024) These advanced materials offer significant potential for improving oil spill cleanup technologies, with various applications summarized in Table 7.1.

Pickering emulsions are stabilized by solid particles, such as nanoparticles or microorganisms, which adsorb at the oil–water interface, preventing droplet coalescence. The stability of these emulsions primarily depends on the wettability of the stabilizing particles, though other factors like particle size, shape, surface area, concentration, and interfacial coverage also play important roles. A high particle coverage at the interface forms a strong barrier that resists droplet aggregation, thereby enhancing emulsion stability.

During the formation of Pickering emulsifiers, particles form an irreversible monolayer at the oil–water interface. For a stable oil-in-water Pickering emulsion, the particles should be partially hydrophobic with a contact angle (θ) greater than

Table 7.1 Nanomaterials and their use in oil spill cleanup

Materials	Use in oil spill cleanup	References
Magnetic polymers composite	Incorporation of Fe_3O_4 nanoparticles on polystyrene and polyvinyl fluoride nanofibers for oil in water separation	Jiang et al. (2015)
Nanowire membranes	Anti-fouling TiO_2 nanowires membranes for oil–water separation	Pan et al. (2019)
Nano-TiO_2	Nano-TiO_2 was able to reduce PAH bioavailability significantly	Vannuccini et al. (2024)
Aerogels	Silica aerogels able to adsorb 96% of oil from saline waters	Sorour et al. (2016)
Graphite oxide	Graphene materials providing 90% oil–water separation efficiency	Avornyo and Chrysikopoulos (2024)
Carbon nanotubes and sponges	Carbon nanotubes showing exceptional affinity and absorption efficiency toward oil	Sehrawat et al. (2024)

90°, preventing them from remaining solely in the aqueous phase. Functionalizing nano-emulsifiers can further improve the stability of Pickering emulsions. Natural clay nanoparticles, such as halloysite nanotubes (HNT), offer an environmentally friendly alternative to traditional emulsifiers. Halloysite, a naturally abundant aluminosilicate, can form oil-in-water Pickering emulsions that remain stable for months. While unmodified HNTs have limited emulsification properties, their surface can be easily modified to enhance wettability and improve emulsification performance (Owoseni et al. 2015).

For instance, functionalizing HNTs with chitosan reverses their surface charge from negative to positive, increasing their hydrophobicity and improving their attachment to the oil–water interface. This surface modification enhances the nanotubes' emulsification capability. Additionally, chitosan, a biocompatible cationic polyelectrolyte, provides an eco-friendly coating for HNTs. Similar modifications with amphiphilic polypeptoid polymers further increase the hydrophobicity of HNTs, improving their ability to reduce interfacial tension and strengthening their thermodynamic affinity for the oil–water interface. Natural biopolymers like lignin, when combined with chitosan, also enhance the emulsifying potential of nanomaterials, demonstrating the versatility and effectiveness of functionalized HNTs as Pickering emulsifiers.

Nanoparticle-Based Oil Herders Oil herders, first introduced in the 1970s as a tool for oil spill remediation, assist in in situ burning or the physical removal of oil. These herders consist of amphiphilic molecules or particles that adsorb at the air–water interface. When applied near an oil spill, they lower the surface tension at the water surface, creating a negative spreading coefficient that confines the oil slick.

A nanoparticle-based oil herder was developed by combining an amphiphile with nanoparticles. This herder, consisting of 1-pentanol and lignin nanoparticles, proved effective in confining oil for nearly complete physical removal. The 1-pentanol created a Marangoni flow, driven by surface tension gradients, which thickened and confined the oil slick. As the 1-pentanol dissolved in water, the lignin nanoparticles adsorbed irreversibly onto the oil–water interface. This adsorption was driven by van der Waals forces, hydrophobic interactions, and electrostatic interactions (Lee et al. 2018).

The nanoparticles formed a stable interfacial film, effectively locking the oil in a confined state and providing long-term stability—addressing a major limitation of conventional oil herders. Nanoparticle-based herders show promise as eco-friendly alternatives with enhanced stability compared to silicone-based herders, though limited research makes it challenging to fully evaluate their environmental suitability.

Magnetic Nanosorbents Sorbents are materials that are insoluble in both water and oil, used to recover oil through absorption, adsorption, or a combination of both. Absorbents soak up and retain oil, often swelling to more than one and a half times their original size, while adsorbents, with porous or rough surfaces, enhance adsorption capacity. The term "sorbent" encompasses both types and any method involving oil removal through absorption or adsorption.

The effectiveness of sorbents depends on factors such as sorption rate, oil retention, and ease of application and removal. Historically, sorbents have had low oil retention and lacked selectivity, absorbing both oil and water. However, advanced nanomaterials aim to improve this by creating sorbents with high selectivity for oil. Nanoparticles, with their larger surface area to volume ratio, increase sorption capacity and oil retention. Magnetic iron oxide-based nanosorbents have gained popularity since the 1970s due to their magnetic properties, allowing for easy removal.

One challenge with iron oxide nanoparticles is their tendency to aggregate, requiring a protective shell. Current research focuses on functionalizing these particles to enhance stability, prevent degradation, improve wettability, increase sorption capacity, and allow for reuse. Functionalization methods, such as coating with oleic acid, biopolymers, polymers, carbon, or silica, have been developed, leading to nearly complete oil removal and sorption capacities exceeding three times their weight. Reviews on magnetic sorbents provide further insights into these functionalization techniques.

Functionalized magnetic nanosorbents have shown promise in repelling water while retaining oil and demonstrating reusability, making them a sustainable alternative to traditional sorbents. However, concerns about the biocompatibility of iron oxide nanoparticles exist, with studies suggesting potential adverse health effects, such as developmental toxicity in zebrafish, though research on human health impacts is limited. High doses of iron oxide in contaminated water may lead to cytotoxicity. Other challenges for functionalized iron oxide nanosorbents include their high cost, complex synthesis, and practicality for use in marine environments. Given the growing interest in magnetic nanosorbents, further studies are needed to assess their toxicity and environmental impact (Zhu et al. 2012).

Multi-walled Carbon Nanotubes (MWCNTs) MWCNTs have gained significant attention as adsorption media due to their excellent capacity for hydrocarbon removal, especially after surface functionalization. MWCNT-based adsorbents have been widely used for removing heavy metals from industrial effluents, showing superior sorption performance compared to other adsorbents. In oily sludge, which contains a mix of organic compounds like aliphatic, aromatic, heteroatom-containing compounds, and asphaltenes, MWCNTs have shown effective removal capabilities for contaminants such as polycyclic aromatic hydrocarbons (PAHs) due to their enhanced surface area, high binding affinity, capacity, and thermal stability.

The large-scale transportation of crude oil leads to significant amounts of oily waste in waterways, creating serious environmental issues due to oil leaks and spills. Various techniques, including membrane filtration, electrochemical methods, air flotation, gravity separation, and biochemical methods, are currently used to separate spilled oil from water or emulsified oily wastewater. However, these methods often have drawbacks, such as high operational costs, low efficiency, and low flux rates (Sengupta and Gupta 2017). Functionalized MWCNTs and graphene oxide have also been effective in separating emulsified oil from water in acidic and neutral conditions, thanks to the different functional groups on their surfaces. However, these materials are ineffective in alkaline conditions due to electrostatic repulsion between the oil droplets and demulsifiers.

The extraction and removal of polycyclic aromatic hydrocarbons (PAHs) from wastewater is among the most promising applications of multi-walled carbon nanotubes (MWCNTs), as their hydrophobic sites effectively adsorb aromatic pollutants (Akinpelu et al. 2019). MWCNTs doped with ferric oxide nanoparticles have been used for the de-oiling of wastewater, and the direct removal of tiny oil droplets from oily water has been achieved using MWCNTs synthesized through the chemical vapor deposition (CVD) method. Fresh MWCNTs were produced in a fluidized-bed reactor via catalytic pyrolysis of propylene over $Fe/Mo/Al_2O_3$ catalysts. These composites exhibit exceptional adsorption capabilities, which have been applied to recycle various metal ions and surfactants. Freshly prepared MWCNTs, with their high specific surface area and unsaturated surface atom coordination, demonstrate excellent chemical reactivity and adsorption properties, making them an attractive choice for applications such as sensors (Nudrat et al. 2018).

7.6 Production of Nanoparticles

For production of nanoparticles, two unique methods of production are used, top-down and bottom-up approach. In top-down approach the bulk material is broken down into smaller fine particles with the help of reducing agents. Top-down approach includes various techniques such as grinding, ball milling, laser/thermal ablation, etc. Whereas bottom-up approach is opposite of top-down approach, wherein the nanoparticles are formed by agglomeration of atoms in new nuclei that grows into particles of nano-size (Abid et al. 2024).

Green synthesis is a modern technology which is used for production of stable NP using organic/plant sources, which increases the shelf life and stability of the NP. This method of synthesis of nanomaterial has been preferred compared to chemical and physical methods because former relies on use of chemicals along with the risk and side effects, and the latter is usually time-consuming and expensive. Hence green synthesis has advantage over both chemical and physical methods, as the nanoparticles are produced when the bulk material meets phytochemicals, amino acids, etc. (Figs. 7.2 and 7.3).

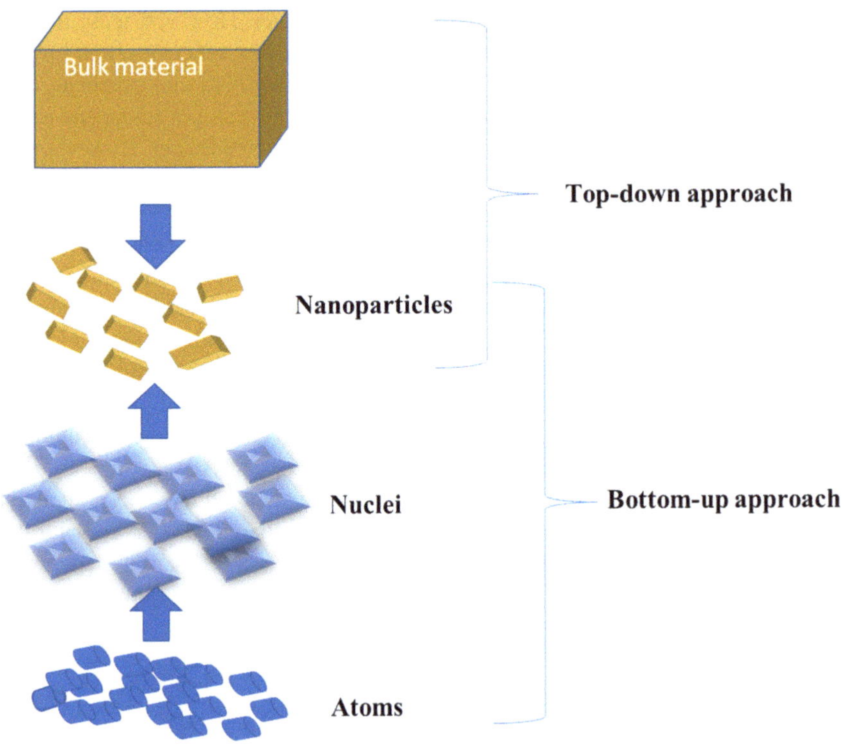

Fig. 7.2 Top-down and bottom-up approach of nanoparticle synthesis (Abid et al. 2024)

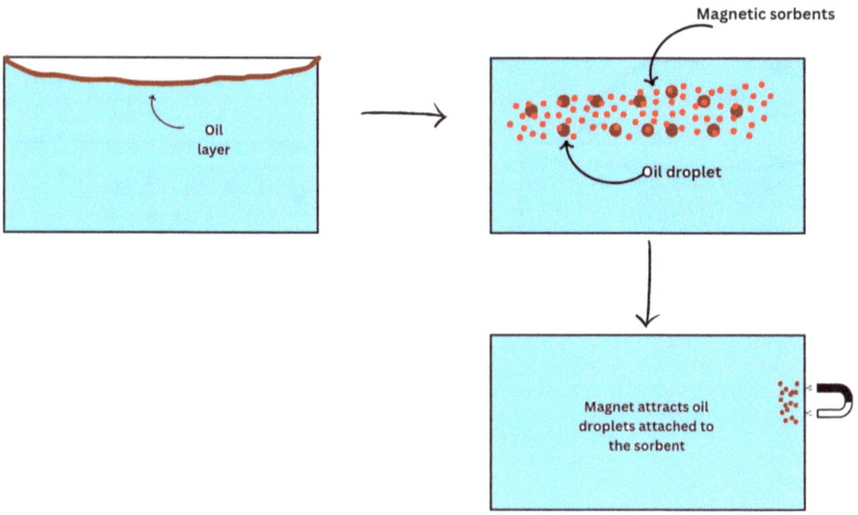

Fig. 7.3 Magnetic remediation of oil spill

7.7 Nano-enhanced Bioremediation of Oil Spills

Even when effective, complete degradation through bioremediation can take months or even years. Although bioremediation has improved over time, integrating nanoscience with bioremediation could speed up progress. In the past five years, new technologies have been developed to enhance bioremediation using nanoparticles. These innovations include utilizing natural Pickering emulsifiers to promote bacterial growth on oil droplets, coating bacteria with magnetic particles for easy removal, and using nutrient-loaded particles for precise nutrient delivery directly at the oil–water interface (Fig. 7.4).

a. Bacterial proliferation: Surfactants are essential for enhancing the biodegradation of oil by increasing its bioavailability to microorganisms. They achieve this by breaking down oil into smaller droplets, expanding the surface area available for microbial activity. This ensures more oil is accessible to bacteria, speeding up the breakdown of hydrocarbons. However, this advantage is balanced by a downside: surfactants lower the oil–water interfacial tension, which can reduce bacterial adhesion to the oil droplets. Since bacterial attachment is crucial for effective degradation, it is important to optimize surfactant use to enhance bioavailability while minimizing interference with bacterial adhesion. One innovative solution to support bacterial growth and activity is the use of clay nanotube Pickering emulsions. These emulsions stabilize oil–water interfaces by forming a protective layer around oil droplets, creating a favorable environment for bacteria to colonize. The porous and surface-active properties of clay nanotubes provide anchoring sites and nutrients for oil-degrading bacteria, enhancing their ability to metabolize hydrocarbons. This strategy not only stabilizes the emulsion but

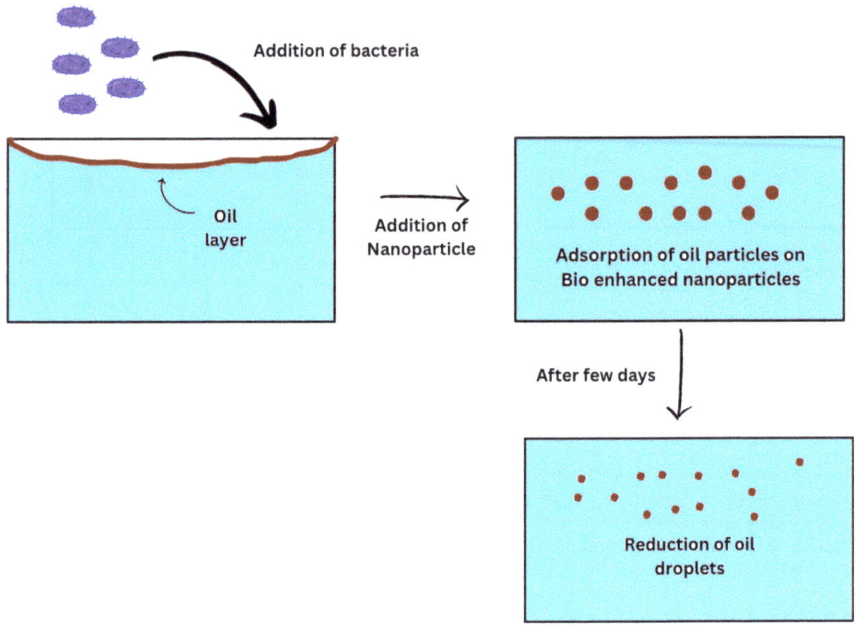

Fig. 7.4 Nano-enhanced bioremediation of oil spill

also sustains microbial activity, improving oil biodegradation efficiency. Additionally, multifunctional magnetic bacteria offer a novel and cost-effective solution for oil spill cleanup. These engineered bacteria act as Pickering emulsifiers, forming stable emulsions while encapsulating oil droplets. Their magnetic properties allow for easy removal of the oil-laden emulsions using external magnetic fields, making them suitable for large-scale and efficient oil spill remediation. Graphene oxide quantum dots (GOQDs) have also shown promise in stimulating indigenous oil-degrading microbial communities. GOQDs interact with bacterial cells, enhancing their metabolic activity and resilience in challenging environments. Their high surface area and unique electronic properties help facilitate electron transfer during microbial respiration, boosting the efficiency of oil degradation. GOQDs can also serve as a carbon source or nutrient carrier, further supporting bacterial growth and activity in contaminated areas. By integrating these advanced techniques—optimized surfactant use, clay nanotube Pickering emulsions, multifunctional magnetic bacteria, and graphene oxide quantum dots—highly effective and sustainable strategies for oil spill remediation can be developed. These methods combine physical, chemical, and biological mechanisms to address the complex challenges of oil contamination in aquatic ecosystems (Pete et al. 2021).

b. Magneto-responsive bacteria: Marine microorganisms have significant potential for oil spill remediation due to their natural ability to degrade oil. Recent innovations aim to enhance their effectiveness by making them magnetically responsive. This modification increases their adhesion to the oil–water interface and allows for their removal using an external magnetic field, improving biodegradation and enabling more efficient cleanup. One such example is the modification of *Alcanivorax borkumensis*, an oil-degrading bacterium, with magnetite nanoparticles coated with cationic polyelectrolytes. This creates a magnetic shell around the bacteria, allowing for nearly complete oil removal when exposed to a magnetic field. However, this modification comes at the cost of reduced bacterial growth rates, likely due to the energy demands or stress caused by the nanoparticle coating. This trade-off emphasizes the need to optimize the process to balance biodegradation efficiency with bacterial health. Another promising method involves using magneto-responsive microbes to stabilize Pickering emulsions at the oil–water interface. These microbes form a physical barrier that keeps oil droplets from coalescing, creating stable emulsions enriched with oil-degrading bacteria. These emulsions can then be removed using magnetic fields, providing both enhanced oil degradation and a physical cleanup mechanism. This approach is considered eco-friendlier and more cost-effective than previous technologies that used standalone magnetic nanoparticles, as it eliminates the need for separate nanoparticle production and distribution. Magneto-responsive bacteria have shown improved oil degradation and removal compared to unmodified bacteria, primarily due to their enhanced adhesion to the oil–water interface, which allows for better access to hydrocarbons and accelerates biodegradation (Pete et al. 2021). Additionally, their removal from the environment after oil degradation reduces the risk of secondary pollution and helps with recovery. Despite these advantages, there are concerns about the ecological impact of introducing magnetic nanoparticles into marine ecosystems. Although these nanoparticles are often deemed biocompatible, their long-term effects on marine organisms, sediment composition, and overall ecosystem health remain unclear. The potential for accumulation or unintended interactions with non-target species requires further investigation. Future research should focus on addressing these concerns, including developing biodegradable or environmentally safe magnetic materials, optimizing bacterial modification to reduce growth inhibition, and conducting thorough ecological risk assessments. With these improvements, magneto-responsive marine microorganisms could play a crucial role in sustainable and efficient oil spill remediation.
c. Target bio stimulation of oil spills: Bio stimulation is a common strategy for oil spill remediation that involves adding nutrients, such as nitrogen and phosphorus, which are essential for the growth of oil-degrading microorganisms. In marine environments affected by oil spills, these nutrients are often in short supply because hydrocarbons provide abundant carbon, but other nutrients are limited. By supplementing these nutrients, bio stimulation can speed up the natural breakdown of hydrocarbons. However, traditional bio stimulation methods face challenges, particularly in open water environments. One significant issue is the

dilution of water-soluble nutrients like nitrates and phosphates, which reduces their availability to oil-degrading microbes and may promote the growth of non-target microorganisms. This dilution reduces the efficiency and sustainability of the process.

To address this limitation, researchers have explored alternative nutrient sources, such as water-insoluble uric acid. This compound can provide a more localized and sustained release of nitrogen at the oil-water interface. Unlike soluble nutrients, uric acid remains near the contamination, making it more accessible to oil-degrading microbes while minimizing nutrient loss due to dilution. This strategy enhances nutrient efficiency and supports microbial growth where it is most needed.

Recent advances in nanotechnology have also contributed innovative solutions for targeted bio stimulation. Nanoparticles can be engineered to deliver nutrients directly to the oil-water interface, reducing nutrient waste and ensuring they are released precisely where they are required. Mesoporous silica nanoparticles (MSNs) are one such platform, designed with hydrophobic properties to interact specifically with oil droplets. MSNs can be loaded with nutrients such as nitrogen, potassium, and phosphorus, and release these nutrients in a controlled manner at the oil-water boundary. This targeted release prevents nutrient dispersion into the surrounding water, reducing the risk of eutrophication and unwanted microbial growth. Additionally, MSNs can degrade after releasing their nutrients, further minimizing potential ecological impacts (Pete et al. 2021).

The combination of nanotechnology with bioremediation has shown promising results in both laboratory and field studies, demonstrating improved growth of oil-degrading microbial communities and more efficient hydrocarbon degradation. Early findings suggest that using nanomaterials for targeted bio stimulation can outperform traditional nutrient supplementation methods, providing a more sustainable and effective approach for oil spill remediation. However, challenges remain, such as ensuring the environmental safety and long-term fate of nanoparticles in marine ecosystems. Ongoing research aims to develop biodegradable and environmentally friendly nanomaterials to minimize risks to marine organisms and habitats. Additionally, the scalability and cost-effectiveness of these technologies need further evaluation for large-scale application in oil spill incidents. Overall, integrating nanotechnology into bio stimulation represents a transformative advancement in oil spill remediation, offering a more efficient, targeted, and eco-friendly approach to enhancing natural biodegradation processes.

7.8 Recent Studies

A study conducted by Sun et al. (2023) explored the use of slow-release microencapsulated fertilizers to enhance the biodegradation of petroleum hydrocarbons after marine oil spills. The fertilizers, made with alginate and bentonite as a base and coated with ethyl cellulose and vegetable wax, provided nitrogen and phosphorus for bio stimulation. The material's macroporous, honeycomb-like structure was confirmed through scanning electron microscopy, element mapping, and spectroscopy techniques. Increased bentonite content improved the slow-release behavior, which followed the Fickian diffusion model. Bio stimulation experiments demonstrated that the fertilizers enhanced the degradation of petroleum hydrocarbons, especially high molecular weight polycyclic aromatic hydrocarbons (PAHs). Analysis of the microbial community revealed that the treatment promoted the prevalence of *Cycloclasticus*, a microorganism effective at degrading PAHs. The alginate/bentonite composite aerogels showed promise as biocompatible carriers for delivering microbial nutrients in environmental restoration efforts.

Banet et al. (2021) conducted a study which compared oil spill remediation using bioaugmentation and bio stimulation. This study investigated two terrestrial oil spills in a hyper-arid region of Israel, one from 1975 and the other from 2014, to examine the effects of bio stimulation on crude oil degradation and the potential for bioaugmentation using bio-stimulated sediments from the older spill. Soil hydrophobicity (measured as Water Drop Penetration Time, WDPT) and hydrocarbon content (Gasoline Range Organics, GRO, and Diesel Range Organics, DRO) were analyzed in one-month ex situ experiments.

Bio stimulation with mineral fertilization and maintaining 50% water saturation significantly reduced hydrophobicity and GRO + DRO content in both spills, while non-bio-stimulated controls showed no significant changes. However, bioaugmentation using bio-stimulated sediments from the 1975 spill did not enhance degradation in the 2014 spill. Interestingly, bio-stimulated sediments from the 2014 spill exhibited a 36% higher degradation rate than those from the 1975 spill. Crude oil degradation in both spills resulted in an increased GRO fraction, and WDPT proved to be a reliable indicator of oil content within a certain range. The study concluded that even in hyper-arid conditions, oil biodegradation capabilities can develop relatively quickly, with bio stimulation being effective, though bioaugmentation with older spill sediments was not.

Guidi et al. (2022) conducted a study which evaluated the use of anatase titanium nanoparticles (n-TiO_2) to remediate the toxic effects of crude oil pollution in marine environments. Petroleum hydrocarbons pose significant risks to ecosystems and human health, with some being carcinogenic and mutagenic. The study focused on the water-soluble fraction of crude oil, which induces severe cellular and molecular damage.

In vivo experiments using *Dicentrarchus labrax* (European sea bass) exposed to crude oil's water-accommodated fraction demonstrated that pure anatase n-TiO_2 nanoparticles were harmless in terms of DNA primary damage. Additionally, n-TiO_2 significantly reduced DNA damage caused by genotoxic pollutants. These findings suggest that anatase n-TiO_2 has potential as a nanotechnology tool for mitigating the harmful cellular effects of oil spills in marine environments. Further research is recommended to validate these results on a larger scale.

Osadebe et al. (2022), explored nano bioremediation to enhance natural attenuation of petroleum-polluted soil using a nanocomposite made from green-synthesized iron oxide nanoparticles (from *Solanum torvum* leaf extracts) decorated on cow bone-derived biochar. The biochar was produced via slow pyrolysis at 500 °C under low oxygen. The iron oxide-biochar nanocomposite (BCNP) was applied to soil microcosms at 10 and 15% w/w amendment levels, and the system was monitored for 60 days.

BCNP-amended soils showed significantly increased bacterial abundance (7.37–8.92 times higher) and total petroleum hydrocarbon (TPH) removal (28.4% and 26.2% higher for 10% and 15% BCNP, respectively) compared to biochar-only controls. Predominant bacteria included *Klebsiella pneumoniae*, *Pseudomonas spp.*, and *Bacillus spp.*, with hydrocarbon-utilizing bacteria dominated by *Pseudomonas spp.*, *Acinetobacter spp.*, and *Enterobacter spp.*. Metagenomic analysis revealed a shift from Proteobacteria and Acidobacteria dominance at the start to a predominance of Proteobacteria, especially *Pseudomonas spp.*, as the study progressed.

The findings highlight the efficacy of iron oxide-biochar nanocomposites in enhancing microbial activity and petroleum hydrocarbon degradation, making it a promising tool for remediating petroleum-contaminated soils.

7.9 Future Perspectives

Developing a consortium leveraging nanomaterials can revolutionize the remediation process by significantly accelerating oil degradation while ensuring minimal environmental impact and avoiding the generation of toxic waste. Such a consortium would integrate advanced nanotechnology with biological and physical approaches to create a comprehensive, efficient, and eco-friendly solution for oil spill cleanup.

Advancements in Nanostructure Design for Oil Sorption

As oil spill remediation often depends on sorption techniques, the future of nano-enhanced bioremediation will likely see innovations in nanostructure design to enhance oil sorption capacity. This could involve the development of new forms of carbon-based nanomaterials, such as graphene oxide or carbon nanotube composites, with tailored surface properties to selectively absorb different oil types or even specific hydrocarbon fractions, leading to more efficient cleanup strategies.

Nanotechnology in Offshore and Remote Oil Spill Remediation

As oil spills in remote or offshore environments become more common, nano-enhanced bioremediation technologies will likely be designed for ease of deployment and use in these challenging locations. This could involve developing nanoparticles that can be delivered via drones, unmanned aerial vehicles (UAVs), or other automated systems to remote areas, thus improving the scalability and efficiency of cleanup efforts in hard-to-reach regions (Fig. 7.5).

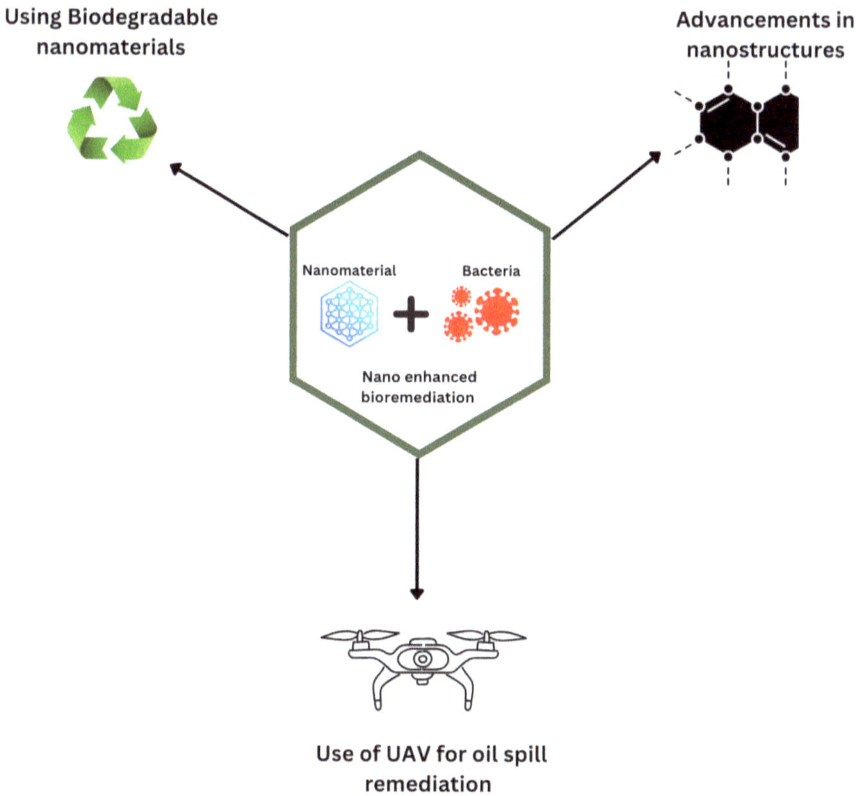

Fig. 7.5 Future perspectives for nano-enhanced bioremediation of oil spill

7.10 Conclusions

The persistent challenge of oil pollution in marine environments calls for innovative and effective remediation strategies. Traditional methods have proven inadequate in addressing the long-term effects of oil spills, emphasizing the need for technological advancements. Emerging solutions utilizing nanotechnology offer promising potential for improving oil spill cleanup. By incorporating nanomaterials like nanoporous sorbents, magnetic nanoparticles, and advanced emulsifiers, we can enhance both physical and biological remediation techniques, boosting oil recovery and accelerating biodegradation.

The future of oil spill remediation lies in the development of enhanced oil absorbing nanomaterials and the scalable nanosynthesis technologies. As these technologies evolve—such as by targeting nutrient delivery with mesoporous silica nanoparticles or using magneto-responsive bacteria—we move closer to more effective, environmentally responsible cleanup methods. Additionally, the ability to deploy these advanced solutions in remote or offshore oil spill scenarios highlights the need for adaptable and scalable approaches to tackle this global issue. By promoting collaboration between scientists, engineers, and environmental policymakers, we can build a comprehensive strategy for oil spill management that prioritizes ecological health and technological progress. Ultimately, a combined approach of physical, chemical, and biological techniques, supported by cutting-edge nanotechnology, will be key to protecting marine ecosystems and the communities that rely on them for the future.

Acknowledgements The authors are grateful to Director, CSIR-National Institute of Oceanography (CSIR-NIO), Goa, India and Scientist-in-Charge, CSIR-NIO, Regional Centre, Mumbai and Director, CSIR-National Chemical Laboratory (CSIR-NCL), Pune, India and Academy of Scientific and Innovative Research (AcSIR) for their encouragement and support.

References

Anani, OA, Jeevanandam J, Adetunji CO, Inobeme A, Oloke JK, Yerima MB, Thangadurai D, Islam S, Oyawoye OM, Olaniyan OT (2021) Application of biosurfactant as a noninvasive stimulant to enhance the degradation activities of indigenous hydrocarbon degraders in the soil. In: Green sustainable process for chemical and environmental engineering and science. Elsevier, pp 69–87

Abid J, Khalil FMA, Saeed S, Khan SU, Iqbal I, Anthony S (2024) Nano revolution in cardiovascular health: nanoparticles (NPs) as tiny titans for diagnosis and therapeutics. Curr Problems Cardiol 102466

Acosta-González A, Marqués S (2016) Bacterial diversity in oil-polluted marine coastal sediments. Curr Opin Biotechnol 38:24–32

Adetunji CO, Anani OA (2021) Bioaugmentation: a powerful biotechnological techniques for sustainable Ecorestoration of soil and groundwater contaminants. Microbial Rejuven Pollut Environ 1:373–398

Akinpelu AA, Ali MdE, Johan MR, Saidur R, Qurban MA, Saleh TA (2019) Polycyclic aromatic hydrocarbons extraction and removal from wastewater by carbon nanotubes: a review of the current technologies, challenges and prospects. Process Safety Environ Protect 122:68–82

Avornyo A, Chrysikopoulos CV (2024) Applications of graphene oxide (GO) in oily wastewater treatment: recent developments, challenges, and opportunities. J Environ Manage 353:120178

Banet G, Turaani AK, Farber R, Armoza-Zvuloni R, Rotem N, Stavi I, Cahan R (2021) The effects of biostimulation and bioaugmentation on crude oil biodegradation in two adjacent terrestrial oil spills of different age, in a hyper-arid region. J Environ Manage 286:112248

Basarkar V, Govardhane S, Shende P (2022) Multifaceted applications of genetically modified micro-organisms: a biotechnological revolution. Curr Pharm des 28(22):1833–1842

Dell'Anno F, Rastelli E, Sansone C, Brunet C, Ianora A, Dell'Anno A (2021) Bacteria, fungi and microalgae for the bioremediation of marine sediments contaminated by petroleum hydrocarbons in the omics era. Microorganisms 9(8):1695

Dhaka A, Chattopadhyay P (2021) A review on physical remediation techniques for treatment of marine oil spills. J Environ Manage 288:112428

Djønne K (2023) Amoco Cadiz and the early days of EC maritime regulation. In: The political economy of maritime safety: EU regulation, ship classification, and the international regime. Springer, Cham, pp 33–66

Gao L, Gu J-D (2021) A new unified conceptual framework involving maintenance energy, metabolism and toxicity for research on degradation of organic pollutants. Int Biodeterior Biodegradation 162:105253

Guidi P, Bernardeschi M, Scarcelli V, Lucchesi P, Palumbo M, Corsi I, Frenzilli G (20220) Nanoparticled titanium dioxide to remediate crude oil exposure. An in vivo approach in dicentrarchus labrax. Toxics 10(3):111

Jiang Z, Tijing LD, Amarjargal A, Park CH, An K-J, Shon HK, Kim CS (2015) Removal of oil from water using magnetic bicomponent composite nanofibers fabricated by electrospinning. Compos Part B: Eng 77:311–318

Joeckel J (2023) The Ashland oil spill. In: Above ground storage tank oil spills. Gulf Professional Publishing, pp 263–290

Lee JG, Larive LL, Valsaraj KT, Bharti B (2018) Binding of lignin nanoparticles at oil–water interfaces: an ecofriendly alternative to oil spill recovery. ACS Appl Mater Interfaces 10(49):43282–43289

Nikolaivits E, Dimarogona M, Fokialakis N, Topakas E (2017) Marine-derived biocatalysts: importance, accessing, and application in aromatic pollutant bioremediation. Front Microbiol 8:265

Nudrat S, Archana Y, Hasan SS (2018) A review study of advances in the science and technology of carbon nanotubes. Res Rev J Phys 5:1–6

Osadebe AU, Akinrodoye TI, Ogugbue CJ, Okpokwasili GC (2022) Green synthesised iron oxide nanoparticles decorated on biochar for enhanced natural attenuation in simulated petroleum compromised soil. Nanotechnol Environ Eng 7(2):517–528

Owoseni O, Zhang Y, Su Y, He J, McPherson GL, Bose A, John VT (2015) Tuning the wettability of halloysite clay nanotubes by surface carbonization for optimal emulsion stabilization. Langmuir 31(51):13700–13707

Pan Z, Cao S, Li J, Du Z, Cheng F (2019) Anti-fouling TiO_2 nanowires membrane for oil/water separation: synergetic effects of wettability and pore size. J Membr Sci 572:596–606

Patowary R, Devi A, Mukherjee AK (2023) Advanced bioremediation by an amalgamation of nanotechnology and modern artificial intelligence for efficient restoration of crude petroleum oil-contaminated sites: a prospective study. Environ Sci Pollut Res 30(30):74459–74484

Pete AJ, Bharti B, Benton MG (2021) Nano-enhanced bioremediation for oil spills: a review. ACS ES&T Eng 1(6):928–946

Rehman SA, Hussain N, Shahbaz A, Smułek W, Jesionowski T (2024) Nanotechnology for remediation of oilfield and refineries wastewater. In: Nanotechnology for oil-water separation. Elsevier, pp 83–104

Saha RC, Reza A, Hasan MS, Saha P (2019) A review-bioremediation of oil sludge contaminated soil. In: E3S web of conferences, vol 96. EDP Sciences, p 01004

Sehrawat M, Singh V, Rani M, Kalra C, Bharadwaj S, Rani R, Bisht A, Singh BP (2024) Nano-welded carbon nanotube sponges for efficient oil spill remediation. J Clean Prod 142841

Sengupta A, Gupta NK (2017) MWCNTs based sorbents for nuclear waste management: a review. J Environ Chem Eng 5(5):5099–5114

Sharma V, Ramish A, Sahu O (2021) Oil spill recovery techniques in petroleum industry: a review on treatment process. J Oil Gas Petrochem Sci 3(1):1–5

Shyam B, Phukan R, Das C (2022) Bioremediation of oil spill cleanup: case studies of two major oil spills. In: North-east research conclave Springer, Singapore, pp 117–133

Sorour MH, Hani HA, Al-Bazedi GA, El-Rafei AM (2016) Hydrophobic silica aerogels for oil spills clean-up, synthesis, characterization and preliminary performance evaluation. J Porous Mater 23:1401–1409

Sun X, Fu H, Bao M, Zhang F, Liu W, Li Y, Li Y, Lu J (2023) Preparation of slow-release microencapsulated fertilizer-Biostimulation remediation of marine oil spill pollution. J Environ Chem Eng 11(2):109283

Upadhyay S (2024) Bioremediation: a sustainable solution (2024)

Vannuccini ML, Della Torre C, Grassi G, Zambonin C, Cotugno P, Leaver MJ, Corsi I (2024) Nano-TiO_2 reduces bioavailability and biotransformation responses to crude oil WAF-associated PAHs in the European sea bass Dicentrachus labrax. Marine Pollut Bull 209:117265

Zamparas M, Tzivras D, Dracopoulos V, Ioannides T (2020) Application of sorbents for oil spill cleanup focusing on natural-based modified materials: a review. Molecules 25(19):4522

Zhu X, Tian S, Cai Z (2012) Toxicity assessment of iron oxide nanoparticles in zebrafish (Danio rerio) early life stages. e46286

Chapter 8
Chitosan and Cellulose Smart Bionanocomposite for Cardiovascular Devices

Amrita Palai and Amarendra Nayak

Abstract This manuscript boons wide ranging appraisal of cellulose–chitosan-based bionanocomposite biomaterial in regulation of cardiovascular disease. This biocomposite gained attention in current epoch due to their biocompatibility, biodegradability, anti-microbial, anti-fungal characteristic features. This paper comprises contemporary fabrication technique intended to enhance mechanical, chemical and biological properties of the bionanocomposite. In biomedicine sector, these bionanocomposite are extensively applicable for drug delivery as well as tissue engineering usage, aid in discharge of active material in controlled manner. This paper encapsulates application of innovative polymeric bio materials for designing diverse cardiovascular devices essential for diagnosis and regulation of cardiovascular disease. The biomaterial is used for the fabrication and analysis of artificial blood vessels, coronary artery stents with biosensors for coronary artery disease management, cardiac patches for heart failure management. This also evaluates enhancement, novelty and sustainable treatment preferences. Finally, this article represents the excellency of the novel polymeric bionanomaterials for different diagnostic and treatment modalities for cardiovascular disease. The progressive evolvement in cardiovascular medicine area paves the way for the development of treatment and diagnostic technique of cardiovascular disease.

Keywords Chitosan · Cellulose · Cardiovascular devices · Biological properties

A. Palai (✉)
Department of Chemistry, Kujang College, Kujang, Odisha, India
e-mail: amritapalai6@gmail.com

A. Nayak
Department of Chemistry, Ravenshaw University, Cuttack, Odisha, India

S. K. Panda and V. P.M. (eds.), *Bionanocomposites from Marine Origin*, Advanced Structured Materials 247, https://doi.org/10.1007/978-3-032-10749-7_8

8.1 Introduction

The current progress in biomedicine region generates prospects for enhancement of diagnosis and administration of cardiovascular disease. In past decades, the application of novel polymeric bionanomaterial is considered as promising material in dealing with cardiovascular disease. In present epoch, lot of effort are deployed for the fabrication of sustainable bionanocomposite because of its renewability, cost-effective, biodegradability, non-petroleum based, low carbon dioxide release, abundance and unique physic-chemical properties. Keeping this in the mind, two most commonly natural polysaccharide, namely chitosan and cellulose are considered as significant biomaterial. Chitosan is generally found in the cell wall of fungi, exoskeletons of insects, crustaceans. It is a straight chain polymer constituting of 1,4 N-acetylglucosamine. It is mainly insoluble in general solvents due to its superior crystalline structure, and hydrogen bonding takes place in between carbonyl, hydroxyl and acetamide groups (Moon et al. 2011; Yen et al. 2009, 2008; Szymanska 2015; Jayakumar et al. 2010a). Chitosan has widely used for diverse applications due to its biocompatibility, biodegradability, less expensive, abundantly available in nature and muco adhesiveness nature. It is found to have significant position in the biomedicine field as it holds anti-microbial activity, wound healing properties and haemostatic activity and thickener in food industry. It has deprived of good mechanical properties which reduces its applicability in the medicine. So, it is reinforced with nano cellulose to enhance its applicability. Cellulose was found to be straight chained homo polymer constituting D 1,4 sugar substrate joined by glycoside bond. Constitute nano sized micro fibrils enclosed by lignin and hemi cellulose. Nano cellulose is broadly two types cellulose nano whisker and cellulose nanofiber. It has wide range usage in biomedicine because of low density, cost-effective, abundantly available in nature, superior mechanical properties, high surface area, considerable flexibility, definite surface barrier feature and small thermal expansion. Nano cellulose conveys rigidity to the nanocomposite even in reinforced in polymer matrix at low temperature because of its excellent optical, rheological and characteristic features and capable of creating network structure by hydrogen bonding. The chitosan cellulose composite has a superior mechanical property which helps it in making promising tool in the drug delivery as well as tissue engineering application (Fernandes et al. 2011; Habibi et al. 2010; Helbert et al. 1996; Liu et al. 2011, 2013). The fabricated bionanopolymers are used in development of artificial blood vessel, coronary artery stents with biosensor for coronary artery disease regulation and cardiac patches for the management of heart failure. The ample number of visions are made to generate novel, sustainable treatment route with biomaterial. The wide spectrum of exploration was done to enhance the novelty of bionanocomposite used for diagnosis and treatment of various cardiovascular diseases.

8.2 Introduction to Chitosan and Cellulose Bionanocomposite

Chitosan and cellulose bionanocomposite are utmost natural abundantly found bio material on earth crust. These are linear, structural polysaccharide found in nature. Both biomaterials are immensely vital in the expansion of material due to its derivatization, extraction and solubilization process. This biomaterial is pertinent to all aspects of human existence, namely food industry, health issue, cloth fibre, medical equipment, pharmaceutical companies, purification equipment, filtration material and construction stuff (Klemm et al. 2005; Crini 2019; Hasegawa et al. 1992; Isogai and Atalla 1992; Rogovina et al. 1999; Holmberg et al. 1997; Myllytie et al. 2009; Vikhoreva et al. 2000; Yang et al. 2002; Shimahara et al. 1989). Amalgamation of two polymers create rigid, straight chained, superior functional molecule having same arrangement with diverse characteristic features. Cellulose mainly represents hydrophillicity nature, structural and mechanical characteristic features but chitosan exhibit ability of electron donor and anti-bacterial activity due to presence of amino group. The development of cellulose-chitosan molecular structure vital for the production of nanofilm, aerogel, hydrogel, nano membrane and nanofibres. The biocompatible and biodegradable polymer are extensively used in the place of the petroleum-based bio polymer as it is eco-friendly and helps in reducing consumption of carbon (Huq et al. 2022; Madni et al. 2021; Casadidio et al. 2019; Jaworska et al. 2018; Lim and Hudson 2003; RaviKumar 1999; Nikolov et al. 2011; Khattak et al. 2019; Usman et al. 2016; Santos et al. 2020). The linkage of two biocomposite, namely cellulose and chitosan leads to the wide spectrum of application in this current age as it exhibits sustainability, renewability, non-toxicity nature, biodegradability properties. The development and analysis of different molecule give rise to the formation of porous molecular structure, namely aerogels, hydrogels, nano membrane, foams, nanofilms, nanofibers (Peter et al. 2021; Tao et al. 2020; Rinaudo 2006; Muzzarelli 1974). It was found that fabrication of different composite requires coating and generation of thin films on molecular structure.

Chitosan was naturally found in the cell wall of the fungi. It is made up of copolymer of N-acetyl-D-glucosamine and glucosamine. It is well known for arrangements of microfibril embedded on the protein matrix which helps in the development of the hierarchical composite for skeletal structure. The chief functional group of the chitosan is hydroxyl group, amino group and N-acetyl groups, which comprises the intermolecular and intra molecular interaction on chitosan structure. It has found ample application in various field, namely pharmaceutical companies, medicine, food packaging, agriculture, biochemistry and waste management. Cellulose was abundantly found polymer on nature which present in the cell wall of plant cell. The cellulose mainly found in plant as cotton fibres, wood, annual plants, weeds, bamboos and animal sources include sea organism, tunicates and micro-organism (Roberts 1992; Sivashankari and Prabaharan 2017; Anwar et al. 2017). Cellulose was found to be main structural component in the plant cell wall which makes its mechanical strength. The properties like hydrophilicity, chirality, biodegradability and biocompatibility are responsible for its extensive usage in the medicinal field. By combining two biopolymer enhancement in the diverse properties occur which leads to the increasing efficiency of biomaterial in various application.

8.3 Fabrication of Chitosan Cellulose Bionanocomposite

Generally, distinctive procedure is extensively used to fabricate chitosan-based bionanocomposite which is described along with their fabrication method, characteristics features and applications.

Solution-Casting Method.

The solution-casting method is known to be quickest method and it involves three phases to fabricate bionanocomposite. Firstly, dispersion of the nanofiller in an appropriate solvent medium through the mechanical agitation and sonication process. The solvent is required to melt the produced nanocomposite. Finally, the polymer solution is blended with the nano filler solution at suitable given temperature. This mixture undergoes casting or precipitation to produce bionanocomposite. Chitosan–gold bionanocomposite is produced by using this technique (Amaral et al. 2005; Gogoi et al. 2020; Kaur et al. 2020; Aranaz et al. 2010).

In situ technique.

This method was found to be most efficient method as it produces nanocomposite in homogeneity dispersed constituents. In this method, nano filler components uniformly distributed in the monomer in the presence or absence of the solvent. The accurate polymerization undergoes incorporation of therapeutic component to the monomer at suitable given temperature. Chitosan-grafted poly acrylo nitrile silver nanocomposite are generated by the chemical reduction of silver ions in graft copolymerization of acrylonitrile onto chitosan film (Li et al. 2019; Jayakumar et al. 2007; Mourya et al. 2010; Manimohan et al. 2020; Wang et al. 2020b).

Electrospinning technique

The electrospinning technique can produce very thin nanofibres. This instrument comprises three main compartment, namely high-voltage electricity source, multineedle containing spinneret and grounded conductor as a collector. The liquid used in this process was filled into syringe at specific rate regulated by syringe pump. The presence of immobilizing charges on the liquid droplet surface helps in generating continuous jet. Fast tracked whipping and stretching of the fluid filament deploys the generation of nanofiber. This method is used to synthesize chitosan-polyethene oxide-based fibre by incorporating chitin nanocomposite (Xia et al. 2020).

Freeze drying technique

This process involves pouring of homogeneous mixture of polymer and filler solution to the copper mould which leads to the quenching of liquid nitrogen at exceptionally low temperature. The polymer substrates are then freeze dried to remove the excess solvent. This polymer structure is widely applicable in tissue engineering. The chitosan based on nanocomposite with halloysite nanotube are fabricated by combining freeze-drying as well as solution-casting technique (Shukla et al. 2013).

8.4 Chitosan and Cellulose Properties and Modifications

Chitosan was found in ample amount in nature. It originates from exo skeletons of shrimps, crabs, cuticles of insects and cell wall of fungi. It is used in the biomedicine in the form of deaceylated derivative by the action of chemical or enzymatic hydrolysis. This polymer was approved by food and drugs administration with little molecular weight and crystallinity value was found to be more than chitin. The structural representation is analogous to cellulose structure where hydroxyl group on two position carbon atom is replaced by amino group. Chitosan is an eco-friendly, biocompatible polymer used as effective and safe product in diet accepted by science protocol in agreement with good manufacturing practice regulation.

Physiochemical properties

Chitosan was found to have array of diverse characteristic feature such as chelation, viscosity, solubility in various medium, formation of nano film, poly electrolytic behaviour and formation of muco adhesively. The morphology of chitosan plays vital role in current time as it influences physiochemical properties. The diverse feature of chitosan constitutes high content of acetylation and low amount of crystalline surface. This property helps in making it potential candidature for different application such as in pharmaceutical industries and biotechnology field. This characteristics property was affected by various reason likely the content of crystallinity, molecular weight, degree of acetylation (DD), degradation procedure (Ashrafi et al. 2020). According to the presence of molecular weight, high molecular weight was found to be 190 to 375 kDa with DD and low molecular weight—have range from 20 to 190 kDa with DD. It was examined that rate of degradation of chitosan increases with decrease in DD and it is reliable on the order, distribution of acetyl group present on the molecule (Shih et al. 2019; Yang et al. 2008; Abdel-Mohdy et al. 2016). It comprises of presence of three chemically reactive group such as amino group and two hydroxyl group in individual glycosidic chain. Therefore, amino group was found to have superior importance since it helps in pH regulation, accountable to the cationic characteristics of chitosan species and administration of different physiochemical properties. The solubility of chitosan reliable on the presence of amino group at low pH level allowing chitosan molecule to dissolve and create bond as soluble cationic polysaccharide while amino group having PH greater than 6, undergo deprotnation helps in making it insoluble. 16 The solubility property influenced by arrangement of acetyl group on the chain, deacetylation procedure and strength of ion. At acidic Ph it is more capable towards showing chelating feature for different metal ions 36–41. This process need independent amino group created by electrostatic force of attraction on acidic solution.

Biological properties

Chitosan was found to be non-toxic, biocompatible, biodegradable polymer showing unique biological characteristic feature such as anti-inflammatory, anti-bacterial, anti-tumour, anti-fungal, haemostatic, analgesic (Ghosh and Ali 2012; Dash et al. 2011; Sarmento and Das Neves 2012; Kumirska et al. 2011). It was found to have excellent anti-bacterial effect on diverse range of bacteria causing diseases because of its cationic nature which helps in the communication between negatively charged lipid and protein present in the cell wall of the bacteria. The diffusion of chitosan into the cell membrane helps in elongation and disruption of the permeability of the membrane which finally encourages leakage in cytoplasm and death of bacterial cell. The ability to have bonded with DNA when it comes in contact with bacteria nuclei and prevents RNA synthesis, due to its anti-bacterial property. It shows anti-fungal property as its electrostatic reaction takes place phospholipid present in the cell membrane. The DD and molecular weight play a vital role in monitoring anti-fungal properties of chitosan. The anti-inflammatory action of chitosan found to be lessening the release of interleukin-8 and tumour 834, necrosis factor from mast cell (Zhang et al. 2014a; Shirvan et al. 2018; Aizat and Aziz 2018; Kumari et al. 2016; Younes and Rinaudo 2015). It has advantages over NSAIDs as it does not produce gastric side effect due to the presence of free amino group helps in the making defensive shield over the stomach. It has ability to cure connective tissue as it endures acid hydrolysis, generation of glucosamine, monosaccharide, and production of proteoglycan structure of connective tissue and cartilage which helps in tissue repairing (Bakshi et al. 2020; Knidri et al. 2020; Moreno et al. 2018; Husain et al. 2017). Chit oligosaccharides were to have anti-tumour effects because of enhanced action of lymphocyte. It increases immunity response and alters the function of immune competent cell. Chitosan containing positive charge endorse adhesion of erythrocyte, adsorption of fibrinogen and activation and adhesion of platelet. This property of chitosan is known as haemostatic property because of amount of polyatomic ion presence and binding of nonspecific plasma membrane. The unique characteristic feature and anti-microbial action has helped in development in diverse area like food industry, cosmetic material, pharmaceutical industry and biomedical sphere. Modified chitosan was produced by chemical and enzymatic reaction to generate novel depolymerize derivative to overcome solubility in many solvents and water. Several aspects influence modification of chitosan such as molecular weight of chitosan, viscosity property and reaction condition. Chitosan is conjugated with other biomolecules and polymers due to the presence of amino groups and hydroxyl groups, various methods such as polymer blending and graft copolymerization are used for chitosan modification. Blending is the process in which new biomaterial is formed by joining two polymers. Grafting process involves covalently bonding of monomers and polymerization of side chain occurs and this process conveys different functional group to the polymer.

8.5 Modification of Chitosan and Cellulose Bionanocomposite

Physical Modification

This is the quickest technique for the alteration of polymer which is done by amalgamation of two polymer generating original substance with diverse physical properties. This method is used to generate composites with optimal, chemical, mechanical, structural, biological and morphological characteristic features. This procedure is effective, reasonable and small time-consuming than other existing method. Chitosan is well known for blending with certain hydrophilic polymer such as polyvinyl alcohol, polyethyl oxide and polyvinyl pyridine. Chitosan and synthetic PVA are well known for blending motive to increase the mechanical, barrier features of chitosan and PVA mixing film. This blending enhances mechanical properties of medicine and helps in regulating drug delivery system (Azmana et al. 2020; Kim 2018; Chang et al. 2019).

Chemical modification

The chemical modification of chitosan-based nanocomposite was done by the different chemical reaction occur by the presence of reactive amino group and hydroxyl group are described below.

Phosphorylation

This technique is helpful in increasing drug loading efficacy, higher degree of swelling and enhancing discharge rate of pH. This process involves phosphorylation of chitosan surface membrane by triethyl phosphate, butanol and phosphoric acid which increase ammonium concentration making it amphoteric in nature. This characteristic feature is considered to have application in anti-cancer agent (Ardila et al. 2017; Wang et al. 2020a; Zhang et al. 2014b; Taher et al. 2019).

Thiolation

This process involves the linkage of covalently bonded thiol group with skeleton of chitosan. This polymer enhances regulation of hydrophilic molecule and higher capabilities of penetration, superior permeation and enhanced cytotoxicity (Ashrafi et al. 2020; Shih et al. 2019).

Hydroxyl alkylation

This process involves reaction of chitosan with ethylene oxide butylene oxide and propylene oxide and glycidol. This modification increases solubility of water in the polymer. The final hydroxybutyl chitosan molecule exhibits excellent hygroscopic nature and retention of moisture, enhances immunocompetence activity and shows superior influence on anti-bacterial activity against gram-positive and gram-negative bacteria (Abdel-Mohdy et al. 2016; Bektas et al. 2020).

Sulphation

According to the process, the skeleton of carbohydrate is made with sulphamate groups and sulphate ester groups. This anionic derivative is water soluble in nature which blocks malicious human melanoma cells. This derivative also helps in the absorption of metal ions, drug delivery, anticoagulation of blood and resistance of haemaglutination. This polymer shows anti-coagulant activity because of similar structure with heparin (Bano et al. 2017; Elsabee and Abdou 2013; Kong et al. 2010).

Carboxylation

This modification produces different car boxy polymers namely carboxymethyl chitosan and car boxy ethyl chitosan. Macro molecule comprising of superior solubility and growth characteristics features is extensively used in the drug delivery system (Verlee et al. 2017; Park and Kim 2010).

Alkylation

This process has represented an alkyl group in the chitosan-based nano skeleton making it alkyl derivative of chitosan-based nanocomposite. The presence of alkyl group in the chitosan chain well known as amphiphillic derivative which weakens the intermolecular hydrogen bonding in the polymer enhancing its solubility features. Alkylated chitosan is applicable in paper making additives. It is applicable in the wound healing as it possesses coagulation and anti-bacterial properties. It is also used water purification system to absorb anionic surfactants (Xia et al. 2011; Lodhi et al. 2014; Singh et al. 2017; Wang 2019; Okamoto et al. 2002; Sherazi 2016; Eidelman and Simon 2004).

Quaternization

This modification process involves the incorporation of quaternary ammonium group into hydroxyl or amino group of the chitosan. Quaternary ammonium chitosan salt exhibits superior solubility and anti-bacterial properties which employs the function ability in nano biomedicine. Hydroxypropyltrimethyl ammonium chloride chitosan is extensively used in the bio engineering as it alter respiratory activity of mitochondria (Meredith et al. 2003; Abolhasani et al. 2010; Khor and Lim 2003; Khoo et al. 2003; Park et al. 2001; Kumari and Singh 2012; Amaral et al. 2005).

Oligomerization

This process undergoes chemical and enzymatic hydrolysis to produce water soluble, low viscous polymer named chitosan oligomer. Chemical hydrolysis comprises deamination of nitrous acid, while enzymatic hydrolysis is done in the presence of chitin deacetylase enzyme. The synthesized material performs superior anti-fungal features which increases with lessening of the polymerization of chitosan oligomer (Kaur et al. 2020).

Graft copolymerization

This technique is used to synthesis graft copolymer by the free radical polymerization in water acetic medium of chitosan in the presence of the initiator. This procedure produces grafted chitosan derivative by modifying polymer skeleton structure which has applicability in tissue engineering, gene transfer and drug delivery system (Aranaz et al. 2010; Li et al. 2019; Jayakumar et al. 2007; Mourya et al. 2010; Manimohan et al. 2020).

Enzymatic Modification

In comparison with the other modification method, this method gained attraction of researchers due its specificity nature and eco-friendly and sustainability. The final fabricated product shows enhancement in anti-oxidant, anti-microbial and thermal degradability features (Manimohan et al. 2020; Wang et al. 2020b; Bobu et al. 2011; Xia et al. 2020; Jeon et al. 2000; Shukla et al. 2013).

8.6 Smart Functionalities in Bionanocomposite for Cardiovascular Applications

Bionanocomposite plays a crucial role in the diagnosis and administration of cardiovascular disorders. Bionanocomposite is group of material having at least one dimension of biological origin and one constituent particle exist in nanometer range. It proposes ample amount of benefits in cardiovascular application comprising improvement in the mechanical and surface characteristic features of implantation, encouraging haemocompatibility, calcification and decreasing possible rate of thrombosis. This material can be used in fabrication of different cardiovascular devices namely heart valves, bypass grafts and stents.

The major usage of bionanocomposite in the application of cardiovascular condition is discussed intensely by acknowledging its need and accountability in present epoch.

Implant Enhancement

Implant enhancement is well known devices employed surgically in the blood vessels and heart used for treatment of different cardiovascular illness. The main aim of this instrument is to enhance efficacy and safety of cardiac implantation. This encompasses improvement in designing devices, materials, techniques, approaches in reducing infection and enhancing long lasting effects. The various type of cardiovascular implantation includes Implantable Cardioverter Defibrillators (ICDs), Pacemakers, Cardiac Resynchronization Therapy (CRT) Devices, Implantable Loop Recorders (ILRs), Ventricular Assist Devices (VADs), Coronary stents, Heart Valves and Stent Grafts.

Advantages of this procedure

- It enhances procedures that increase heart functionality, lessening complication which enriches patient's quality of life.
- It helps in minimizing risk of infection and malfunctioning of devices.
- It helps in maintaining longevity of the devices used.
- It helps in offering personalised treatment and medication considering needs of patients.

Haemocompatibilty

This condition states the capability of the employed devices to communicate with blood cell without causing unfavourable effects.

Main features of haemocompatibility are stated below:

- The different test is conducted to haemocompatibility that means the interaction of blood with device to know capability to prevent blood clotting, destruction of red blood cell and immunity.
- Haemocompatibility is used to measure safety and effectiveness of the instrument used in the medicinal field.
- Surface morphology likely structure of material, chemical reaction determines hem compatibility.

- Haemocompatibility test can be difficult due to diversified characteristic features of the red blood cell.
- Numerous hostile effects are seen likely platelet activation and thrombus formation by the communication of blood cell with poly biomaterials.

Antithrombogenicity

Antithrombogenicity is a property of biomaterial which helps in the minimizing the clotting of the blood within the blood vessel. This is one of the characteristic properties of the blood cell interacting biomaterial as it plays vital in the prevention of risk of thrombosis by safeguarding its biocompatibility property.

Main characteristic features of Antithrombogenicity

- Anthrombogenic biomaterial is used to reduce blood cell clotting which leads to the smooth flow of blood cell within blood vessel by minimizing complications like embolism.
- These materials are necessary for the enduring biocompatibility nature by preventing thrombosis.
- It is widely applicable for the improving performance of the medical instrument and methods by focusing in the reduction of the risk of thrombosis.

Tissue engineering

Tissue engineering seems to be an interdisciplinary area in biomedicine which uses principle of the life science as well as engineering for development of biologically substitutes of injured and unhealthy tissues and organs. This technique is used for linking cell and biomaterial to produce potent organs and effective tissues. The main components of this method are:

Cells

The tissue formation is done by using patient's individual cell or other cell by tissue engineering process.

Scaffolds

These structures give supportive system for the cell growth as well as differentiation and imitating the natural extracellular matrix.

Growth factor

This technique depends on some characteristics, namely cell growth, differentiation and arrangement of tissue.

Applications

Tissue engineering is helpful in substituting unhealthy organs and tissues such as skin, bone, cartilage and blood vessel

This method is helpful in organ transplantation by replacing organ.

This technique is used to assess the effectiveness of the modelled drugs.

Drug Delivery

Drug delivery is the term which aims on delivery of drug to specific targeted tissue with efficacy for the administration of the cardiac illness. This method focusses on the improvement of the drug efficiency, reducing side effects and enhancing compliance of the patient. This method uses specific strategies and procedure to achieve targeted drug delivery, namely drug-eluting stents and drug-coated balloons, nanoparticles, implantable devices, theranostics, extracellular vesicles and magnetic properties.

Smart Biomaterials

Smart material termed as biomaterial which alter vigorously alter the characteristic properties by responding to the stimulus by improving applicability of heart related issues. These materials are explored for diverse application namely cardiac tissue regeneration, drug delivery and implantable devices. The different types of the smart biomaterial are shape memory alloy, shape memory polymer, bioconjugates and hydrogels.

8.7 Application in Cardiovascular System

It can be used to personalized individuals heart valve which can imitate structure as well as application of natural valve.

These materials are used to produce cardiac patch which can be implanted in the place of damaged tissue of the heart.

These materials are applied to transfer drugs directly to the heart for the generation of new blood vessel.

Atherosclerosis treatment

This is a term which signifies the blockage of blood to different body parts including brain and heart organ. This obstructs with usual working of the body organ and condition related to limited flow of blood which was termed as carotid artery disease. Biomaterial named statins are used for the treatment of atherosclerosis.

Cardiac tissue engineering

Cardiac tissue engineering emphasis on the substituting and repairing affected heart with the help of engineered biomaterial and cell. This aims at producing new heart tissue which resembles structure and function of the present myocardium. This area of nano science aims at generating therapeutics treatment for heart failure and cardiovascular conditions.

Application in cardiac system

- This system used to replace damaged tissue with new functional myocardium.
- This method is applicable for the prevention of the heart failure.
- The new formed tissue is helpful for restoring and improving the pumping capacity of the heart.
- This system reduces the possibility of heart transplantation.

8.8 Bionanocomposite for Designing Cardiovascular Devices

The linkage of biopolymer and nanomaterial is used to generate cardiovascular device which provide enriched characteristic feature and biocompatibility nature. These renewable biomaterials are extensively applicable in biomedical function encompassing drug delivery, tissue engineering and wound healing properties.

Rapid progress in the field of nanomedicine has led to formation of wide array of unique pair of materials to confront as solutions to diverse cardiac repair compliance. Amidst this, vascular graft is broadly used for the regulation of cardiac conditions.

Vascular grafts

Vascular grafts are fabricated artificial blood vessels by using bio nanomaterial which straight supply blood to the ischaemic portion of the heart and repairing and substituting infected blood vessels. In case of myocardial ischaemia, a medical term for the widening and blockage of coronary artery, vascular grafts are used to construct passage to avoid the diseased vessels and restore blood supply to the ischaemic portion. The two types of material are used for the fabrication of the vascular grafts, namely natural vascular tissue and synthetic polymers. The natural vascular tissue comprises arteries and veins fragments obtained from animal and human having excellent biocompatibility and biological application. This process needs surgical procedure which leads to the risk of injury and related to issue compiled with the donor compatibility during transplantation process. To surpass these challenges, unique procedure and strategies undergo examination and it was found that formation of thrombosis, infection risk, long inadequacy of long-term stability. The artificially generated bionanomaterial containing biodegradable polymer gelatin, gelatin-PEG, gelatin-PLGA copolymer gained lot of attraction of the researchers. Naturally produced polymer, namely collagen, chitosan, sodium alginate has intrinsic biodegradability and biocompatibility nature. The development of vascular graft leads to generate prothrombotic atmosphere within the circulatory system increasing the risk of thrombotic complications leading to the vascular luminal occlusion. So, specific strategies are used to create vascular grafts from biomaterial which embedded with bioactive agents which minimizes the risk of restenosis and diminishes the negative effects of therapeutic procedures.

Cardiac Patch

A cardiac patch is a specific structured biomaterial focussed at myocardial defects, improving cell adhesion, proliferation and giving suitable mechanical support to regulate generation of new tissue in improving the function of heart organ. One of the vital features of cardiac patch is its customized structure which helps it imitate or act accurately on the defective portion of the individual's heart depending on its morphology and function. The materials are used in the synthesis of the cardiac patch is commonly divided into natural and synthetic materials having their own benefits and limitation with the characteristics features of superior biodegradability which encourages the prevention of refusal of implantation, incorporation of bioactive molecule enhances adhesion and regeneration of cardiac cell. The bioactive substance used in this procedure comprises biodegradable polymers such as gelatin, collagen, PLA, biodegradable scaffold materials likely biodegradable polymer nanofiber scaffolds, bioactive substances such as growth factors, cytokines and cellular materials, namely CMs, stem cells. These materials are used to fabricate cardiac patch to deliver suitable mechanical support system, enrich repair process, increase adhesion and proliferation which regulates new generation of tissue muscle. These patches generate appropriate physical and biochemical atmosphere for the growth and restructure of the myocardial cells. Recent progress in the field of the research leads to the evolvement of biomaterial having improved characteristics namely biocompatibility, degradability, mechanical properties constructed on conductive nanofiber yarn network aligned within the hydrogel shell resembles usual cardiac structure of natural heart. The customized structure, developed fabrication method and optimizing usage value, has helped it to become significant tool for enhancing patient recovery with cardiac disorders.

Hydrogel

Hydrogel is biomaterial which has capability of forming gel-like structure. It comprises biocompatible polymer and bioactive molecule shows attributes as adsorbent construction and good amount of water content. Ample number of research was done related to the fabrication method of the hydrogel by taking good polymeric materials, naturally derived polymer and gelation conditions leads to the formation of potent hydrogel structure. The biomaterial undergoes gelation process after the injection was done leads to the formation of three-dimensional structure with enhanced stability provide supports to the growth and regeneration of the tissue organ and regenerative medicine. The naturally produced polymer, namely silk fibroin and chitosan are extensively investigated for the tissue regeneration process owing to its biocompatibility and biodegradability features. The growth of the myocardial fibre in the cardiac tissue where injection of hydrogel was done specify the efficacy of the hydrogel which enhances neovascularization and cardiac regeneration which convey significant improvement in cardiac condition.

8.9 Future Prospects and Challenges

Challenges

In spite of having unique biological and physiochemical characteristics features, certain disadvantages make the biomaterial inefficient in application in few noticeable area of medicine. The major drawbacks which should be addressed immediately are recorded below.

- The physical properties like pH have low solubility which has disadvantage in medicinal sector.
- The safety and hygiene of the prepared biomaterial was not considered by European Food Safety Authority.

- The colloidal stability of the chitosan-based nanocomposite is low which delay the large-scale production in drug delivery system.
- The chitosan-based bionanocomposite has high elasticity features which hinder its application in the biomedicine.
- The chitosan-based bionanocomposite face lots of hurdles in the drug delivery purposes, namely efficacy of drug release, capacity of drug loading, degradation rate of used bionanocomposite and time period of drug delivery.
- Lastly, industries and processing unit have economic issue which hinders commercial opportunities of eco-friendly biopolymer globally.

Future Prospects

Nanochitosan can be used in agricultural lands for controlling pest, growing crops, preservation of seeds, generating innate immune system crops, drought resistance crops and salinity resistance crops. So, chitosan-based bionanocomposite is used instead of synthetic chemical as it protects natural land, environment as well as agricultural farm.

Chitosan-based nano carrier incorporated with antibodies has shown long-term non-toxicity which helps in the drug delivery to the brain organ. Further studies should be done to support this application.

Wound healing process was hindered by the irradiation of UV light as it inhibits growth of epithelial cell which should be acknowledged.

Chitosan-based bionanocomposite was found to be potent heterogeneous bio catalyst with encouraging results in the pharmacological as well as therapeutic treatment.

These biomaterial with colloidal instability act as major drawback in the large-scale drug delivery system which should be taken care and managed.

The focus should be given to produce biomaterial flexible elasticity.

The main source of the chitosan was nature. In order to meet needs of population, research should be done to produce alternative source for the chitosan.

The cost-effective and eco-friendly procedures should be deployed for the extraction other than use of chemical solvent in the industries.

8.10 Conclusion

In current era, bionanocomposite has gained attention in the medicinal research field due to its intrinsic attribute's likely biocompatibility, biodegradability features with enhanced structural and functional characteristics. Chitosan-based bionanocomposite was found to be promising tool in the biomedicinal field due to its unique properties. This review explained natural sources of chitosan, modification procedure, fabrication methods, usage in cardiovascular application and designing of cardiovascular devices. It also describes application in cardiac condition, advantages and disadvantages of the biomaterial. This paper also indicates the specific region where further research is required to improve physiochemical characteristics features of the chitosan-based bionanocomposite. The chitosan-based bionanocomposite having excellent characteristics features, namely mechanical, thermal, barrier properties, biocompatibility and biodegradability nature behaves as potent biomaterial in biomedicine field in sustainable manner.

References

Abdel-Mohdy FA, El-Bisi MK, Abou-Okeil A, Sleem AA, El-Sabbagh S, El-Shafei K, El-Sayed HS (2016) El Sawy SM Antimicrobial and haemostatic effect of chitosan/polyacrylic acid hybrid membranes. Egypt J Chem 59(1):45–57

Abolhasani MM, Arefazar A, Mozdianfard M (2010) Effect of dispersed phase composition on morphological and mechanical properties of PET/EVA/PP ternary blends. J Polym Sci B Polym Phys 48(3):251–259

Aizat MA, Aziz F (2018) 12 -chitosan nanocomposite application in wastewater treatments. Nanotechnol. Water Wastewater Treat Theory Appl

Amaral IF, Granja PL, Barbosa MA (2005) Chemical modification of chitosan by phosphorylation: an XPS, FT-IR and SEM study. J Biomater Sci Polym Ed 16(12):1575–1593

Anwar M, Anggraeni AS, Amin MHA (2017) Comparison of green method for chitin deacetylation. AIP Conf Proc 1823:020071

Aranaz I, Harris R, Heras A (2010) Chitosan amphiphilic derivatives. Chem Appl Curr Org Chem 14(3):308–330

Ardila N, Daigle F, Heuzey MC, Ajji A (2017) Effect of chitosan physical form on its antibacterial activity against pathogenic bacteria. J Food Sci 82(3):679–686

Ashrafi M, Bayat M, Mortazavi P, Hashemi SJ, Meimandipour A (2020) Antimicrobial effect of chitosan–silver–copper nanocomposite on Candida albicans. J. Nanostructure Chem. 10:87–95

Azmana M, Mahmood S, Hilles AR, Mandal UK, Al-Japairai KAS, Raman S (2020) Transdermal drug delivery system through polymeric microneedle: a recent update. J Drug Deliv Sci Technol 60:101877

Bakshi PS, Selvakumar D, Kadirvelu K, Kumar NS (2020) Chitosan as an environment friendly biomaterial—a review on recent modifications and applications. Int J Biol Macromol 150:1072–1083

Bano I, Arshad M, Yasin T, Ghauri MA, Younus M (2017) Chitosan: a potential biopolymer for wound management. Int J Biol Macromol 102:380–383

Bektas N, Senel B, Yenilmez E, Özatik O, Arslan R (2020) Evaluation of wound healing effect of chitosan-based gel formulation containing vitexin. Saudi Pharm J 28(1):87–94

Bobu E, Raluca N, Lupei M, Ciolacu FL, Desbri' eres J (2011) Synthesis and characterization of n-alkyl chitosan for papermaking applications. Cellul Chem Technol 619–625

Casadidio C, Peregrina DV, Gigliobianco MR, Deng S, Censi R, Di Martino P (2019) Chitin and chitosans: characteristics, eco-friendly processes, and applications in cosmetic science. Mar Drugs 17:369

Chang SH, Lin YY, Wu GJ, Huang CH, Tsai GJ (2019) Effect of chitosan molecular weight on anti-inflammatory activity in the RAW 264.7 macrophage model. Int. J. Biol. Macromol. 131:167–75

Crini G (2019) Historical review on chitin and chitosan biopolymers. Environ Chem Lett 17:1623–1643

Dash M, Chiellini F, Ottenbrite RM, Chiellini E (2011) Chitosan—a versatile semi-synthetic polymer in biomedical applications. Prog Polym Sci 36(8):981–1014

Eidelman N, Simon CG (2004) Characterization of combinatorial polymer blend composition gradients by FTIR microspectroscopy. J Res Natl Inst Stand Technol 109(2):219

ElKnidri H, Dahmani J, Addaou A, Laajeb A, Lahsini A (2019) Rapid and efficient extraction of chitin and chitosan for scale-up production: effect of process parameters on deacetylation degree and molecular weight. Int J Biol Macromol 139:1092–1102
Elsabee MZ, Abdou ES (2013) Chitosan based edible films and coatings: a review. Mater Sci Eng C 33(4):1819–1841
Fernandes SC, Freire CS, Silvestre AJ, Pascoal Neto C, Gandini A (2011) Novel materials based on chitosan and cellulose. Polym Int 60:875–882
Ghosh A, Ali MA (2012) Studies on physicochemical characteristics of chitosan derivatives with dicarboxylic acids. J Mater Sci 47(3):1196–1204
Gogoi P, Dutta A, Ramteke A, Maji TK (2020) Preparation, characterization and cytotoxic applications of curcumin-(±) a-lipoic acid coloaded phosphorylated chitosan nanoparticles in MDA MB 231 breast cancer cell line. Polym Adv Technol 31(11):2827–2841
Habibi Y, Lucia LA, Rojas OJ (2010) Cellulose nanocrystals: chemistry, self-assembly, and applications. Chem Rev 3479–3500
Hasegawa M, Isogai A, Onabe F, Usuda M, Atalla RH (1992) Characterization of cellulose-chitosan blend films. J Appl Polym Sci 45:1873–1879
Helbert W, Cavaille JY, Dufresne A (1996) Thermoplastic nanocomposites filled with wheat straw cellulose whiskers. Part i: Process Mech Behav Polym Compos 17:604–611
Holmberg M, Berg J, Stemme S, Ödberg L, Rasmusson J, Claesson P (1997) Surface force studies of langmuir-blodgett cellulose films. J Colloid Interface Sci 186:369–381
Huq T, Khan A, Brown D, Dhayagude N, He Z, Ni Y (2022) Sources, production and commercial applications of fungal chitosan: Areview. J Bioresour Bioprod 7:85–98
Husain S, Al-Samadani KH, Najeeb S, Zafar MS, Khurshid Z, Zohaib S, Qasim SB (2017) Chitosan biomaterials for current and potential dental applications. Mater (Basel) 10(6):602
Isogai A, Atalla RH (1992) Preparation of cellulose-chitosan polymer blends. Carbohydr Polym 19:25–28
Jaworska MM, St̨epniak I, Gali´ nski M, Kasprzak D, Binia´ s D, Górak A (2018) Modification of chitin structure with tailored ionic liquids. Carbohydr Polym 202:397–403
Jayakumar R, Nwe N, Tokura S, Tamura H (2007) Sulfated chitin and chitosan as novel biomaterials. Int J Biol Macromol 40(3):175–181
Jayakumar R, Menon D, Manzoor K, Nair SV, Tamura H (2010a) Biomedical applications of chitin and chitosan based nanomaterials—a short review. Carbohydr Polym 82:227–232
Jeon YJ, Shahidi F, Kim SK (2000) Preparation of chitin and chitosan oligomers and their applications in physiological functional foods. Food Rev Int
Kaur A, Kumar P, Kaur L, Sharma R, Kush P (2020) Thiolated chitosan nanoparticles for augmented oral bioavailability of gemcitabine: preparation, optimization, in vitro and in vivo study. J Drug Deliv Sci Technol 61:102169
Khattak S, Wahid F, Liu L-P, Jia S-R, Chu L-Q, Xie Y-Y, Li Z-X, Zhong C (2019) Applications of cellulose and chitin/chitosan derivatives and composites as antibacterial materials: Current state and perspectives. Appl Microbiol Biotechnol 103:1989–2006
Khoo CGL, Frantzich S, Rosinski A, Sj¨ostr¨om M, Hoogstraate J (2003) Oral gingival delivery systems from chitosan blends with hydrophilic polymers. Eur. J. Pharm. Biopharm 55(1):47–56
Khor E, Lim LY (2003) Implantable applications of chitin and chitosan. Biomaterials 24(13):2339–2349
Kim S (2018) Competitive biological activities of chitosan and its derivatives: antimicrobial, antioxidant, anticancer, and anti-inflammatory activities. Int J Polym Sci 2018(1):1708172
Klemm D, Heublein B, Fink H-P, Bohn A (2005) Cellulose: fascinating biopolymer and sustainable raw material. Angew Chem 44:3358–3393
El Knidri H, Laajeb A, Lahsini A (2020) Chitin and chitosan: chemistry, solubility, fiber formation, and their potential applications. In: Handbook of chitin and chitosan. Elsevier, pp 35–57
Kong M, Chen XG, Xing K, Park HJ (2010) Antimicrobial properties of chitosan and mode of action: a state of the art review. Int J Food Microbiol 144(1):51–63

Kumari S, Singh RP, Chavan NN, Annamalai PK (2016) Chitosan-based bionanocomposites for biomedical application. Bioinspired Biomim Nanobiomater
Kumari S, Singh RP (2012) Glycolic acid-g-chitosan-gold nanoflower nanocomposite scaffolds for drug delivery and tissue engineering. Int J Biol Macromol 50(3):878–883
Kumirska J, Weinhold MX, Thöming J, Stepnowski P (2011) Biomedical activity of chitin/chitosan based materials—influence of physicochemical properties apart from molecular weight and degree of N-acetylation. Polym (Basel) 3:1875–1901
Li J, Wu X, Shi Q, Li C, Chen X (2019) Effects of hydroxybutyl chitosan on improving immunocompetence and antibacterial activities. Mater Sci Eng C 105:110086
Lim S-H, Hudson SM (2003) Review of chitosan and its derivatives as antimicrobial agents and their uses as textile chemicals. J Macromol Sci 43:223–269
Liu D, Chen X, Yue Y, Chen M, Wu Q (2011) Structure and rheology of nanocrystalline cellulose. Carbohydr Polym 84:316–322
Liu T, Yang X, Wang ZL, Yan X (2013) Enhanced chitosan beads-supported Fe0-nanoparticles for removal of heavy metals from electroplating wastewater in permeable reactive barriers. Water Res 47:6691–6700
Lodhi G, Kim YS, Hwang JW, Kim SK, Jeon YJ, Je JY, Ahn CB, Moon SH, Jeon BT, Park PJ (2014) Chitooligosaccharide and its derivatives: preparation and biological applications. Biomed. Res. Int. 2014(1):654913
Madni A, Kousar R, Naeem N, Wahid F (2021) Recent advancements in applications of chitosan-based biomaterials for skin tissue engineering. J Bioresour Bioprod 6:11–25
Manimohan M, Pugalmani S, Sithique MA (2020) Biologically active water soluble novel-biopolymer/hydrazide based O-carboxymethyl chitosan schiff bases: synthesis and characterisation. J Inorg Organomet Polym Mater 30(9):1–19
Meredith JC, Sormana JL, Keselowsky BG, García AJ, Tona A, Karim A, Amis EJ (2003) Combinatorial characterization of cell interactions with polymer surfaces. J Biomed Mater Res A 66(3):483–490
Moon RJ, Martini A, Nairn J, Simonsen J, Youngblood J (2011) Cellulose nanomaterials review: structure: properties and nanocomposites. Chem Soc Rev 40:3941–4399
Moreno JAS, Mendes AC, Stephansen K, Engwer C, Goycoolea FM, Boisen A, Nielsen LH, Chronakis IS (2018) Development of electrosprayed mucoadhesive chitosan microparticles. Carbohydr Polym 190:240–247
Mourya VK, Inamdar NN, Tiwari A (2010) Carboxymethyl chitosan and its applications. Adv Mater Lett 1(1):11–33
Muzzarelli RAA (1974) Natural chelating polymers. Pergamon Press, Oxford, UK, p 254–27
Myllytie P, Salmi J, Laine J (2009) The influence of pH on the adsorption and interaction of chitosan with cellulose. BioResources 4:1647–1662
Nikolov S, Fabritius H, Petrov M, Friák M, Lymperakis L, Sachs C, Raabe D, Neugebauer J (2011) Robustness and optimal use of design principles of arthropod exoskeletons studied by initio-based multiscale simulations. J Mech Behav Biomed Mater 4:129–145
Okamoto Y, Kawakami K, Miyatake K, Morimoto M, Shigemasa Y, Minami S (2002) Analgesic effects of chitin and chitosan,. Carbohydr Polym
Park BK, Kim MM (2010) Applications of chitin and its derivatives in biological medicine. Int J Mol Sci 11(12):5152–5164
Park SY, Jun ST, Marsh KS (2001) Physical properties of PVOH/chitosan-blended films cast from different solvents. Food Hydrocoll 15:499–502
Peter S, Lyczko N, Gopakumar D, Maria HJ, Nzihou A, Thomas S (2021) Chitin and chitosan based composites for energy and environmental applications: a review. Waste Biomass Valorization. 12:4777–4804
RaviKumar MNV (1999) Chitin and chitosan fibres: a review. Bull Mater Sci 22:905–915
Rinaudo M (2006) Chitin and chitosan: properties and applications. Prog Polym Sci 31:603–632
Roberts GAF (1992) Chitin Chemistry. Macmillan, New York

Rogovina SZ, Vikhoreva GA, Akopova TA, Erina NA (1999) Properties of films made from cellulose-chitosan blends. Polym Sci Ser B 41:335–337

Santos VP, Marques NSS, Maia PCSV, Lima MABD, Franco LDO, de Campos-Takaki GM (2020) Seafood waste as attractive source of chitin and chitosan production and their applications. Int J Mol Sci 21:4290

Sarmento B, Das Neves J (2012) Chitosan-based systems for biopharmaceuticals: delivery. Target Polym Ther

Sherazi TA (2016) In: Drioli E, Giorno L (eds) Graft polymerization BT—encyclopedia of membranes. Springer, Berlin, Heidelberg

Shih P-Y, Liao Y-T, Tseng Y-K, Deng F-S, Lin C-H (2019) A potential antifungal effect of chitosan against Candida albicans is mediated via the inhibition of SAGA complex component expression and the subsequent alteration of cell surface integrity. Front Microbiol 10:602

Shimahara K, Takiguchi Y, Kobayashi T, Uda K, Sannan T (1989) Screening of mucoraceae strains suitable for chitosan production. In: Skjak-Braek G, Anthonsen T, Sandford P (eds) Chitin and chitosan: source, chemistry, biochemistry, physical properties and application. Elsevier, pp 319–332

Shirvan AR, Shakeri M, Bashari A (2018) Recent advances in application of chitosan and its derivatives in functional finishing of textiles. Impact Prospect Green Chem Text Technol 1:107–133

Shukla SK, Mishra AK, Arotiba OA, Mamba BB (2013) Chitosan-based nanomaterials: a state-of-the-art review. Int J Biol Macromol

Singh R, Shitiz K, Singh A (2017) Chitin and chitosan: biopolymers for wound management. Int Wound J 14(6):1276–1289

Sivashankari PR, Prabaharan M (2017) Deacetylation modification techniques of chitin and chitosan. In: Chitosan based biomaterials, vol 1. Elsevier, pp 117–133

Szymanska E, Winnicka K (2015) Stability of chitosan—a challenge for pharmaceutical and biomedical applications. Mar Drugs 13:1819–1846

Taher FA, Ibrahim SA, Abd El-Aziz A, Abou El-Nour MF, El-Sheikh MA, El-Husseiny N (2019) Mohamed MM Anti-proliferative effect of chitosan nanoparticles (extracted from crayfish Procambarus clarkii, crustacea: Cambaridae) against MDA-MB-231 and SK-BR-3 human breast cancer cell lines. Int J Biol Macromol 12:478–487

Tao F, Cheng Y, Shi X, Zheng H, Du Y, Xiang W, Deng H (2020) Applications of chitin and chitosan nanofibers in bone regenerative engineering. Carbohydr Polym 230:115658

Usman A, Zia KM, Zuber M, Tabasum S, Rehman S, Zia F (2016) Chitin and chitosan based polyurethanes: A review of recent advances and prospective biomedical applications. Int J Biol Macromol 86:630–645

Verlee A, Mincke S, Stevens CV (2017) Recent developments in antibacterial and antifungal chitosan and its derivatives. Carbohydr Polym 164:268–283

Vikhoreva GA, Kil'deeva NR, Gorbacheva IN, Shablykova EA, Rogovina SZ, Akopova TA (2000) Study of cellulose—chitosan composites, Solid-phase modification, rheology, films. Fibre Chem 32:402–406

Wang YW, Liu CC, Cherng JH, Lin CS, Chang SJ, Hong ZJ, Liu CC, Chiu YK, Hsu SD, Chang H (2019) biological effects of chitosan-based dressing on hemostasis mechanism. Polymers 11(11):1906

Wang N, Ji Y, Zhu Y, Wu X, Mei L, Zhang H, Deng J, Wang S (2020a) Antibacterial effect of chitosan and its derivative on enterococcus faecalis associated with endodontic infection. Exp Ther Med 19(6):3805–3813

Wang W, Meng Q, Li Q, Liu J, Zhou M, Jin Z, Zhao K (2020b) Chitosan derivatives and their application in biomedicine. Int J Mol Sci 21:487

Xia W, Liu P, Zhang J, Chen J (2011) Biological activities of chitosan and chitooligosaccharides. Food Hydrocoll 25(2):170–179

Xia C, Fu B, Zhang X, Qin C, Jin JC (2020) Chitosan quaternary ammonium salt induced mitochondrial membrane permeability transition pore opening study in a spectroscopic perspective. Int J Biol Macromol

Yang L, Hsiao WW, Chen P (2002) Chitosan–cellulose composite membrane for affinity purification of biopolymers and immunoad sorption. J Membr Sci 197:185–197

Yang J, Tian F, Wang Z, Wang Q, Zeng YJ, Chen SQ (2008) Effect of chitosan molecular weight and deacetylation degree on hemostasis. J Biomed Mater Res B Appl Biomater 84(1):131–137

Yen MT, Yang JH, Mau J (2008) L, Antioxidant properties of chitosan from crab shells. Carbohyd Polym 74:840–844

Yen MT, Yang JH, Mau J (2009) L, Physicochemical characterization of chitin and chitosan from crab shells. Carbohyd Polym 75:15–21

Younes I, Rinaudo M (2015) Chitin and chitosan preparation from marine sources. Structure, properties and applications. Mar Drugs 13(3):1133–74

Zhang L, Li Y, Yu JC (2014b) Chemical modification of inorganic nanostructures for targeted and controlled drug delivery in cancer treatment. J Mater Chem B 2(5):452–470

Zhang Z, Ortiz O, Goyal R, Kohn J (2014a) Biodegradable polymers. In: Principles of Tissue Engineering. Elsevier, pp 441–473

Chapter 9
Applications of Bionanocomposites from Marine Waste

Hoang Ngoc Cuong, Nguyen Cong Minh, Pham Viet Nam, and Nguyen Van Hoa

9.1 Introduction

The marine environment is rich in bioresources, some of which are often regarded as biowaste. This includes waste produced by the seafood-processing industry and seaweed that washes up on beaches. The seafood-processing industry typically produces around 75% of the total catch weight, which is considered as by-products, also known as "seafood waste" or "rest raw materials" (Lal 2023; Liu et al. 2024; Rawdkuen et al. 2019; Venugopal 2021). These wastes have been known to contain many valuable components, including chitin and chitosan extracted from crustacean shells, collagen and gelatin from fish skin, hydroxyapatite (HA) from fish bones, and alginate from brown algae (Elkhenany 2024; Santos et al. 2020; Zhan et al. 2024). Moreover, bioactive molecules, including marine-derived lipids, unsaturated fatty acids, proteins, polyphenols, and carotenoids, have been reported (Nieri et al. 2023; Menaa 2021). Historically, marine waste was primarily viewed as an issue of treatment, minimization, and prevention due to the environmental challenges associated with its disposal. As a result, these wastes were often processed and utilized as poultry feed; however, such products were typically low in value and price. Nowadays, marine wastes are increasingly recognized as significant sources of valuable nutraceuticals, food, and other materials, which have various applications in biomedicine, packaging,

H. N. Cuong
Faculty of Biotechnology, Binh Duong University, Ho Chi Minh City, Vietnam

N. C. Minh
Insitute for Biotechnology and Environment, Nha Trang University, Nha Trang, Vietnam

P. V. Nam
Faculty of Food Science and Technology, Ho Chi Minh City University of Industry and Trade, Ho Chi Minh City, Vietnam

N. Van Hoa (✉)
Faculty of Food Technology, Nha Trang University, Nha Trang, Vietnam
e-mail: hoanv@ntu.edu.vn

S. K. Panda and V. P.M. (eds.), *Bionanocomposites from Marine Origin*, Advanced Structured Materials 247, https://doi.org/10.1007/978-3-032-10749-7_9

food preservation, and environmental engineering (Venugopal, 2021; Zhan et al. 2024; Rudovica et al. 2021; Šimat et al. 2020). Utilizing marine-derived resources to develop value-added products aligns well with the principles of a circular economy, which focuses on waste reduction and improving resource recovery.

Recently, the fabrication and use of nanocomposites made from marine waste have emerged as sustainable solutions to tackle environmental and resource challenges. By converting abundant biopolymers and minerals found in seafood by-products and natural marine waste into high-performance materials, these nanocomposites offered significant potential much professional literature has reported the recovery, properties, and applications of value-added products derived from marine waste (Ranganadhareddy and Chandrasekhar, 2023; Santulli et al. 2023; Gamage et al. 2023; Gomaa and Danial 2023). However, few published summaries presented the applications of bioactive nanocomposites sourced from marine waste. This chapter aims to provide an overview of the nanocomposites derived from marine biowaste sources, emphasizing current and potential applications across various fields.

9.2 Fabrication and Properties of Bionanocomposites from Marine Waste

Bionanocomposites were created from marine waste using various methods, as summarized in Fig. 9.1 and Table 9.1. These methods could be categorized into chemical, biological, and physical approaches, each offering unique advantages in controlling material structure and functionality. Chemical methods, such as precipitation, co-precipitation, sol–gel processes, and hydrothermal synthesis, were commonly used due to their ability to precisely control reaction conditions. This control was found to contribute to the porosity, crystallinity, and stability of the resulting nanocomposites (Gobena and Woldeyonnes 2024; Azadi et al. 2024). For example, HA derived from fish bones was typically produced through hydrothermal synthesis, a high-temperature, and high-pressure technique. This process generated a material with porosity and crystallinity that closely resembled human bone minerals, making it suitable for bone grafting and orthopedic applications (Alsharif et al. 2023; Pal et al. 2017).

Sol–gel methods were often used to produce composites made from chitosan, SiO_2, and alginate. These composites enhance antimicrobial and antioxidative properties, making them suitable for applications in food packaging and biomedicine (Dong et al. 2022; Tian et al. 2021). Additionally, electrochemical synthesis allowed for the creation of chitosan/Ag nanocomposites, which possessed strong antimicrobial properties and were ideal for wound care and infection prevention (Nešović et al. 2019; Canama et al. 2023).

Biological methods have emerged as an eco-friendly approach for producing nanocomposites, utilizing bacteria, fungi, or enzymes instead of harsh chemicals. This makes them particularly suitable for applications in the medical and food sectors.

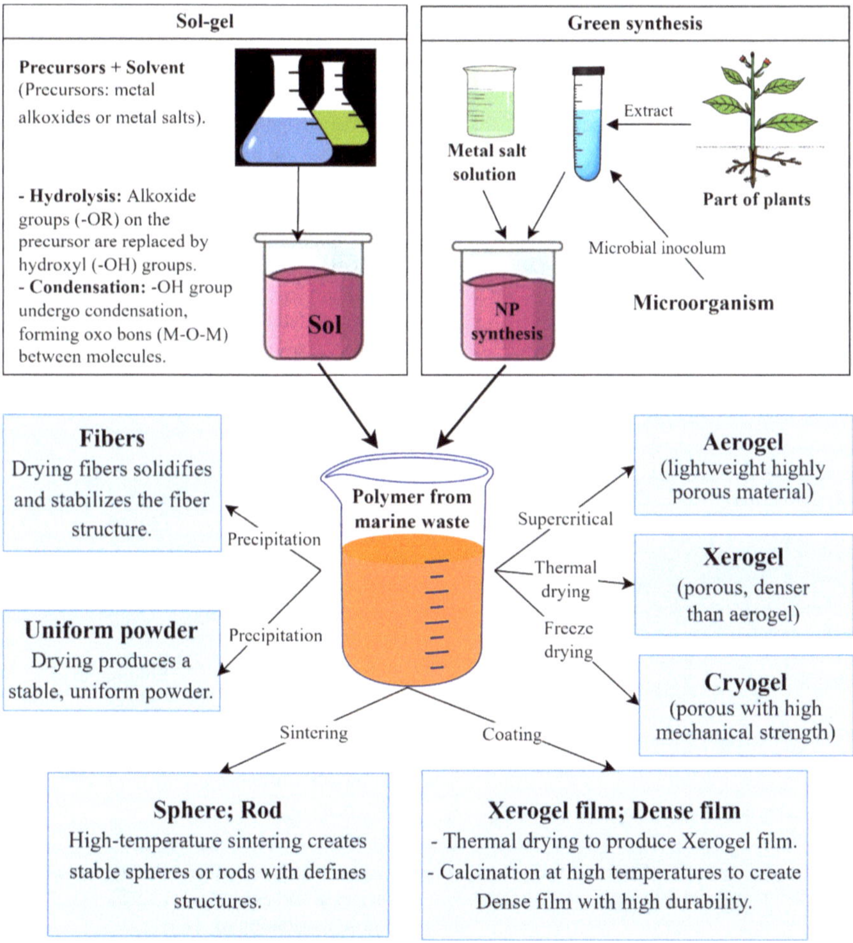

Fig. 9.1 Outline of different steps involved in preparing a nanocomposite using a sol–gel method and green synthesis

For instance, enzymatic processes have been employed to synthesize nanocomposites from chitosan and alginate, which helps preserve the bioactivity of the source materials while minimizing toxicity. This is ideal for food packaging and biomedical uses (Niculescu and Grumezescu 2022; Khubiev et al. 2023). Additionally, Ag/chitosan nanocomposites have been synthesized through enzymatic methods and utilized as antibacterial films for food preservation, effectively preventing microbial growth and extending shelf life (Canama et al. 2023; Kyung and Ko 2014). Similarly, microbial processes had been used to prepare metal nanoparticles, such as silver or gold, combined with collagen or chitosan. These composites were produced using *Bacillus* or *Aspergillus* and demonstrate enhanced antibacterial effects, making them suitable for wound healing and medical device applications (El-Zawawy, et al. 2023; Arif and

Table 9.1 Different synthetic methods and applications of various marine-based nanocomposites

Synthetic method	Nanocomposites	Remark properties of nanocomposites	References
Precipitation	Chitosan/SiO_2	10–100 nm. High surface area, thermal stability, biocompatibility, strong adsorption capacity	Zhong et al. (2022)
	Alginate/calcium phosphate	50–200 nm. Porosity, bioactivity, biocompatibility, mechanical stability	Zahorán et al. (2022)
	Collagen/HA	20–150 nm (HA crystals). Enhanced mechanical strength, bioactivity, osteoconductivity, biodegradability	Shen et al. (2011)
Hydrothermal	Chitosan/Ag	20–50 nm. High antibacterial, thermal stability, uniform size	Hileuskaya et al. (2021)
	Alginate/calcium phosphate	50–200 nm. Good porosity, biocompatibility, mechanical stability	Wang et al. (2015)
Sol–gel	ZnO/alginate,	20–50 nm. Enhanced photocatalytic activity. Improved antibacterial properties. Biocompatibility suitable for biomedical applications	Siddiqui et al. (2022)
	Chitosan/SiO_2	10–100 nm. High specific surface area and porosity. Uniform distribution of SiO_2 nanoparticles within the chitosan matrix	Budnyak et al. (2015)
	TiO_2/chitosan	15–60 nm. Enhanced photocatalytic degradation of pollutants. Antimicrobial activity. Biodegradability and biocompatibility. Improved mechanical properties	Bui et al. (2017)
Co-precipitation	Fe_3O_4/chitosan	10–20 nm. Superparamagnetic behavior. Biocompatibility and biodegradability. Enhanced thermal stability. Potential for targeted drug delivery	Wang et al. (2015)

(continued)

Table 9.1 (continued)

Synthetic method	Nanocomposites	Remark properties of nanocomposites	References
	Fucoidan/chitosan	100–150 nm. Targeted binding to P-selectin for atherosclerosis therapy. Anti-inflammatory and antioxidative properties. Positive surface charge ensuring stability	Liu et al. (2022)
Electrochemical	Chitosan/Ag	10–15 nm. Enhanced antibacterial activity. Biocompatibility suitable for biomedical applications. Improved stability due to chitosan matrix	Tran et al. (2016)
	Alginate/Cu	20–30 nm. High electrical conductivity. Biodegradability and non-toxicity. Enhanced mechanical strength	Zhang et al. (2024)
	Collagen/Au	15–25 nm. Excellent biocompatibility. Enhanced optical properties due to gold nanoparticles. Improved mechanical properties	Do et al. (2025)
Microemulsion	Alginate/Fe_3O_4	10–20 nm. Superparamagnetic behavior. High surface area. Biocompatibility	Zadeh et al. (2024)
	Collagen/Au	15–25 nm. Enhanced biocompatibility. Improved mechanical strength. Optical properties suitable for biosensing	Sun et al. (2024)
	Chitosan/Ag	5–15 nm. High antibacterial activity. Biodegradability. Enhanced mechanical properties	Ramesh et al. (2017)

(continued)

Table 9.1 (continued)

Synthetic method	Nanocomposites	Remark properties of nanocomposites	References
	Fucoidan/alginate	0.5–1.7 nm. Formation of water-soluble interpolymer complexes. Pronounced inhibitory effect against Gram (+) and Gram (−) bacteria. Biocompatibility suitable for pharmaceutical and medical applications	Brovko (2024)
Solution polymerization	PVA/ZnO	10–20 nm. Enhanced thermal stability. Improved mechanical properties. Good antimicrobial activity	Barman et al. (2020)
	Gelatin/SiO_2	Not specified. Enhanced mechanical properties. Improved barrier properties against water vapor and oxygen. Increased UV light absorbance	Tabatabaei et al. (2018)
Grafting	Chitosan/Au	10–15 nm. Enhanced catalytic activity. Biocompatibility. Stability in aqueous solutions	Bibi et al. (2016)
	Collagen/Ag	5–20 nm. Antibacterial properties. Biocompatibility. Stability in biological environments	Cardoso et al. (2014)
	Alginate/Cu	10–30 nm. Good antibacterial activity and biodegradability. Enhanced mechanical properties	Reddy et al. (2014)
Liquid-phase chemical	Ag/chitosan	10–20 nm. Excellent antimicrobial activity and biocompatibility. Enhanced mechanical properties	Trivedi et al. (2014)
	Fe_3O_4/gelatin	15–25 nm. Superparamagnetic behavior, biodegradability and enhanced stability	Mofatehnia et al. (2023)

(continued)

Table 9.1 (continued)

Synthetic method	Nanocomposites	Remark properties of nanocomposites	References
Direct precipitation	HA/chitosan	20–50 nm. Enhanced mechanical strength. Improved biocompatibility. Increased osteoconductivity	Attar Nosrati et al. (2020)
	Alginate/Fe_3O_4	15–25 nm. Superparamagnetic behavior. Biodegradability. Enhanced stability	Mofatehnia et al. (2023)
Spray pyrolysis	Alginate/CuO	15–25 nm. Good antibacterial properties. Biodegradability. Improved mechanical strength	Leonardi et al. (2021)
	Gelatin/TiO_2	20–30 nm. Photocatalytic activity. Biocompatibility. Enhanced thermal stability	Cheaburu-Yilmaz (2015)
Reverse micelle	Alginate/chitosan	20–100 nm. Biocompatibility. Biodegradability. Enhanced stability	Kafshgari et al. (2012)
Solvent evaporation	Chitosan/ Poly(lactic-co-glycolic acid)	200–300 nm. Enhanced drug encapsulation efficiency. Controlled drug release profiles. Biodegradability and biocompatibility	Ahmed and Aljaeid (2016)
	Chitosan/ Polycaprolactone	100–200 nm. Improved mechanical strength. Enhanced thermal stability. Biodegradability	Arabahmadi et al. (2019)
Emulsion cross-linking	Chitosan/HA microspheres	100–200 μm. Enhanced biocompatibility. Improved mechanical strength. Controlled drug release	Wu et al. (2014)
Green synthesis	Chitosan-Ag/Au	10–30 nm. Dual antibacterial activity from silver and gold. Enhanced biocompatibility. High stability in biological environments	El-Sherbiny and Sedki (2019)

(continued)

Table 9.1 (continued)

Synthetic method	Nanocomposites	Remark properties of nanocomposites	References
	Chitosan/ZnO	20–50 nm. Strong antibacterial, UV protective, biodegradable, high mechanical stability	Fais et al. (2024)
	Alginate/Ag	10–30 nm. High antibacterial activity, conductive, excellent stability in aqueous environments	Anbalagan (2024)
	Chitin/Fe_3O_4	12–25 nm. Magnetic, strong adsorption capacity, biocompatible	Yosri et al. (2024)
	Collagen/TiO_2	Antibacterial, UV protective, biocompatible, enhanced mechanical strength	Borah et al. (2024)
Biological synthesis	Chitosan/SiO_2	Antibacterial, high mechanical strength, excellent biocompatibility	Kamala et al. (2024)
	Collagen/TiO_2	30–80 nm. Antibacterial, UV protective, enhanced mechanical stability	Santos et al. (2024)
	Chitin/magnetite	Magnetic, high adsorption capacity, biocompatible	Nallasamy and Natarajan (2024)
Algal, fungal, plant-based	Chitosan/alginate	High mechanical strength, biodegradable, excellent adsorption properties	Almutairi et al. (2021)
	GO-based nanocomposites from seaweed	Antibacterial, anti-algae, and strong adsorption properties	Soleimani et al. (2023)
	Seaweed polysaccharide nanocomposites	Biodegradable, flexible, durable, high oxygen barrier properties	Bukhari et al. (2023)
	Ag nanocomposites from red seaweed	High antimicrobial and antifungal activities, biodegradable	Roseline et al. (2019)
Polyelectrolyte complex	Chitosan/alginate	50–100 nm. High mechanical strength, excellent adsorption properties, and biocompatibility	Almutairi et al. (2021)

(continued)

Table 9.1 (continued)

Synthetic method	Nanocomposites	Remark properties of nanocomposites	References
	Marine polysaccharide/clay	High structural integrity, thermal stability, and improved mechanical properties	da Costa et al. (2016)
	Fucoidan/chitosan	Antibacterial, anti-inflammatory, and biodegradable properties	Venkatesan et al. (2018)
	Chitosan/alginate Polyelectrolyte	Antibacterial, enhanced biocompatibility, and improved adsorption for environmental use	Hardy et al. (2018)
Coacervation	Carrageenan/gelatin	High mechanical stability, biodegradable, excellent encapsulation efficiency	Bakry et al. (2019)
	Chitosan/gum arabic	Excellent encapsulation efficiency, antioxidant properties, and biocompatibility	Cui et al. (2025)
	Marine protein/polysaccharide	High structural integrity, biocompatible, and durable	Vahdati et al. (2020)
Ultrasonication	Chitosan/ZnO	Antioxidant, antibacterial, high thermal stability, and biocompatible	Diab et al. (2024)
	Chitosan/fucoidan	Biodegradable, high antibacterial activity, excellent adhesion under wet conditions	Galland et al. (2024)
	SiO_2-polyurethane/Graphene	Superhydrophobic, non-toxic, and antifouling	Selim et al. (2025)
Ionic gelation	Chitosan/alginate	100–200 nm. Biodegradable, strong antibacterial activity, and high encapsulation efficiency	Venkatesan et al. (2016)
	Chitosan/CuO	100–150 nm. Biodegradable, enhanced mechanical stability, and strong adsorption properties	Khoerunnisa et al. (2023)
	Fucoidan/chitosan	80–100 nm. Excellent biocompatibility, antioxidant, and antimicrobial properties	Ganesan et al. (2022)

(continued)

Table 9.1 (continued)

Synthetic method	Nanocomposites	Remark properties of nanocomposites	References
Centrifugation	Ag/seaweed	Anti-inflammatory, strong antibacterial properties, biodegradable	Kavishri et al. (2024)
Mechanical milling	Marine-derived polymer	Biodegradable, excellent thermal and mechanical stability	Yosri et al. (2024)
Sputtering	Cu-coated cellulose/alginate	High antimicrobial activity, biodegradable, environmentally friendly	Świerczyńska et al. (2024)
Thermal evaporation	Chitosan/ZnO	150–200 nm. Strong antibacterial properties, biodegradable, high mechanical strength	Pandey et al. (2024)
	Graphene/marine shell-based	Strong corrosion resistance, high conductivity, and lightweight properties	Vaseashta (2015)
Physical vapor deposition	Marine polymer/metal oxide	High corrosion resistance, strong adhesion, and thermal stability	Wang et al. (2023)
	Seaweed-derived HA	Biocompatible, excellent structural properties, and high thermal stability	Thongjamroon et al. (2023)
Electrospinning	Marine-derived bio-based nanofibers	150–300 nm. Biodegradable, high mechanical strength, and environmentally friendly	Oliver-Cuenca et al. (2024)
Homogenization	Chitosan/ZnO	200–300 nm. Antioxidant, antidiabetic, antibacterial, and biodegradable	Diab et al. (2024)
	Calcium/alginate	200–300 nm. Biocompatible, controlled release properties, and biodegradable	Bae et al. (2019)
Microwave-assisted synthesis	Sodium alginate/chitosan	Biodegradable, excellent encapsulation properties, and sustained release of active agents	Ristić et al. (2024)

(continued)

Table 9.1 (continued)

Synthetic method	Nanocomposites	Remark properties of nanocomposites	References
Spray drying	Chitosan/sodium alginate	100–200 nm. Biodegradable, excellent encapsulation efficiency, and controlled release properties	Kumar et al. (2017)
	Collagen/HA	200–500 nm. Strong mechanical properties, biocompatible, and environmentally friendly	Kumar Balu et al. (2021)

Uddin 2021). Furthermore, green synthesis methods had been employed to create nanocomposites like collagen-gold and alginate-copper, which possess antioxidative properties. These composites were formulated using plant extracts and enzymes and are applied in cosmetics and dermatological products (Akturk et al. 2016; Anbalagan 2024).

Physical synthesis methods, such as electrospinning, freeze-drying, and mechanical milling, have been utilized to create nanocomposites with improved mechanical properties and specific structures, thereby expanding their applications in various fields. For instance, chitosan/HA nanofibers were produced using the electrospinning technique. These products demonstrated high mechanical strength and surface area, making them ideal for use as scaffolds in tissue engineering and bone regeneration (Frohbergh et al. 2012; Hong Thien et al. 2013). Additionally, the freeze-drying method helped maintain the porous structure of chitosan, alginate, or gelatin-based nanocomposites, which were well-suited for wound dressings that required high biocompatibility and structural integrity. Mechanical milling and other physical techniques were reported to further enhance particle distribution and uniformity within these composites, broadening their applications in areas like environmental filtration and food packaging (Afjoul et al. 2020; Gorrasi and Sorrentino 2015).

9.3 Applications of Bio-Nanocomposites from Marine Wastes

Typically, the characteristics of a composite are largely dependent on the properties of each component and their interactions. Given the valuable bioactive components, the applicability of marine-based nanocomposites spans a wide range of industries. Examples of bionanocomposites from marine wastes are presented in Table 9.1. These bionanocomposites are primarily divided into three groups: (i) nanocomposites consisting of metal nanoparticles and biopolymers, (ii) hybrids of metal oxides, hydroxides, carbon-based materials (graphene oxide (GO), graphene, carbon nanotubes (CNT)) or salt nanoparticles combined with biopolymers, and (iii) blends of two or more biopolymers at the nanoscale. The prepared marine-based nanocomposites demonstrated unique bioactive properties, including antioxidative, antimicrobial, biocompatible, and biodegradable activities, as well as effects related to their nanosize. Consequently, these nanocomposites were studied and applied in various fields, including biomedical and nutritional applications, functional food packaging, controlled drug delivery systems, cosmetics for anti-aging products, food preservation, adjuncts in cancer therapies, tissue engineering scaffolds, wound dressings, and water pollution treatment.

9.3.1 Bio-Based Applications

Marine-derived nanocomposites are extensively utilized in various bio-based applications due to their renewable, biodegradable, and biocompatible properties. Besides, the combination of nanomaterials and biopolymers significantly improves their mechanical properties, including tensile strength, stiffness, and toughness. Moreover, the processing of biopolymers-based composites generally has a lower environmental impact than traditional plastics, as they produce fewer greenhouse gases (Sadasivuni et al. 2020; Vinod, et al. 2020). The most commonly used biopolymers from marine wastes include chitosan, alginate, collagen, HA, and gelatin. These biomaterials play a crucial role in developing sustainable nanocompotites that minimize environmental impact while enhancing functionality (Silva et al. 2023, 2020; Iliou et al. 2022). Advanced synthetic methods such as hydrothermal synthesis, sol–gel processing, co-precipitation, electrospinning, and green synthesis are employed to customize these composites for specific applications, optimizing characteristics like particle size, stability, and bioactivity (Mane et al. 2023; Lin and Zhang 2023). Marine-derived nanocomposites have been found in applications across multiple sectors, including agriculture, food preservation, nutraceuticals, and environmental management, showcasing unique advantages that align with the needs of the bio-based industry. Table 9.2 illustrates bio-based applications of marine-derived nanocomposites.

9.3.2 Engineering Applications

Marine-derived nanocomposites are gaining significant attention in engineering applications due to the increasing demand for sustainable, high-performance materials that offer versatility, durability, and environmental compatibility. These nanocomposites are created from renewable marine resources, using biomaterials such as chitosan, alginate, collagen, and HA as matrix materials. Advanced synthesis

Table 9.2 Bio-based applications of marine-derived nanocomposites

Nanocomposite	Potential applications	References
Chitosan/alginate	A carrier for slow-release fertilizers and improves plant growth under abiotic stress	Leonardi et al. (2021)
Chitosan/collagen	Biostimulatory effects on plants and enhances nutrient delivery for crop development	Kumar and Madihally (2022)
Chitosan/fucoidan	Boosts seed germination and enhances plant resistance to fungal infections	Kumar and Madihally (2022)
Chitosan/gelatin	Provides controlled nutrient release for plants and improves soil nutrient availability	Khan et al. (2021)
Chitosan/calcium phosphate	Aids in soil remediation and enhances plant nutrient uptake	Niranjana et al. (2022)
Chitosan/HA	Stimulates plant growth and protects against abiotic stressors like drought and salinity	Pighinelli and Kucharska (2013)
Chitosan/Ag	Acts as an antimicrobial agent to protect crops from bacterial and fungal infections	Katas et al. (2018)
Alginate/HA	Bone tissue regeneration, dental applications, and wound healing	Costa et al. (2024)
Alginate/chitosan	Drug delivery, wound healing, and tissue engineering	Ofudje et al. (2023)
Alginate/gelatin	Tissue regeneration, wound healing, and controlled drug release	Fernando et al. (2021)
Alginate/fucoidan	Wound healing, tissue regeneration, and drug delivery	Lu et al. (2019)
Alginate/chitin	Bone tissue regeneration and wound healing	Karimipour-Fard et al. (2021)
Collagen/HA	Bone tissue regeneration, dental applications, and wound healing	Alkaron et al. (2024)
Collagen/chitosan	Tissue engineering, wound healing, and drug delivery	Fernando et al. (2021)
Collagen/alginate	Tissue engineering and drug delivery	Costa et al. (2024)
Collagen/gelatin	Wound healing, bone regeneration, and tissue scaffolds	Fernando et al. (2021)
Collagen/fucoidan	Wound healing and tissue regeneration	Lu et al. (2019)
Collagen/Chitin	Bone tissue regeneration and wound healing	Karimipour-Fard et al. (2021)

(continued)

Table 9.2 (continued)

Nanocomposite	Potential applications	References
Collagen/carrageenan	Cartilage repair and tissue regeneration	Swain et al. (2021)
Collagen/pectin	Bone implants, tissue regeneration, and wound healing	Costa et al. (2024)
HA/chitosan	Biomedicine, especially in tissue engineering and bone implants	Costa et al. (2024)
HA/Alginate	Tissue engineering and other biomedical applications	Costa et al. (2024)
HA/Collagen	Bone tissue regeneration and dental applications	Alkaron et al. (2024)
HA/Gelatin	Tissue engineering and other biomedical applications	Fernando et al. (2021)
HA/Fucoidan	Orthopedic implants due to improved biocompatibility and bioactivity	Young et al. (2016)
HA/Chitin	Enhances osteogenesis and bone mineralization due to its hierarchical structure	Duan et al. (2017)
HA/Carrageenan	Provides controlled drug release capabilities when loaded with active agents such as hesperidin nanoparticles	Taymouri et al. (2024)
HA/Carboxymethyl cellulose	Tissue engineering and other biomedical applications	Hussin et al. (2022)
HA/Pectin	Promotes osteogenesis and supports neo-bone formation for tissue regeneration	Sumathra et al. (2017)
HA/Clay	Improves bioactivity and supports in vivo tissue formation for bone repair and regeneration	Šupová (2024)

techniques are employed to develop nanocomposites that meet the stringent requirements of various engineering fields, including water treatment, construction, electronics, and energy storage (Yosri et al. 2024; Unal 2019). For example, bionanocomposites offer a sustainable method for water treatment due to their renewable nature, biodegradability, and low toxicity (Spoială et al. 2021; Khandegar et al. 2021; Ahmad 2019; Yarahmadi et al. 2024; Rana et al. 2024). Additionally, biocomposites are effective at removing a wide variety of contaminants from water, such as suspended solids, heavy metals, organic pollutants, and microorganisms. They can be utilized in different water treatment processes, including coagulation, flocculation, sedimentation, filtration, and bioremediation (Udayakumar et al. 2021; Pachaiappan et al.

2022). Table 9.3 provides details on multiple marine-derived nanocomposites specifically designed for different engineering applications, such as water purification, structural reinforcement, electronics, and renewable energy (Fig. 9.2).

9.3.3 *Medical Applications*

Marine-derived nanocomposites are becoming increasingly important in medical applications due to their unique bioactivity, biocompatibility, and structural adaptability. It is well-known that the addition of nanomaterials to biopolymers greatly improves their mechanical properties, including tensile strength, stiffness, and toughness. This enhancement is vital for applications such as bone implants and tissue engineering scaffolds, which must endure physiological loads (Montaser et al. 2016; Chae et al. 2013; Purohit et al. 2019; Kikuchi and Bulletin 2013). In addition, bionanocomposites are biocompatible, meaning they are well tolerated by the human body and do not trigger an adverse immune response. Moreover, the properties of bionanocomposites can be accurately adjusted by modifying the type, concentration, and distribution of nanomaterials within the biopolymer matrix. This capability enables the development of materials with specific characteristics that are tailored to meet the requirements of particular medical applications (Shaikh et al. 2024; Khane et al. 2024). These nanocomposites are made from various marine-derived components, including chitosan, alginate, collagen, HA, gelatin, fucoidan, and laminarin. Researchers have employed several synthesis techniques, such as electrospinning, sol–gel processing, green synthesis, and co-precipitation, to develop a range of nanocomposites for medical uses (Kumar et al. 2024). Table 9.4 presents different marine-derived nanocomposites tailored for various medical applications, including wound care and regeneration, tissue engineering scaffolds, orthopedic implants, biosensors, controlled drug delivery, and antimicrobial coatings.

9.3.4 *Packaging Applications*

The increasing demand for sustainable packaging materials is driving research into marine-derived nanocomposites. These materials provide biodegradable, biocompatible, and functional alternatives to traditional plastics (Yosri et al. 2024; Sarkar et al. 2023). It noted that the addition of nanomaterials to biopolymers greatly improved their barrier properties against gases, such as oxygen and carbon dioxide, as well as moisture (Vilvert et al. 2023; Cesur et al. 2018; Viacava et al. 2024). This enhancement was crucial for prolonging the shelf life of packaged food and other perishable items. Besides, bionanocomposites were demonstrated greater tensile strength, stiffness, and toughness than neat biopolymers, making them more resistant to damage during handling, transportation, and storage (Martí et al. 2019; Kazemi and Rezaei 2015; Prakash et al. 2020). Marine-sourced materials such as chitosan, alginate,

Table 9.3 Engineering applications of various marine-derived nanocomposites

Nanocomposite	Potential applications	References
Chitosan/HA	Water treatment, especially for removing heavy metals and toxic compounds	Spoială et al. (2021)
Chitosan/GO	Water treatment, with high adsorption capabilities for organic and inorganic pollutants	Khandegar et al. (2021)
Chitosan/Alginate	Water treatment, particularly for removing heavy metal ions and organic pollutants	Ahmad (2019)
Chitosan/Gelatin	Water treatment, with adsorption capabilities for organic and inorganic pollutants	Yarahmadi et al. (2024)
Chitosan/Chitin	Water treatment, with adsorption capabilities for organic and inorganic pollutants	Rana et al. (2024)
Alginate/HA	Tissue engineering, particularly for bone regeneration	Farshidfar et al. (2023)
Alginate/GO	Tissue engineering, with enhanced electrical and mechanical properties	Alves et al. (2021)
Alginate/Carbon	Enhances chitosan/alginate bonding, improves mechanical properties and water absorption	Khachatryan et al. (2023)
Alginate/Fe_3O_4	Tissue engineering, with magnetic properties assisting in cell and molecular guidance	Muraca et al. (2020)
Alginate/SiO_2	Enhanced mechanical and thermal properties of alginate films, food packaging and biomedical applications	Yang and Xia (2017)
Alginate/MgO	Enhanced mechanical properties and biodegradability of alginate films, used in tissue engineering	Silva et al. (2017)
Alginate/polymer	Enhances mechanical properties and biodegradability of alginate, used in drug delivery and tissue engineering	Işıklan et al. (2024)
Collagen/GO	Flexible electronics and wearable sensors, with enhanced conductivity and flexibility	Martín et al. (2023)
Collagen/Ag	High antimicrobial coatings and electrochemical sensing platforms	Peng et al. (2022)
Collagen/Au	Optical and electrochemical biosensors for medical diagnostics	Bai et al. (2023)
Collagen/ZnO	Piezoelectric sensing devices and wearable health monitoring systems	Zhang et al. (2023a)
Collagen/TiO_2	Photocatalytic sensors and environmental monitoring applications	Zhu et al. (2023)
HAp/CNT	Enhances mechanical strength and crack resistance in bio/concrete	Siddiqui et al. (2018)
HA/GO	Improves compressive strength and durability of cementitious materials	Raval et al. (2019)
HA/SiO_2	Provides enhanced thermal stability and mechanical properties in construction materials	Kareem et al. (2024)

(continued)

Table 9.3 (continued)

Nanocomposite	Potential applications	References
Gelatin/GO	High-performance supercapacitors and energy storage systems	Wan et al. (2011)
Gelatin/TiO_2	Dye-sensitized solar cells to improve photocatalytic efficiency	Hasan et al. (2020)
Gelatin/SiO_2	Explored for solid electrolytes in lithium-ion batteries	Sachithanadam and Joshi (2016)
Gelatin/ZnO	Piezoelectric nanogenerators for mechanical energy harvesting	Zare et al. (2021)

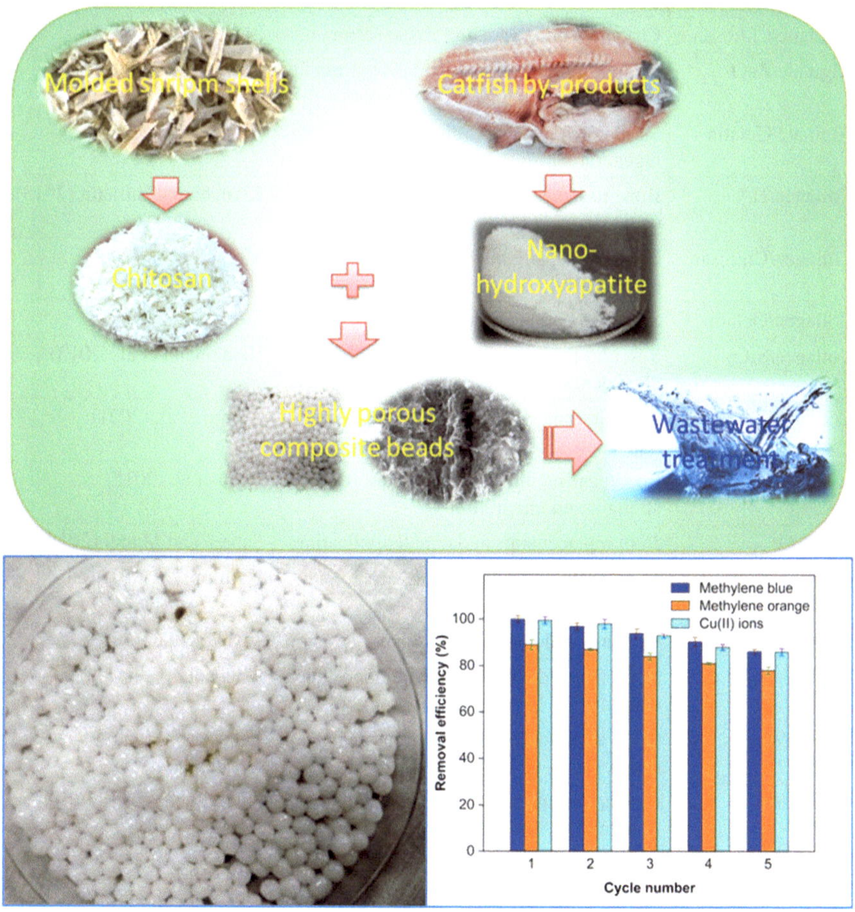

Fig. 9.2 Schematic diagram of the preparation, appearance, and adsorption capacity of the nano-HA/chitosan composites for wastewater treatment from the cheap waste sources (Trung et al. 2022)

Table 9.4 Medical applications of various marine-derived nanocomposites

Nanocomposite	Potential applications	References
Chitosan/HA	Used for targeted bone drug delivery	Said et al. (2023)
Chitosan/GO	Cancer therapy with pH-sensitive drug release	Pieklarz et al. (2018)
Chitosan/Au	Photothermal therapy and controlled drug release	Lee et al. (2010)
Chitosan/ZnO	Antimicrobial drug delivery to treat infections	Toiserkani (2016)
Alginate/Ag	Antimicrobial wound dressings	Montaser et al. (2016)
Alginate/HA	Bone tissue engineering and wound repair	Chae et al. (2013)
Alginate/Chitosan	Chronic wound care and tissue repair	Zhang et al. (2023b)
Alginate/ZnO	Address infections and oxidative stress in wound environments	Belhachem et al. (2024)
Alginate/Gelatin	Scaffolds for tissue engineering and regeneration	Purohit et al. (2019)
Collagen/HA	Bone tissue engineering for rigidity and osteoconductivity	Kikuchi and Bulletin (2013)
Collagen/Chitosan	Cartilage repair and skin tissue regeneration	Moreira et al. (2016)
Collagen/Gelatin	Soft tissue regeneration scaffolds	Monroy et al. (2018)
Collagen/SiO_2	Vascular tissue engineering to improve angiogenesis	Heinemann et al. (2013)
Collagen/ZnO	Antimicrobial wound healing and tissue repair	Araujo et al. (2021)
HA/TiO_2	Load/bearing implants for enhanced strength and integration	Chen et al. (2015)
HA/ZrO	Joint replacements and dental applications	Sabree and Mahdi (2019)
HA/GO	Good antibacterial properties and implant durability	Hashim et al. (2019)
HA/Chitosan	Coatings and bone defect repair	Tavakol et al. (2013)
HA/SiO_2	Bone scaffolds for thermal and mechanical stability	Noviyanti et al. (2023)
Gelatin/Ag	Good antimicrobial coatings for medical devices	Yang et al. (2022)
Gelatin/Au	Cancer therapy and targeted drug delivery systems	Khodashenas et al. (2021)
Gelatin/GO	Delivering hydrophobic drugs with controlled release	Kalkhoran et al. (2018)
Gelatin/ZnO	Ideal for wound healing and antimicrobial coatings	Yang et al. (2021)
Gelatin/Chitosan	Employed in localized and sustained drug release systems	Sharmeen et al. (2018)

collagen, and other bioactive compounds like fucoidan and laminarin are being increasingly utilized in packaging applications due to their structural integrity, antimicrobial properties, and ability to preserve freshness. Research has shown that these prepared nanocomposites can effectively extend shelf life, prevent microbial contamination, and support eco-friendly packaging solutions (Sharma et al. 2023). A variety of marine-derived nanocomposites specifically developed for packaging applications are listed in Table 9.5, each offering unique benefits for sustainable food preservation, protective coatings, and active packaging.

9.4 Conclusions and Future Trends

The development of bionanocomposites derived from marine waste has significantly impacted various industries by combining sustainability with functionality. These materials not only address environmental concerns related to waste management but also provide nanocomposites with enhanced properties such as biocompatibility, biodegradability, and functional versatility. Marine-derived biocomponents, including chitosan, alginate, gelatin, collagen, and HA, serve as excellent materials for creating advanced nanocomposites. Their applications extend to food packaging, biomedical engineering, structural reinforcement, and water purification. Reports highlight a shift toward sustainable practices, where marine waste is seen as a valuable resource rather than a disposal problem. However, challenges still exist regarding the scalability, cost-effectiveness, and stability of these materials in various environmental conditions. To overcome these challenges, it is essential for material scientists, biologists, and engineers to collaborate across disciplines. This collaboration can help optimize synthesis methods and enhance the functionality of these composites. Emerging technologies, such as 3D printing, electrospinning, and green synthesis, present promising opportunities for creating customized materials with improved performance. Additionally, it is important to establish regulatory standards to ensure the safe use of marine-derived nanocomposites in food and medical applications. On the other hand, the integration of these nanocomposites into circular economy models is essential for promoting sustainable industrial practices. Innovations in the recycling and upcycling of marine waste into high-value nanomaterials can significantly improve resource efficiency. Furthermore, these materials can be customized for advanced applications, including responsive packaging systems and intelligent biomedical devices. The future of marine-derived bionanocomposites hinges on their potential to tackle urgent global challenges such as food security, environmental conservation, and healthcare sustainability. By harnessing the untapped potential of marine resources, nanocomposites are set to play a vital role in sustainable development, in line with international efforts to reduce waste and encourage innovation in green technologies.

Table 9.5 Packing applications of various marine-derived nanocomposites

Nanocomposite	Applications	References
Chitosan/Ag	Good antimicrobial food wraps and coatings	Tripathi et al. (2011)
Chitosan/ZnO	Food films for bacterial growth inhibition	Sanuja et al. (2015)
Chitosan/GO	Provides oxygen barrier and antimicrobial properties	Vilvert et al. (2023)
Chitosan/Nanoclay	Improves water vapor and gas barrier properties	Cesur et al. (2018)
Chitosan/CuO	Antibacterial and antioxidant properties for packaging	Laanoja (2024)
Alginate/Nanoclay	Improved water vapor and gas barrier properties	Viacava et al. (2024)
Alginate/Ag	Used as antimicrobial films for perishable foods	Kanagaraj et al. (2023)
Alginate/ZnO	Provides antimicrobial and mechanical enhancements	Bakil et al. (2020)
Alginate/GO	Offers superior oxygen barrier and mechanical flexibility	Martí et al. (2019)
Alginate/Gelatin	Enhances flexibility and water resistance in films	Kazemi and Rezaei, (2015)
Collagen/Ag	Good antimicrobial active packaging	She et al. (2024)
Collagen/ZnO	Applied in antimicrobial and UV-blocking films	Albu et al. (2016)
Collagen/GO	Used for oxygen barrier and intelligent freshness monitoring	Zine et al. (2017)
Collagen/Au	Employed in smart packaging systems for spoilage detection	Akturk et al. (2016)
HA/Nanoclay	Enhanced water vapor and gas barrier properties in films	Uddin et al. (2024)
HA/GO	Excellent oxygen barrier and mechanical strength	Prakash et al. (2020)
HA/Ag	Offers antimicrobial activity for food packaging	Silva-Holguín and Reyes-López (2020)
HA/ZnO	Good antimicrobial and UV/blocking properties	Wang et al. (2024)
HA/Gelatin	Biodegradable films for sustainable packaging	Türe (2019)
Gelatin/Ag	Antimicrobial edible films for food safety	Menezes et al. (2019)
Gelatin/ZnO	Good UV/blocking and antibacterial properties	Alizadeh-Sani et al. (2021)

(continued)

Table 9.5 (continued)

Nanocomposite	Applications	References
Gelatin/Nanoclay	Improved water vapor barrier properties	Villalba-Rodríguez et al. (2021)
Gelatin/GO	Enhanced oxygen barrier properties for food spoilage prevention	Adilah et al. (2021)
Gelatin/Alginate	Biodegradable and edible coatings for fruits and vegetables	Li et al. (2022)

References

Adilah AN et al (2021) Incorporation of graphene oxide to enhance fish gelatin as bio-packaging material. Food Packag Shelf Life 28:100679

Afjoul H et al (2020) Freeze-gelled alginate/gelatin scaffolds for wound healing applications: an in vitro, in vivo study. Mater Sci Eng, C 113:110957

Ahmad N et al (2019) Chitosan based nanocomposites as efficient adsorbents for water treatment. Modern age waste water problems: solutions using applied nanotechnology, pp 69–83

Ahmed TA, Aljaeid BM (2016) Preparation, characterization, and potential application of chitosan, chitosan derivatives, and chitosan metal nanoparticles in pharmaceutical drug delivery. Drug Des Dev Therapy 483–507

Akturk O et al (2016) Collagen/gold nanoparticle nanocomposites: a potential skin wound healing biomaterial. J Biomater Appl 31(2):283–301

Albu MG et al (2016) Preparation and biological activity of new collagen composites, part I: collagen/zinc titanate nanocomposites. Appl Biochem Biotechnol 180:177–193

Alizadeh-Sani M et al (2021) Preparation of active nanocomposite film consisting of sodium caseinate, ZnO nanoparticles and rosemary essential oil for food packaging applications. J Polym Environ 29(2):588–598

Alkaron W et al (2024) Hydroxyapatite-based natural biopolymer composite for tissue regeneration. Materials 17(16):4117

Almutairi FM et al (2021) Application of chitosan/alginate nanocomposite incorporated with phycosynthesized iron nanoparticles for efficient remediation of chromium. Polymers 13(15):2481

Alsharif SA et al (2023) Hydrothermal extraction and physicochemical characterization of biogenic hydroxyapatite nanoparticles from buffalo waste bones for in vivo xenograft in experimental rats. Sci Rep 13(1):17490

Alves Z et al (2021) Design of alginate-based bionanocomposites with electrical conductivity for active food packaging. Int J Mol Sci 22(18):9943

Anbalagan G et al (2024) Green synthesis of copper and copper oxide nanoparticles from brown algae turbinaria species' aqueous extract and its antibacterial properties. Cureus 16(4)

Andonegui San Martín M et al (2023) Sustainable collagen composites with graphene oxide for bending resistive sensing. Polymers 15(19):3855

Arabahmadi S et al (2019) Electrospun biocompatible gelatin-chitosan/polycaprolactone/hydroxyapatite nanocomposite scaffold for bone tissue engineering. Int J Nano Dimens 10(2):169–179

Araujo TAT et al (2021) Collagen membranes for skin wound repair: a systematic review. J Biomater Appl 36(1):95–112

Arif R, Uddin R (2021) A review on recent developments in the biosynthesis of silver nanoparticles and its biomedical applications. Med Dev Sens 4(1):e10158

Attar Nosrati S et al (2020) Optimized precipitation process for efficient and size-controlled synthesis of hydroxyapatite–chitosan nanocomposite. J Korean Ceram Soc 57:632–644

Azadi E et al (2024) Sources and extraction of biopolymers and manufacturing of bio-based nanocomposites for different applications. Molecules 29(18):4406

Bae SB, Nam HC, Park WH (2019) Electrospraying of environmentally sustainable alginate microbeads for cosmetic additives. Int J Biol Macromol 133:278–283

Bai Z et al (2023) Versatile nano–micro collagen fiber-based wearable electronics for health monitoring and thermal management. J Mater Chem A 11(2):726–741

Bakil SNA et al (2020) Sodium alginate-zinc oxide nanocomposite film for antibacterial wound healing applications. Biointerface Res Appl Chem 10(2):6245–6252

Bakry AM et al (2019) Myofibrillar protein with κ-or λ-carrageenans as novel shell materials for microencapsulation of tuna oil through complex coacervation. Food Hydrocolloids 96:43–53

Barman A et al (2020) Stabilization and dispersion of ZnO nanoparticles in PVA matrix. J Inorg Organomet Polym Mater 30:2248–2257

Belhachem A et al (2024) Study of the antibacterial activity of hybrid nanocomposites "metal oxide/alginate" synthesized for therapeutic purposes. Discov Med 1(1):1–19

Bibi S et al (2016) Eco-friendly synthesis and catalytic application of chitosan/gold/carbon nanotube nanocomposite films. RSC Adv 6(65):60180–60186

Borah D et al (2024) Sustainable green synthesis of SnO_2 quantum dots: a stable, phase-pure and highly efficient photocatalyst for degradation of toxic dyes. Mater Today Sustain 26:100770

Brovko O et al (2024) Investigation of interpolymer complexes of fucoidan with sodium alginate in solutions and films. J Appl Phycol 37:539–551

Budnyak TM et al (2015) Synthesis and adsorption properties of chitosan-silica nanocomposite prepared by sol-gel method. Nanoscale Res Lett 10:87

Bui VKH, Park D, Lee Y-C (2017) Chitosan combined with ZnO, TiO_2 and Ag nanoparticles for antimicrobial wound healing applications: a mini review of the research trends. Polymers 9(1):21

Bukhari NTM et al (2023) Seaweed polysaccharide nanocomposite films: a review. Int J Biol Macromol 245:125486

Canama GJC et al (2023) Synthesis of chitosan-silver nanocomposite and its evaluation as an antibacterial coating for mobile phone glass protectors. ACS Omega 8(20):17699–17711

Cardoso VS et al (2014) Collagen-based silver nanoparticles for biological applications: synthesis and characterization. J Nanobiotechnol 12:1–9

Cesur S et al (2018) Antimicrobial and biodegradable food packaging applications of polycaprolactone/organo nanoclay/chitosan polymeric composite films. J Vinyl Add Tech 24(4):376–387

Chae T et al (2013) Novel biomimetic hydroxyapatite/alginate nanocomposite fibrous scaffolds for bone tissue regeneration. J Mater Sci: Mater Med 24(8):1885–1894

Cheaburu-Yilmaz CN et al (2015) Eco-friendly chitosan-based nanocomposites: chemistry and applications. In: Eco-friendly polymer nanocomposites: chemistry and applications, pp 341–386

Chen H-T et al (2015) Assessment of bone morphogenic protein and hydroxyapatite–titanium dioxide composites for bone implant materials. Surf Coat Technol 276:168–174

Costa WB et al (2024) Polysaccharide hydroxyapatite (nano) composites and their biomedical applications: an overview of recent years. ACS Omega 9(28):30035–30070

Cui Y et al (2025) The complex coacervation of gum Arabic and krill protein isolate and their application for Antarctic krill oil encapsulation. Carbohyd Polym 348:122831

da Costa MPM, de Mello Ferreira IL, de Macedo Cruz MT (2016) New polyelectrolyte complex from pectin/chitosan and montmorillonite clay. Carbohyd Polym 146:123–130

De Silva R et al (2017) Magnesium oxide nanoparticles reinforced electrospun alginate-based nanofibrous scaffolds with improved physical properties. Int J Biomater 2017(1):1391298

Diab AS et al (2024) Antioxidant and antidiabetic potential of Cystoseira myrica-mediated green metallic zinc nanoparticles using the algal extract: synthesis and characterization. Mater Res Exp 11(11):115402

Do MH et al (2025) In situ synthesis of gold nanoparticles embedded in magnetic nanocomposite of glucosamine/alginate for enhancing recyclable catalysis performance of nitrophenol reduction. Nanoscale Adv 7:886–898

Dong W et al (2022) Characterization and antioxidant properties of chitosan film incorporated with modified silica nanoparticles as an active food packaging. Food Chem 373:131414
Duan B et al (2017) Hierarchical microspheres constructed from chitin nanofibers penetrated hydroxyapatite crystals for bone regeneration. Biomacromol 18(7):2080–2089
Elkhenany H et al (2024) Innovative marine-sourced hydroxyapatite, chitosan, collagen, and gelatin for eco-friendly bone and cartilage regeneration. J Biomed Mater Res, Part A 113:e37833
El-Sherbiny IM, Sedki M (2019) Green synthesis of chitosan-silver/gold hybrid nanoparticles for biomedical applications. In: Pharmaceutical nanotechnology: basic protocols. Springer New York, New York, NY, pp 79–84
El-Zawawy NA et al (2023) Mycosynthesis of silver nanoparticles from endophytic Aspergillus flavipes AUMC 15772: ovat-statistical optimization, characterization and biological activities. Microb Cell Fact 22(1):228
Fais G et al (2024) Cytotoxic effects of ZnO and Ag nanoparticles synthesized in microalgae extracts on PC12 cells. Mar Drugs 22(12):549
Farshidfar N, Iravani S, Varma RS (2023) Alginate-based biomaterials in tissue engineering and regenerative medicine. Mar Drugs 21(3):189
Fernando MS et al (2021) Biopolymer-based nanohydroxyapatite composites for the removal of fluoride, lead, cadmium, and arsenic from water. ACS Omega 6(12):8517–8530
Frohbergh ME et al (2012) Electrospun hydroxyapatite-containing chitosan nanofibers crosslinked with genipin for bone tissue engineering. Biomaterials 33(36):9167–9178
Galland P et al (2024) Tuning the underwater adhesiveness of antibacterial polysaccharides complex coacervates. J Colloid Interface Sci 661:196–206
Gamage A et al (2023) Recent application prospects of chitosan based composites for the metal contaminated wastewater treatment. Polymers 15(6):1453
Ganesan S et al (2022) Preparation and characterization of salsalate-loaded chitosan nanoparticles: in vitro release and antibacterial and antibiofilm activity. Mar Drugs 20(12):733
Gobena ST, Woldeyonnes AD (2024) A review of synthesis methods, and characterization techniques of polymer nanocomposites for diverse applications. Discover Materials 4(1):52
Gomaa M, Danial AW (2023) Seaweed-based alginate/hydroxyapatite composite for the effective removal of bacteria, cyanobacteria, algae, and crystal violet from water. J Biol Eng 17(1):69
Gorrasi G, Sorrentino A (2015) Mechanical milling as a technology to produce structural and functional bio-nanocomposites. Green Chem 17(5):2610–2625
Hardy A et al (2018) β-Cyclodextrin-functionalized chitosan/alginate compact polyelectrolyte complexes (CoPECs) as functional biomaterials with anti-inflammatory properties. ACS Appl Mater Interfaces 10(35):29347–29356
Hasan MM et al (2020) Biopolymer-based electrolytes for dye-sensitized solar cells: a critical review. Energy Fuels 34(12):15634–15671
Hashim NC, Frankel D, Nordin D (2019) Graphene oxide-modified hydroxyapatite nanocomposites in biomedical applications: a review. Ceram-Silik 63(4):426–448
Heinemann S, Coradin T, Desimone MF (2013) Bio-inspired silica–collagen materials: applications and perspectives in the medical field. Biomater Sci 1(7):688–702
Hileuskaya K et al (2021) Hydrothermal synthesis and properties of chitosan–silver nanocomposites. Russ J Inorg Chem 66(8):1128–1134
Hussin MSF et al (2022) Extraction of natural hydroxyapatite for biomedical applications—a review. Heliyon 8(8)
Iliou K et al (2022) Marine biopolymers as bioactive functional ingredients of electrospun nanofibrous scaffolds for biomedical applications. Mar Drugs 20(5):314
Işıklan N, Geyik G, Güncüm E (2024) Alginate-based bio-nanocomposite reinforced with poly (2-hydroxypropyl methacrylamide) and magnetite graphene oxide for delivery of etoposide and photothermal therapy. Mater Today Chem 41:102323
Kafshgari MH et al (2012) Preparation of alginate and chitosan nanoparticles using a new reverse micellar system. Iran Polym J 21:99–107

Kalkhoran AHZ et al (2018) Synthesis and characterization of graphene-grafted gelatin nanocomposite hydrogels as emerging drug delivery systems. Biomed Phys Eng Exp 4(5):055017
Kamala K et al (2024) Therapeutic potential of selenium nanoparticles synthesized by mangrove plant: combatting oral pathogens and exploring additional biological properties. J Trace Elements Miner 9:00167
Kanagaraj SSP et al (2023) Antimicrobial activity of green synthesized biodegradable alginate–silver (Alg-Ag) nanocomposite films against selected foodborne pathogens. Appl Nanosci 13(1):651–662
Kareem RO, Bulut N, Kaygili O (2024) Hydroxyapatite biomaterials: a comprehensive review of their properties, structures, medical applications, and fabrication methods. J Chem Rev 6(1):1–26
Karimipour-Fard P et al (2021) Development, processing and characterization of polycaprolactone/nano-hydroxyapatite/chitin-nano-whisker nanocomposite filaments for additive manufacturing of bone tissue scaffolds. J Mech Behav Biomed Mater 120:104583
Katas H et al (2018) Biosynthesis and potential applications of silver and gold nanoparticles and their chitosan-based nanocomposites in nanomedicine. J Nanotechnol 2018:4290705
Kavishri S et al (2024) Facile synthesis of silver nanoparticles from sustainable Sargassum sp. seaweed material and its anti-inflammatory application. Cureus 16(4)
Kazemi SM, Rezaei M (2015) Antimicrobial effectiveness of gelatin–alginate film containing oregano essential oil for fish preservation. J Food Saf 35(4):482–490
Khachatryan G et al (2023) Design of carbon nanocomposites based on sodium alginate/chitosan reinforced with graphene oxide and carbon nanotubes. Polymers 15(4):925
Khan O et al (2021) Green synthesis and evaluation of calcium-based nanocomposites fertilizers: a way forward to sustainable agricultural. J Saudi Soc Agric Sci 20(8):519–529
Khandegar V, Kaur PJ, Chanana PJB (2021) Chitosan and graphene oxide-based nanocomposites for water purification and medical applications: a review. BioResources 16(4):8525
Khane Y et al (2024) Fabrication, characterization and application of biocompatible nanocomposites: a review. Eur Polymer J 214:113187
Khodashenas B et al (2021) Gelatin-coated gold nanoparticles as an effective pH-sensitive methotrexate drug delivery system for breast cancer treatment. Mater Today Chem 20:100474
Khoerunnisa F et al (2023) Design and application of chitosan-CuO nanocomposites synthesized via novel hybrid ionic gelation-ultrasonication methods for water disinfection. J Water Process Eng 52:103556
Khubiev OM et al (2023) Chitosan-based antibacterial films for biomedical and food applications. Int J Mol Sci 24(13):10738
Kikuchi MJB, Bulletin P (2013) Hydroxyapatite/collagen bone-like nanocomposite. Biol Pharm Bull 36(11):1666–1669
Kumar S, Madihally SV (2022) Role of chitosan and chitosan-based nanomaterials in plant sciences. Academic Press
Kumar Balu S et al (2021) Emerging marine derived nanohydroxyapatite and their composites for implant and biomedical applications. J Mech Behav Biomed Mater 119:104523
Kumar LR et al (2017) Evaluation of chitosan as a wall material for microencapsulation of squalene by spray drying: characterization and oxidative stability studies. Int J Biol Macromol 104:1986–1995
Kumar B, Singh N, Kumar P (2024) A review on sources, modification techniques, properties and potential applications of alginate-based modified polymers. Eur Polymer J 213:113078
Kyung GS, Ko SH (2014) Antimicrobial chitosan-silver nanocomposite film prepared by green synthesis for food packaging. Korean J Food Sci Technol 46(3):347–351
Laanoja J et al (2024) Synthesis and synergistic antibacterial efficiency of chitosan-copper oxide nanocomposites. Heliyon 10(15)
Lal J et al (2023) Diverse uses of valuable seafood processing industry waste for sustainability: a review. Environ Sci Pollut Res 31:62249–62263
Lee S-C et al (2010) Preparation and characterization of chitosan–gold nanocomposites for drug delivery application. Surf Rev Lett 17(02):165–172

Leonardi M et al (2021) Smart nanocomposites of chitosan/alginate nanoparticles loaded with copper oxide as alternative nanofertilizers. Environ Sci Nano 8(1):174–187
Li H et al (2022) Preparation and characterization of sodium alginate/gelatin/Ag nanocomposite antibacterial film and its application in the preservation of tangerine. Food Packag Shelf Life 33:100928
Lin S, Zhang T (2023) Research progress in preparation and application of spinel-type metallic oxides (M$\geq$ 2). J Alloy Compd 962:171117
Liu M et al (2022) Synthesis and characterization of fucoidan-chitosan nanoparticles targeting p-selectin for effective atherosclerosis therapy. Oxid Med Cell Longev 2022:8006642
Liu C et al (2024) Value-added utilization technologies for seaweed processing waste in a circular economy: developing a sustainable modern seaweed industry. Compr Rev Food Sci Food Safety 23(6):e70027
Lu H-T et al (2019) Development of genipin-crosslinked and fucoidan-adsorbed nano-hydroxyapatite/hydroxypropyl chitosan composite scaffolds for bone tissue engineering. Int J Biol Macromol 128:973–984
Mane RS, Jadhav V, Al-Enizi AM (2023) Solution methods for metal oxide nanostructures. Elsevier
Martí M et al (2019) Calcium alginate/graphene oxide films: Reinforced composites able to prevent Staphylococcus aureus and methicillin-resistant Staphylococcus epidermidis infections with no cytotoxicity for human keratinocyte HaCaT cells. Eur Polymer J 110:14–21
Maia FR, Oliveira JM, Reis RL (eds) (2024) Handbook of the extracellular matrix: biologically-derived materials. Springer. https://doi.org/10.1007/978-3-031-56363-8
Menaa F et al (2021) Marine algae-derived bioactive compounds: a new wave of nanodrugs? Mar Drugs 19(9):484
Menezes MDLLR et al (2019) Effect of tannic acid as crosslinking agent on fish skin gelatin-silver nanocomposite film. Food Packaging Shelf Life 19:7–15
Mofatehnia P, Elhamifar D, Ziarani GM (2023) Cu-containing core–shell structured Fe_3O_4@ Gelatin nanocomposite: a novel catalyst for the preparation of hexahydroquinolines. Res Chem Intermed 49(3):967–978
Monroy DAP et al (2018) Gelatin and collagen nanofiber scaffolds for tissue engineering. INTECH. https://doi.org/10.5772/intechopen.73316
Montaser A et al (2016) Preparation and characterization of alginate/silver/nicotinamide nanocomposites for treating diabetic wounds. Int J Biol Macromol 92:739–747
Moreira CD et al (2016) Thermogelling chitosan–collagen–bioactive glass nanoparticle hybrids as potential injectable systems for tissue engineering. Mater Sci Eng, C 58:1207–1216
Muraca D, Moscoso Londono O, Pirota KR (2020) Alginate based nanocomposites with magnetic properties. Compos A Appl Sci Manuf 135:105936
Nallasamy P, Natarajan S (2024) Organic fertilizer integrated with marine waste derived $CaCO_3$ nanocarrier system: a focus on enhanced yield and quality in tomato cultivation. Sci Rep 14(1):25299
Nešović K et al (2019) Chitosan-based hydrogel wound dressings with electrochemically incorporated silver nanoparticles–in vitro study. Eur Polymer J 121:109257
Niculescu A, Grumezescu A (2022) Applications of chitosan-alginate-based nanoparticles—an up-to-date review. Nanomaterials 12(2):186
Nieri P et al (2023) Bioactive molecules from marine diatoms and their value for the nutraceutical industry. Nutrients 15(2):464
Niranjana L et al (2022) Characterization of calcium phosphate chitosan nanocomposite as plant growth promoter. J Exp Biol Agricul Sci 3:567–574
Noviyanti AR et al (2023) Hydrothermally synthesized hydroxyapatite-silica composites with enhanced mechanical properties for bone graft applications. Chemistry 5(3):1645–1655
Ofudje EA et al (2023) Bioactivity properties of hydroxyapatite/clay nanocomposites. Sci Rep 13(1):19896
Oliver-Cuenca V et al (2024) Bio-based and biodegradable polymeric materials for a circular economy. Polymers 16(21):3015

Pachaiappan R et al (2022) A review on biofiltration techniques: recent advancements in the removal of volatile organic compounds and heavy metals in the treatment of polluted water. Bioengineered 13(4):8432–8477
Pal A et al (2017) Synthesis of hydroxyapatite from lates calcarifer fish bone for biomedical applications. Mater Lett 203:89–92
Pandey SK et al (2024) Synthesis, fabrication and tribo-mechanical studies of microcapsules filled and Al_2O_3/glass fibre reinforced UHMWPE based self-lubricating and self-healing composites. Proc Inst Mech Eng Part L: J Mater Des Appl 239(8):1569–1578
Peng Z et al (2022) Flexible wearable pressure sensor based on collagen fiber material. Micromachines 13(5):694
Pieklarz K et al (2018) Applications of chitosan–graphene oxide nanocomposites in medical science: a review. Progress Chem Appl Chitin Deriv 23:5–24
Pighinelli L, Kucharska MJCP (2013) Chitosan–hydroxyapatite composites. Carbohyd Polym 93(1):256–262
Prakash J et al (2020) Nanocomposite chitosan film containing graphene oxide/hydroxyapatite/gold for bone tissue engineering. Int J Biol Macromol 154:62–71
Purohit SD et al (2019) Development of a nanocomposite scaffold of gelatin–alginate–graphene oxide for bone tissue engineering. Int J Biol Macromol 133:592–602
Ramesh R et al (2017) Synthesis and characterization of the superparamagnetic Fe_3O_4/Ag nanocomposites. J Cluster Sci 28:963–969
Rana AK et al (2024) Sustainable MXene-chitosan/chitin composites for Interdisciplinary applications in water purification, bio-medical, bio-sensing and electronic fields. Mater Today Sustain 25:100671
Ranganadhareddy A, Chandrasekhar C (2023) Biopolymer from marine waste biomass and its applications-a review. J Biochem Technol 14(2–2023):87–93
Raval JP (2019) Fabrication and applications of hydroxyapatite-based nanocomposites coating for bone tissue engineering. In: Applications of nanocomposite materials in orthopedics. Woodhead Publishing, pp 71–82
Rawdkuen S, Kaewprachu P (2019) Valorization of food processing by-products as smart food. Food Preservation and Waste Exploitation, 107
Reddy PRS et al (2014) Synthesis of alginate based silver nanocomposite hydrogels for biomedical applications. Macromol Res 22:832–842
Ristić I et al (2024) Eco-friendly microwave synthesis of sodium alginate-chitosan hydrogels for effective curcumin delivery and controlled release. Gels 10(10):637
Roseline TA et al (2019) Nanopesticidal potential of silver nanocomposites synthesized from the aqueous extracts of red seaweeds. Environ Technol Innov 13:82–93
Rudovica V et al (2021) Valorization of marine waste: use of industrial by-products and beach wrack towards the production of high added-value products. Front Mar Sci 8:723333
Sabree IK, Mahdi OS (2019) Characterization of zirconia-hydroxyapatite nanocomposites for orthopedic and dental applications. Int J Eng Technol 7(4.19):926–930
Sachithanadam M, Joshi SC (2016) Effect of granule sizes on acoustic properties of protein-based silica aerogel composites via novel inferential transmission loss method. Gels 2(1):11
Sadasivuni KK et al (2020) Recent advances in mechanical properties of biopolymer composites: a review. Polym Compos 41(1):32–59
Said HA et al (2023) Manufacturing methods, properties, and potential applications in bone tissue regeneration of hydroxyapatite-chitosan biocomposites: a review. Int J Biol Macromol 243:125150
Santos VP et al (2020) Seafood waste as attractive source of chitin and chitosan production and their applications. Int J Mol Sci 21(12):4290
dos Santos VCT et al (2024) Metallic nanoparticles synthesized by marine microorganisms and its application against pathogenic microorganisms: challenges and opportunities in marine nanotechnology. Desalin Water Treat 317:100283

Santulli C et al (2023) Use of sea waste to enhance sustainability in composite materials: a review. J Marine Sci Eng 11(4):855
Sanuja S, Agalya A, Umapathy MJ (2015) Synthesis and characterization of zinc oxide–neem oil–chitosan bionanocomposite for food packaging application. Int J Biol Macromol 74:76–84
Sarkar S et al (2023) Current status of marine animal derived polysaccharides in sustainable food packaging. ACS Food Sci Technol 3(11):1877–1889
Selim MS et al (2025) Non-toxic silicone-polyurethane nanocomposites filled with graphene/tungsten disulphide nanorods as hierarchical superhydrophobic marine antifouling surfaces. Chem Eng Sci 200:108987
Shaikh AA et al (2024) Biopolymer-based nanocomposites for application in biomedicine: a review. J Polym Eng 44(2):83–116
Sharma A et al (2023) A propitious role of marine sourced polysaccharides: drug delivery and biomedical applications. Carbohyd Polym 308:120448
Sharmeen S et al (2018) Polyethylene glycol functionalized carbon nanotubes/gelatin-chitosan nanocomposite: an approach for significant drug release. Bioact Mater 3(3):236–244
She J et al (2024) Recent advances in collagen-based hydrogels: materials, preparation and applications. React Funct Polym React Funct Polym 207:106136
Shen X et al (2011) A novel method for the fabrication of homogeneous hydroxyapatite/collagen nanocomposite and nanocomposite scaffold with hierarchical porosity. J Mater Sci: Mater Med 22:299–305
Siddiqui HA, Pickering KL, Mucalo MR (2018) A review on the use of hydroxyapatite-carbonaceous structure composites in bone replacement materials for strengthening purposes. Materials 11(10):1813
Siddiqui VU et al (2022) Fabrication of a zinc oxide/alginate (ZnO/Alg) bionanocomposite for enhanced dye degradation and its optimization study. RSC Adv 12(12):7210–7228
Silva SS et al (2020) Marine-derived polymers in ionic liquids: architectures development and biomedical applications. Mar Drugs 18(7):346
Silva SS et al (2023) Antimicrobial marine-derived materials. Handbook of the extracellular matrix: biologically-derived materials. Springer, pp 1–36
Silva-Holguín PN, Reyes-López SY (2020) Synthesis of hydroxyapatite-Ag composite as antimicrobial agent. Dose-Response 2020(1):1–14
Šimat V et al (2020) Recent advances in marine-based nutraceuticals and their health benefits. Mar Drugs 18(12):627
Soleimani S et al (2023) Green synthesis of graphene oxide-based nanocomposite by Polycladia myrica: antibacterial, anti-algae, and acute zooplanktonic responses. J Appl Phycol 35(3):1417–1429
Spoială A et al (2021) Chitosan-based nanocomposite polymeric membranes for water purification—a review. Materials 14(9):2091
Sumathra M et al (2017) Sustainable pectin fascinating hydroxyapatite nanocomposite scaffolds to enhance tissue regeneration. Sustain Chem Pharm 5:46–53
Sun L et al (2024) Preparation and modification of collagen/sodium alginate-based biomedical materials and their characteristics. Polymers 16(2):171
Šupová MJM (2024) Calcium orthophosphate–clay composites—preparation, characterisation, and applications: a review. Minerals 14(2):169
Swain S, Kwon TY, Rautray TR (2021) Fabrication of silver doped nano hydroxyapatite-carrageenan hydrogels for articular cartilage applications. BioRxiv 31(424664):2020–2112
Świerczyńska M et al (2024) Preparation and biochemical activity of copper-coated cellulose nonwoven fabric via magnetron sputtering and alginate-calcium ion complexation. Mar Drugs 22(10):436
Tabatabaei RH et al (2018) Preparation and characterization of nano-SiO_2 reinforced gelatin-k-carrageenan biocomposites. Int J Biol Macromol 111:1091–1099
Tavakol S et al (2013) Bone regeneration based on nano-hydroxyapatite and hydroxyapatite/chitosan nanocomposites: an in vitro and in vivo comparative study. J Nanopart Res 15:1–16

Taymouri S et al (2024) Biocompatible nanocomposite scaffolds based on carrageenan incorporating hydroxyapatite and hesperidin loaded nanoparticles for bone tissue regeneration. Polym Adv Technol 35(1):e6284

Thongjamroon S, Wootthikanokkhan J, Poolthong NJC (2023) Photocatalytic performances and antifouling efficacies of alternative marine coatings derived from polymer/metal oxides (WO_3@ TiO_2)-based composites. Catalysts 13(4):649

Tian B et al (2021) Formation chitosan-based hydrogel film containing silicon for hops β-acids release as potential food packaging material. Int J Biol Macromol 191:288–298

Toiserkani HJCI (2016) Fabrication and characterization chitosan/functionalized zinc oxide bionanocomposites and study of their antibacterial activity. Compos Interfaces 23(3):175–189

Tran HV et al (2016) Facile hydrothermal synthesis of silver/chitosan nanocomposite and application in the electrochemical detection of hydrogen peroxide. Sens Lett 14(1):32–38

Tripathi S, Mehrotra G, Dutta PK (2011) Chitosan–silver oxide nanocomposite film: preparation and antimicrobial activity. Bull Mater Sci 34:29–35

Trivedi TJ, Rao KS, Kumar A (2014) Facile preparation of agarose–chitosan hybrid materials and nanocomposite ionogels using an ionic liquid via dissolution, regeneration and sol–gel transition. Green Chem 16(1):320–330

Trung TS et al (2022) Valorization of fish and shrimp wastes to nano-hydroxyapatite/chitosan biocomposite for wastewater treatment. J Sci Adv Mater Dev 7(4):100485

Türe H (2019) Characterization of hydroxyapatite-containing alginate–gelatin composite films as a potential wound dressing. Int J Biol Macromol 123:878–888

Udayakumar GP et al (2021) Ecofriendly biopolymers and composites: preparation and their applications in water-treatment. Biotechnol Adv 52:107815

Uddin MN et al (2024) Research and applications of nanoclays: a review. SPE Polym 5(4):507–535

Unal S et al (2019) Production and characterization of calcium phosphates from marine structures: the fundamentals basics. Springer Ser Biomater Sci Eng 113–135

Vahdati M et al (2020) Coacervate-based underwater adhesives in physiological conditions. ACS Appl Polym Mater 2(8):3397–3410

Van Hong TD et al (2013) Electrospun chitosan/hydroxyapatite nanofibers for bone tissue engineering. J Mater Sci 48:1640–1645

Vaseashta A (2015) Life cycle analysis of nanoparticles: reducing risk and liability. DEStech Publications, Inc.

Venkatesan J et al (2016) Seaweed polysaccharide-based nanoparticles: preparation and applications for drug delivery. Polymers 8(2):30

Venkatesan J et al (2018) Preparation, characterization and biological applications of biosynthesized silver nanoparticles with chitosan-fucoidan coating. Molecules 23(6):1429

Venugopal V (2021) Valorization of seafood processing discards: Bioconversion and bio-refinery approaches. Front Sustain Food Syst 5:611835

Viacava GE, Marcovich NE, Ansorena MR (2024) Studies on nanostructure functionalized materials for active and intelligent food packaging. In: Nanostructured materials for food packaging applications. In: Nanostructured Materials for Food Packaging Applications, pp 477–511

Villalba-Rodríguez AM et al (2021) Nanoclay/polymer-based hydrogels and enzyme-loaded nanostructures for wound healing applications. Gels 7(2):59

Vilvert JC et al (2023) Preservation of quality and bioactive compounds in mangoes using chitosan-Graphene-oxide-based biodegradable packaging. Horticulturae 9(10):1145

Vinod A et al (2020) Renewable and sustainable biobased materials: an assessment on biofibers, biofilms, biopolymers and biocomposites. J Clean Prod 258:120978

Wan C, Frydrych M, Chen BJSM (2011) Strong and bioactive gelatin–graphene oxide nanocomposites. Soft Matter 7(13):6159–6166

Wang Y et al (2015) Alginate-intervened hydrothermal synthesis of hydroxyapatite nanocrystals with nanopores. Cryst Growth Des 15(4):1949–1956

Wang H et al (2023) Improvement in surface performance of stainless steel by nitride and carbon-based coatings prepared via physical vapor deposition for marine application. J Market Res 27:6021–6046
Wang B, Zhang X, Ma X (2024) PVC-based ultraviolet screening films modified by chitosan capped ZnO–hydroxyapatite. J Appl Spectrosc 91(4):921–927
Wang X-Y et al (2015) Preparation Fe_3O_4 @ chitosan magnetic particles for covalent immobilization of lipase from Thermomyces lanuginosus. Int J Biol Macromol 75:44–50
Wu H et al (2014) Emulsion cross-linked chitosan/nanohydroxyapatite microspheres for controlled release of alendronate. J Mater Sci: Mater Med 25:2649–2658
Yang M, Xia Y (2017) Preparation and characterization of nano-SiO_2 reinforced alginate-based nanocomposite films (II). J Appl Polym Sci 134(38):45286
Yang Z, Chaieb S, Hemar YJPR (2021) Gelatin-based nanocomposites: a review. Polym Rev 61(4):765–813
Yang M et al (2022) Silver nanoparticle-loaded gelatin-based nanocomposite films toward enhanced mechanical properties and antibacterial activity. ACS Appl Bio Mater 5(5):2193–2201
Yarahmadi A et al (2024) Materials based on biodegradable polymers chitosan/gelatin: a review of potential applications. Front Bioeng Biotechnol 12:1397668
Yosri N et al (2024) Sustainability in the green engineering of nanocomposites based on marine-derived polysaccharides and collagens: a review. Int J Biol Macromol 274:33249
Young AT et al (2016) Interaction of stem cells with nano hydroxyapatite-fucoidan bionanocomposites for bone tissue regeneration. Int J Biol Macromol 93:1488–1491
Zadeh KK, Jafari D (2024) Activated carbon/alginate/Fe_3O_4 magnetic nanocomposite as a superior functional material for removal of lead from aqueous media. Biomass Convers Biorefinery 14(16):19025–19043
Zahorán R et al (2022) Flow-driven synthesis of calcium phosphate–calcium alginate hybrid chemical gardens. Soft Matter 18(42):8157–8164
Zare H et al (2021) Fabrication of multifunctional nano gelatin/zinc oxide composite fibers. Autex Res J 21(4):403–407
Zhan Z et al (2024) Valorization of seafood waste for food packaging development. Foods 13(13):2122
Zhang X et al (2023) Collagen-based flexible electronic devices for electrochemical energy storage and sensing. Macromol Rapid Commun 44(10):2200977
Zhang M et al (2023) Multifunctional chitosan/alginate hydrogel incorporated with bioactive glass nanocomposites enabling photothermal and nitric oxide release activities for bacteria-infected wound healing. Int J Biol Macromol 232:123445
Zhang X et al (2024) Using electrodeposition of carboxylated chitosan for green preparation of copper nanoclusters and nanocomposite films. J Wuhan Univ Technol-Mater Sci Ed 39(6):1348–1357
Zhong T et al (2022) Chitosan/silica nanocomposite preparation from shrimp shell and its adsorption performance for methylene blue. Sustainability 15(1):47
Zhu J et al (2023) Pathways toward wearable and high-performance sensors based on hydrogels: toughening networks and conductive networks. Nat Sci Rev 10(9):nwad180
Zine R, Sinha M (2017) Nanofibrous poly (3-hydroxybutyrate-co-3-hydroxyvalerate)/collagen/graphene oxide scaffolds for wound coverage. Mater Sci Eng, C 80:129–134

Chapter 10
Applications of Marine Nanoscience and Nanotechnology

Abid H. Banday, Syed Irtiqa Urfi, Meer Mehru Manzoor, Syed Hadi Mashad, and Mohnad Abdalla

Abstract Nanoscience and nanotechnology represent groundbreaking and innovative domains dedicated to the exploration, manipulation, and application of materials at the nanoscale, which ranges from 1 to 100 nm. At this minute scale, materials display distinct physical, chemical, and biological characteristics, facilitating the creation of advanced technologies that surpass conventional approaches. The objective of nanoscience is to uncover the fundamental mechanisms and processes occurring at the nanoscale, while nanotechnology leverages this understanding to engineer new tools, systems, and devices with diverse practical applications. This field has revolutionized the creation of new materials by altering the properties of bulk substances. Notably, materials science has evolved to incorporate nanometer-sized components, leading to the fabrication of materials and structures that are unattainable through traditional techniques. Nanostructures are typically lightweight, robust, and highly adjustable, making them essential for the advancement of sophisticated manufacturing systems and tools. In the context of the marine environment, marine nanoscience investigates natural nanosystems and processes within the oceans, examining their mechanisms and impacts on structure and function. Subsequently, marine nanotechnology applies this knowledge to devise innovative tools, systems, and devices aimed at addressing challenges faced by marine ecosystems. Together, nanoscience and nanotechnology provide sustainable solutions for

A. H. Banday (✉) · S. H. Mashad
Department of Chemistry, Islamia College of Science and Commerce, Hawal, Srinagar, India
e-mail: abidrrl@gmail.com

A. H. Banday
Department of Chemistry and Biochemistry, Auburn University, Auburn, AL, USA

S. I. Urfi
Department of Metallurgical and Materials Engineering, National Institute of Technology, Hazratbal, Srinagar, India

M. M. Manzoor
Department of Chemistry, School of Sciences, Cluster University of Jammu, Jammu, India

M. Abdalla
Pediatric Research Institute, Children's Hospital Affiliated to Shandong University, Jinan, Shandong, China

S. K. Panda and V. P.M. (eds.), *Bionanocomposites from Marine Origin*, Advanced Structured Materials 247, https://doi.org/10.1007/978-3-032-10749-7_10

the future of marine systems, enhancing environmental protection and advancing scientific understanding.

Keywords Marine nanoscience · Marine nanotechnology · Nanocoatings · Biogeochemical cycles · Nanoecology · Blue economy · Marine exploration · Remediation

10.1 Introduction

Nanoscience and nanotechnology are innovative revolutionary fields focused on the study, manipulation and application of materials at the nanoscale (typically between 1 and 100 nm) (Joudeh and Linke 2022; Mekuye and Abera 2023). At this scale, materials exhibit unique physical, chemical, and biological properties, opening doors to advanced and innovative technologies beyond traditional methods (Hassett et al. 2021; Patra and Baek 2014; Munyalo and Zhang 2018). While nanoscience focuses on uncovering the underlying mechanisms to understand phenomena at the nanoscale and advance knowledge about processes, nanotechnology focuses on using this understanding to design and develop new tools, systems, and devices with a wide range of practical applications (Zäch et al. 2006; Bayda et al. 2020; Dowling et al. 2004). Nanotechnology has revolutionized almost every branch of science. It has replaced traditional synthesis methods with greener and more sustainable approaches, attracting the attention of researchers globally (Fig. 10.1) (Alqarni et al. 2022; Khan et al. 2022; Harinisri et al. 2023). Since oceans are vast ecosystems and hotbeds of biodiversity, they offer rich resources for nanoparticle synthesis (Virginia Armbrust and Palumbi 1979; O'Dor et al. 2009). The marine-based synthetic processes are environment friendly, energy efficient, and time-saving (Kiuru et al. 2014).

Over the past decade, nanotechnology has revolutionized several fields and has achieved incredible success through the use of nanoparticles. In agriculture, nanoparticles are used as herbicides, nano-fertilizers, nano-pesticides, and as gene vectors targeted to specific plant organelles to deliver their contents precisely (Mgadi et al. 2024; Pestovsky and Martínez-Antonio 2017). These innovations minimize the loss of agrochemicals through the formation of stable emulsions, use of minimal amounts of pesticide, better coverage of the leaf surface, and precise application (Singh et al. 2021; Salvador-Morales and Grodzinski 2022).

Nanotechnology has also enabled the creation of new materials, tools, and systems by modifying the properties of bulk materials (Salvador-Morales and Grodzinski 2022; Squire et al. 2023; Shams et al. 2022). The field of material science in particular has been transformed by the use of nanometer-sized building blocks. It is now possible to create unique materials and structures that are not achievable through traditional chemical processes. Nanostructures are light, strong and often highly tunable, making them essential for the development of advanced manufacturing tools and systems (Sajid 2022; Wu et al. 2020).

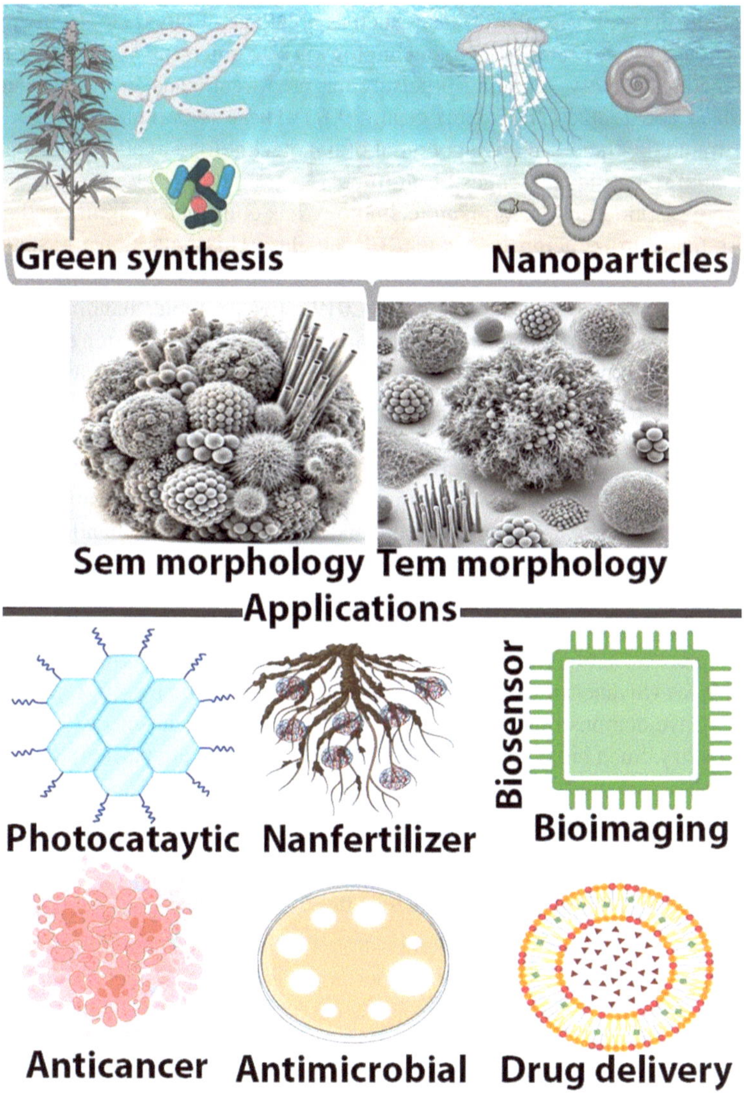

Fig. 10.1 Green synthesis of nanoparticles and their applications

The impact of nanotechnology extends beyond materials science to other fields such as medicine, petrochemicals, electronics, defense, and quantum computing. It has successfully overcome many limitations of traditional science through miniaturization and minimization. A decade ago, the concept of microscopic machines capable of traveling inside the human body to target and eliminate bacteria or pathogens seemed inconceivable. Today, these nanomachines are a reality, with applications in cardiovascular diseases such as breakage of blood clots which potentially increase life expectancy. Nanotechnology is also emerging as a pioneering tool in therapeutic delivery, enabling precisely targeted treatments to the human body (Farokhzad and Langer 2009; Suri et al. 2007). These remarkable advances have demonstrated the transformative potential of nanotechnology to address some of the most crucial scientific challenges of our time.

In the transportation sector, nanotechnology has contributed significantly to the progress made in reducing the weight of vehicles. This reduction translates into increased efficiency, higher speeds, improved safety and lower energy consumption (Ahmed et al. 2019; Mathew et al. 2019). For example, nanoparticle-based steel alloys produce lighter yet stronger materials with higher strength-to-weight ratios than traditional metals. In the automotive industry, nanomaterials are widely used to develop parts with better wear resistance, leading to longer product life. Additionally, nanotechnology has driven innovation in renewable energy and energy storage systems. Nanomaterials are used in solar cells to harness hydrogen as a clean, sustainable fuel, contribute to the development of next-generation batteries, and create corrosion-resistant surfaces and materials (Parmar and Bhande 2022; Zang 2011; Hussein 2015).

Marine biodiversity has the potential to produce cost-effective and sustainable nanoparticles. Various marine organisms, including algae, yeasts, fungi, cyanobacteria, seagrasses, and marine animals, have been used to efficiently synthesize nanoparticles (Sivanesan et al. 2021; Jeong et al. 2022). The resulting green nanomaterials have demonstrated significant promise in biomedical sciences and medicinal chemistry. Such biosynthetically derived nanoparticles are particularly valuable in pharmaceutical, nutraceutical, and cosmeceutical applications, showcasing their versatility and potential in advancing healthcare and sustainable technology (Aithal and Aithal 2021; Cornier et al. 2017; Duan et al. 2023).

Applications of Marine nanoscience and nanotechnology

Nanoscience and nanotechnology demonstrate the theoretical basis and the practical execution of this advanced field for various application, respectively. In the context of marine environment, marine nanoscience explores the study of natural nanoscale systems and processes within oceanic settings, analyzing their mechanisms and their effects on structure and function. At the same time, marine nanotechnology leverages this information to develop groundbreaking tools, systems, and devices designed to tackle challenges within marine ecosystems. Collectively, nanoscience and nanotechnology work in harmony to promote sustainable solutions for the prospective future of marine systems (Chinnappan et al. 2015; Alishahi 2015).

10.2 Applications of Marine Nanoscience

Marine nanoscience mainly concentrates on studying, analyzing, and comprehending nanoscale processes and phenomena occurring within the marine ecosystem and environment. The primary applications and advancements in this area comprise:

10.2.1 Nature and Role of Nanoparticles in the Marine Systems

Nanoscience studies the sources of nanoparticles in ocean waters, which may originate from hydrothermal vents, volcanic activity, or chemical reactions. It also analyzes the chemical characteristics of these nanoscale structures and their interactions within marine ecosystems. Additionally, the transport of natural nanoparticles in the ocean, along with their functions in pollutant binding and nutrient recycling, is evaluated (Barhoum et al. 2022).

10.2.2 Study of Natural Nanostructures in Marine Organisms

Marine nanoscience contributes to understand essential biological processes, including shell formation and biomineralization in marine organisms like mollusks and corals, which utilize nanoscale mineral structures for constructing their shells and skeletons (Maliszewska et al. 2020; Johnston 2010). Likewise, it investigates the function of nanoscale silica structures in photosynthesis, the carbon cycle, and ecosystem stabilization in marine diatoms and phytoplankton. Nanoscience also examines nanosurfaces in marine organisms, such as the anti-fouling and self-cleaning characteristics of shark skin. Furthermore, numerous marine organisms, including fish and crustaceans, possess nanoscale photonic crystals that impart them iridescent colors and camouflage capabilities (Vigneron and Simonis 2012; Vukusic and Sambles 2003). These nanoscale attributes provide significant insights into natural phenomena and motivate innovative technological applications (Kong et al. 2016).

10.2.3 Nanoecology in Marine Systems

Marine nanoscience centers on the intricate relationships between nanoparticles and microbes in ocean ecosystems. Studies assess how microorganisms engage with both naturally occurring and human-made nanoparticles. Additionally, investigations look into microbial adaptations to these particles and their consequences for marine food webs. Marine nanoecotoxicology examines the influence of naturally existing nanoparticles on marine organisms and offers insights into the effects of human activities on ocean ecosystems (Oshiro et al. 2024; Haro 2011; Brander et al. 2011; Wang 2023).

10.2.4 Nanoscience in Biogeochemical Cycles

Nanoscience investigates the manner in which different nanoscale particles found in the ocean aid the development of marine life. For instance, researchers examine the significance of iron nanoparticles in fostering phytoplankton growth, which plays a role in carbon sequestration via the biological pump. The interactions between nanoscale particles in the carbon, nitrogen, and sulfur cycles within marine environments are also analyzed. These processes are further explored for their implications on the stability of marine ecosystems and their possible effects on global climate control (Cusson et al. 2015; Plank and Law 2012).

10.2.5 Nanoscale Processes in Ocean Chemistry

Nanoscience seeks to understand the nanoscale phenomena taking place beneath the ocean's surface. This encompasses the investigation of mineral and salt crystallization from seawater. Furthermore, research delves into the function of nanostructures in gas exchange at the ocean–atmosphere boundary, with a particular emphasis on processes like CO_2 absorption and emission (Babakhani et al. 2022; Carafa 2024).

10.2.6 Marine Sediment and Nanostructures

Marine nanoscience investigates how nanoscale particles play a role in the formation of oceanic sediments, especially their contribution to the entrapment of nutrients, organic matter, and carbon (Wallace et al. 2003). The study also concentrates on the creation of nanoscale minerals at hydrothermal vents and how these nanostructures affect deep-sea ecosystems (Gharekhan and Upasani 2019; Wu et al. 2018).

10.2.7 Marine Nanoscience in Climate Studies

Marine nanoscience is crucial for comprehending the influence of environmental and climate changes on marine life (Maltby et al. 2022; Craig 2012). Research investigates how ocean acidification impacts nanoscale carbonate structures in microorganisms, specifically concerning fluctuations in pH. Moreover, the composition of nanoscale aerosols generated by ocean currents and their role in climate regulation is analyzed.

10.2.8 Marine Nanobiotechnology for Science

Marine nanoscience studies the distinct nanostructures present in marine organisms, offering inspiration for biomimetic designs. It also investigates the role of evolutionary biology in understanding how these nanostructures have evolved over time to facilitate the survival of marine organisms in extreme ocean environments (Singh et al. 2020a; Sangeetha et al. 2013).

10.2.9 Marine Ecology and Nanoscience

Marine nanoscience assesses the ecological impact of nanosized particles, particularly emphasizing nanosized plastics in ocean ecosystems. Research tracks the origins, effects, and breakdown of these particles in marine systems (Mehmood et al. 2021). Additionally, the contribution of nanomaterials to nutrient recycling, especially in the formation of marine snow within deep-sea ecosystems, is examined (Singh et al. 2020b; Page and Lawley 2022).

Evolving Marine Nanoscience Applications

The domain of marine nanoscience shows considerable promise, particularly in enhancing our understanding of oceanographic measurements and marine genomics. As researchers persist in creating more accurate tools for tracking nanoscale variations in salinity, temperature, and pH, we anticipate more precise and timely insights into the health and dynamics of marine ecosystems. Furthermore, studying how nanoscale processes impact the operation of proteins, RNA, and DNA in marine organisms is likely to uncover new biological mechanisms that affect ecosystem dynamics (Mann and Lazier 2005; Fujii et al. 2020).

These innovations might change not just scientific understanding but also our approach to urgent environmental challenges like ocean acidification, climate change, and loss of biodiversity. The ability to view these nanoscale interactions in extraordinary detail could equip us with the resources necessary to gain a deeper insight into the ocean's intricate systems and the delicate ways they are affected by both natural and anthropogenic changes. As the discipline evolves, it is probable that these advancements will create new paths for investigation and applications that will influence our connection with the marine environment.

10.3 Applications of Marine Nanotechnology

10.3.1 Marine Materials and Coatings

Marine nanotechnology is emerging as a transformative solution for various significant challenges encountered by the marine industry, especially in transportation. A key application of nanotechnology is the creation of nanostructured materials that resist corrosion, which are essential for extending the operational life and performance of marine vessels, offshore platforms, equipment, and pipelines. These materials are crafted to form protective surfaces that can endure harsh and corrosive saltwater environments, ensuring the structures remain functional and resilient over prolonged periods.

Moreover, nanotechnology is crucial in addressing biofouling, which happens when marine organisms like algae and barnacles accumulate on the hulls of ships and other submerged infrastructure (Sinha et al. 2023; Krishna Mohan et al. 2024). Anti-fouling nanocoatings are notably effective in preventing the accumulation of these organisms, resulting in decreased maintenance expenses and enhanced operational efficiency. By reducing drag from biofouling, these coatings improve the speed and fuel efficiency of vessels, making them more environmentally friendly and economical.

Another significant benefit of nanotechnology in the marine sector is the creation of nanocomposites (Joseph et al. 2020; Wang et al. 2023). These materials facilitate the development of lightweight yet exceptionally strong structures for shipbuilding and cargo packaging (Youssef and Abdel-Aziz 2013). The improved strength-to-weight ratio of these nanocomposites leads to superior overall performance, including enhanced fuel efficiency and lower emissions, which are vital for the sustainability of the marine transportation industry.

Collectively, these advancements in marine nanotechnology are not just enhancing the functionality and durability of marine transportation systems but also fostering more efficient and sustainable practices in an industry that is crucial for global trade and environmental protection.

10.3.2 Nanotechnology in Marine Energy Harvesting

Marine nanoparticles are crucial for the production of hydrogen from ocean sources, renewable energy generation, and the innovation of next-generation batteries (Bosu and Rajamohan 2024; Liao et al. 2023). Photocatalytic nanomaterials are employed to generate hydrogen fuel by splitting water molecules through solar energy. Nanomaterials also enhance the performance of tidal and wave energy conversion technologies (Rosa-Santos et al. 2022; Roy et al. 2023; Wang et al. 2019). Furthermore, nanotechnology supports the development of high-capacity, durable batteries for underwater drones and marine vessels while improving the durability of underwater energy harvesters and turbines via nanocoatings that lower maintenance expenses (Radhakrishnan et al. 2022; Liu et al. 2023).

10.3.3 Marine Pollution Monitoring and Remediation

Nanotechnology is essential in the monitoring and cleaning up of marine pollution (Fig. 10.2) (Corsi et al. 2021; Saliu et al. 2021; Abouelkheir et al. 2023). Nanosensors are extremely sensitive and capable of detecting pollutants such as oils, heavy metals, and microplastics in real time (Mustafa et al. 2021). They also aid in identifying hazardous toxins and algal blooms (Corsi et al. 2021; Jigyasa Rajput 2022). Nanoparticles, such as iron oxide, are utilized to extract radioactive isotopes and heavy metals from seawater. Additionally, photocatalytic nanoparticles are applied to degrade organic pollutants in the water. Carbon nanotubes and other nanoscale devices play a vital role in the cleanup of oil spills (Arum et al. 2021; Vetrivel et al. 2018).

10.3.4 Robotics in Marine Exploration

Nanoelectronic devices and nanostructured sensors are crucial for mapping, navigation, and environmental assessments, particularly in deep-sea exploration (Katzschmann et al. 2018; Aracri et al. 2021). Autonomous underwater vehicles (AUVs), driven by nanoelectronics, are small, high-performance drones utilized for investigating marine environments (Segovia Ramírez et al. 2021; Ciaccio and Troisi 2021). Nano-enabled optical and acoustic devices improve dependable communication in the difficult underwater conditions.

10.3.5 Ecosystem Protection in Marine Systems

Nanomaterials, especially nanosensors and nanoprobes, play a key role in monitoring and rehabilitating marine species. Nanoparticles are employed specifically to target and manage invasive species while preserving native ones. These materials are also beneficial in the restoration of marine habitats, including aiding coral growth in reef restoration efforts. Nanoprobes and sensors are essential for examining marine microorganisms, plankton, and other minute organisms. Furthermore, they help evaluate the effects of pollution and climate change on these marine microorganisms (Plank and Law 2012; Fujii et al. 2020).

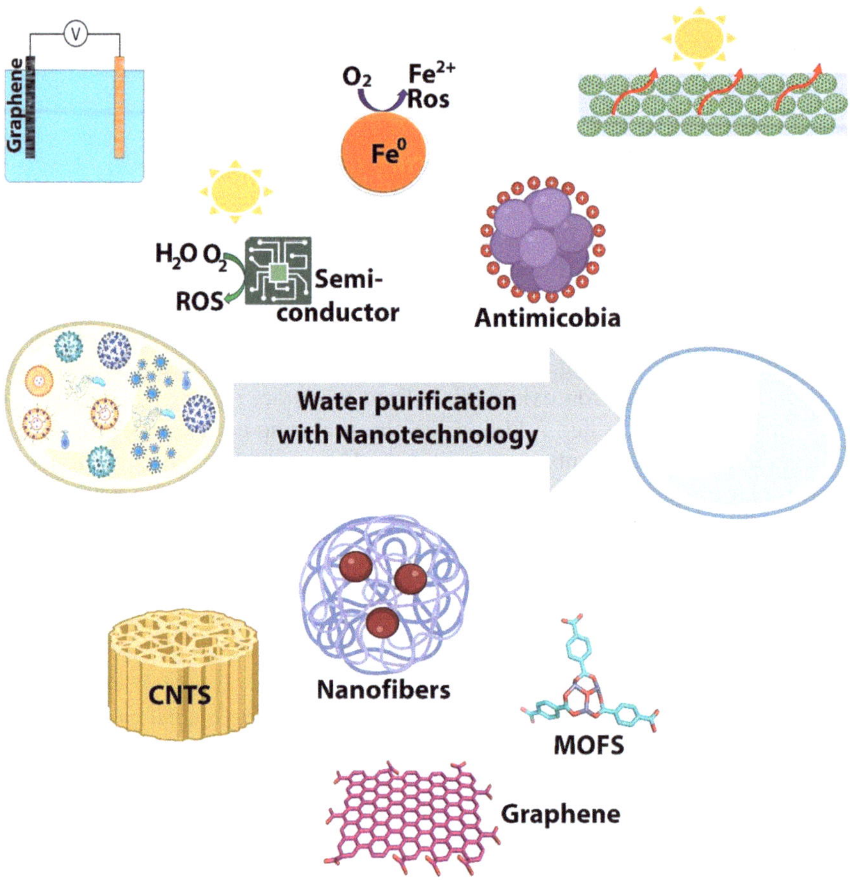

Fig. 10.2 Water treatment through nanoparticles

10.3.6 Marine Food and Fisheries

Nanotechnology greatly influences seafood safety, the enhancement of aquaculture, and the quality of sea-feed (Ogunkalu 2019; Dar et al. 2020). Nano-additives incorporated into fish feed improve its nutritional value and boost its absorption efficiency by marine organisms (Nasr-Eldahan et al. 2021). Nanosensors are utilized to identify pollutants like heavy metals (for instance, arsenic and mercury), indicators of spoilage in seafood items, and pathogens. Moreover, nanostructured water treatment systems aid in maintaining ideal water quality within aquaculture environments (Gehrke et al. 2015; Qu et al. 2013).

10.3.7 Nanotechnology in Carbon Sequestration

A variety of nanostructured materials and devices are employed to capture and neutralize CO_2 from oceanic waters, assisting in the fight against ocean acidification (Chausali et al. 2023; Anwar et al. 2018). Nanotechnology is also utilized in the crafting of artificial nanoscale structures that emulate marine organisms for the efficient absorption of carbon. In addition, nanomaterials offer protection against ice development on offshore infrastructures and maritime vessels.

10.3.8 Marine Medicine and Biotechnology

Nanomaterials and devices inspired by marine life are utilized in creating biocompatible medical treatments and tools. These innovations also assist in extracting rare bioactive substances from marine species (Banday et al. 2024; Beygmoradi and Homaei 2017), which are in high demand within the pharmaceutical and healthcare industries. Nano-based biosensors are used to detect marine pathogens that could pose risks to human health (Kim and Chojnacka 2015; Halvorson and Quezada 2019).

10.3.9 Naval Defense and Security

Nanotechnology is widely applied in the creation of nanocoatings for naval defense, enhancing radar absorption, minimizing visibility, and improving underwater threat detection. It also aids in the advancement of lightweight armor materials for naval ships, reinforcing their protective capabilities.

Nanotechnology and the Blue Economy

The blue economy (Sarangi 2023; Nguyen et al. 2021), which centers on economic activities pertaining to the sea, ocean, and coastal areas, gains substantial benefits from marine nanotechnology (Pace et al. 2023; Martínez-Vázquez et al. 2021). Nanotechnology enhances the effectiveness of extracting valuable minerals from ocean floor sediments. Additionally, marine nanoproducts play a role in developing potent cosmeceuticals, nutraceuticals, and other essential consumer products. Furthermore, nanofiltration technologies improve the efficiency and decrease the energy usage of seawater desalination procedures.

10.4 Future Prospects and Challenges

Marine nanoscience and nanotechnology possess immense potential in transforming various sectors and numerous areas of current significance such as biodiversity conservation, renewable energy, blue economy, and underwater exploration. It holds significant promise for alleviating environmental issues linked to the release of nanomaterials in marine environments. Marine nanotechnology represents a convergence of cutting-edge science and responsible environmental management, providing solutions to contemporary marine problems while creating new pathways for sustainable development. However, the technologies have not yet reached their full potential, necessitating the creation of cost-effective and scalable nanotechnological solutions.

10.5 Conclusion

Marine nanoscience and nanotechnology together constitute an essential and transformative domain, presenting groundbreaking methods for comprehending, preserving, and sustainably managing marine systems and environments. While marine nanoscience focuses on the investigation and evaluation of nanoscale structures, phenomena, processes, and behaviors, marine nanotechnology utilizes this expertise to devise innovative solutions for industry, scientific inquiry, and sustainable development. This collaboration between marine nanoscience and nanotechnology continues to encourage sustainable and meaningful solutions for marine and environmental issues.

Key advancements driven by marine nanoscience and nanotechnology include:

- Development of cutting-edge materials such as anti-fouling surfaces and corrosion-resistant coatings, which reduce economic expenses and environmental contamination.
- Improved monitoring of marine ecosystems through nanosensors that facilitate real-time evaluations of water quality and better ecosystem management.
- Progress in energy, which includes increased efficiency of marine renewable energy systems and breakthroughs in seawater-based hydrogen production.
- Efforts aimed at restoring marine biodiversity and enhancing aquaculture methods.
- Sustainable approaches to tackle marine challenges such as water pollution, biofouling, and oil spills.

Despite notable progress, numerous challenges in marine nanoscience and nanotechnology are still unresolved, including the scalability of these technologies, the environmental ramifications of nanomaterials, and the establishment of strong regulations for their utilization. Tackling these challenges is crucial for the field to persist in providing transformative solutions that safeguard the oceans, unlock their capabilities, and ensure their longevity for future generations. By merging discovery with innovation, marine nanoscience and nanotechnology aid in unraveling the intricacies of marine systems while promoting global environmental resilience and technological advancement.

Acknowledgements AHB thanks USIEF, New Delhi for grant of Nehru-Fulbright Academic and Professional Excellence award at Auburn University, Alabama, USA. All authors thank HED, Government of Jammu and Kashmir for sanction of funds toward establishment of state-of-the-art Interdisciplinary research Center at Islamia College of Science and Commerce, Srinagar, India. The invaluable support and encouragement of Principal Prof. (Dr.) Tehmina Yousuf is sincerely appreciated.

References

Abouelkheir SS, et al (2023) Nanobiotechnology: a sustainable approach for marine environment bioremediation. In: Industrial wastewater reuse: applications, prospects and challenges. https://doi.org/10.1007/978-981-99-2489-9_8
Ahmed I et al (2019) Applications of nanotechnology in transportation engineering. In: Sustainable infrastructure: breakthroughs in research and practice. https://doi.org/10.4018/978-1-7998-0948-7.ch044
Aithal P, Aithal S (2021) Green nanotechnology innovations to realize UN sustainable development goals 2030. Int J Appl Eng Manage Let 5(2):96–105. https://doi.org/10.47992/ijaeml.2581.7000.0106
Alishahi A (2015) Application of nanotechnology in marine-based products: a review. J Aquat Food Prod Technol 24:533–543. https://doi.org/10.1080/10498850.2013.788113
Alqarni LS, Alghamdi MD, Alshahrani AA, Nassar AM (2022) Green nanotechnology: recent research on bioresource-based nanoparticle synthesis and applications. J Chem 2022:4030999. https://doi.org/10.1155/2022/4030999
Anwar MN et al (2018) CO_2 capture and storage: A way forward for sustainable environment. J Environ Manage 226:131–144
Aracri S et al (2021) Soft robots for Ocean exploration and offshore operations: a perspective. Soft Robot 8:625–639
Arum LAS, et al An overview on carbon nanotubes as innovative absorbent for marine oil spill. In: IOP conference series: Earth and environmental science, vol 847
Babakhani P et al (2022) Nature nanotechnology Potential use of engineered nanoparticles in ocean fertilization for large-scale atmospheric carbon dioxide removal. Nat Nanotechnol 17:1342–1351
Banday AH et al (2024) Exploring the potential of marine natural products in drug development: a comprehensive review. Phytochem Lett 59:124–135
Barhoum A, Jeevanandam J, Danquah MK (2022) Fundamentals of bionanomaterials. https://doi.org/10.1016/C2020-0-00039-3
Bayda S, Adeel M, Tuccinardi T, Cordani M, Rizzolio F (2020) The history of nanoscience and nanotechnology: from chemical-physical applications to nanomedicine. Molecules 25:112. https://doi.org/10.3390/molecules25010112
Beygmoradi A, Homaei A (2017) Marine microbes as a valuable resource for brand new industrial biocatalysts. Biocatal Agric Biotechnol 11:131–152. https://doi.org/10.1016/j.bcab.2017.06.013
Bosu S, Rajamohan N (2024) Influence of nanomaterials in biohydrogen production through photo fermentation and photolysis—Review on applications and mechanism. Int J Hydrogen Energy 52:61–79
Brander SM et al (2011) The ecotoxicology of plastic marine debris. Am Biol Teacher 73:474–478

Carafa A (2024) Storing CO_2 in the Ocean: emerging technologies, challenges, and enabling factors. https://doi.org/10.1029/2007GL031131

Chausali N, Saxena J, Prasad R (2023) Nanotechnology as a sustainable approach for combating the environmental effects of climate change. J Agric Food Res 12:100541

Chinnappan RS, Kandasamy K, Sekar A (2015) A review on marine based nanoparticles and their potential applications. Afr J Biotechnol 14:1525–1532

Cornier J, Owen A, Kwade A, Van de Voorde M (2017) Pharmaceutical nanotechnology: innovation and production. pharmaceutical nanotechnology: innovation and production. https://doi.org/10.1002/9783527800681

Corsi I et al (2021) Ecosafe nano-based solutions for pollution monitoring and control in the marine environment. In: 2021 IEEE International workshop on metrology for the sea: learning to measure sea health parameters, MetroSea 2021—Proceedings. https://doi.org/10.1109/MetroSea52177.2021.9611624

Craig RK (2012) Marine biodiversity, climate change, and governance of the oceans. Diversity (Basel) 4:224–238

Cusson M et al (2015) Relationships between biodiversity and the stability of marine ecosystems: comparisons at a European scale using meta-analysis. J Sea Res 98:5–14

Dar AH, Rashid N, Majid I, Hussain S, Dar MA (2020) Nanotechnology interventions in aquaculture and seafood preservation. Crit Rev Food Sci Nutrit 60:1912–1921. https://doi.org/10.1080/10408398.2019.1617232

Di Ciaccio F, Troisi S (2021) Monitoring marine environments with autonomous underwater vehicles: a bibliometric analysis. Results in Engineering 9:100205. https://doi.org/10.1016/j.rineng.2021.100205

Dowling A et al (2004) Nanoscience and nanotechnologies : opportunities and uncertainties. London The Royal Society The Royal Academy of Engineering Report, vol 46

Duan Y, Wu J, Qi W, Su R (2023) Eco-friendly marine antifouling coating consisting of cellulose nanocrystals with bioinspired micromorphology. Carbohydr Polym 304:120504

Farokhzad OC, Langer R (2009) Impact of nanotechnology on drug delivery. ACS Nano 3:16–20

Fujii T, Pondella DJ, Todd VLG, Guerin A (2020) Editorial: seafloor heterogeneity: artificial structures and marine ecosystem dynamics. Front Mar Sci 7:378. https://doi.org/10.3389/fmars.2020.00378

Gehrke I, Geiser A, Somborn-Schulz A (2015) Innovations in nanotechnology for water treatment. Nanotechnol Sci Appl 8:1–7. https://doi.org/10.2147/NSA.S43773

Gharekhan CH, Upasani VN (2019) Diatoms from saline ecosystems and biotechnological applications: an overview. Int J Pharm Biol Sci 9

Halvorson HO, Quezada F (2019) Marine biotechnology. In: Encyclopedia of Ocean sciences, vol 1–5, 3rd edn.

Harinisri K, Jayanthi N, Suresh Kumar R (2023) Diverse application of green nanotechnology—A review. Mater Today Proc. https://doi.org/10.1016/j.matpr.2023.06.085

De Haro L (2011) News in marine toxicology. In: Annales de Toxicologie Analytique, vol 23

Hassett KJ et al (2021) Impact of lipid nanoparticle size on mRNA vaccine immunogenicit0y. J Control Rel 335:237–246

Hussein AK (2015) Applications of nanotechnology in renewable energies—A comprehensive overview and understanding. Renew Sustain Energy Rev 42:460–476. https://doi.org/10.1016/j.rser.2014.10.027

Jeong GJ, Khan S, Tabassum N, Khan F, Kim YM (2022) Marine-bioinspired nanoparticles as potential drugs for multiple biological roles. Mar Drugs 20:527. https://doi.org/10.3390/md20080527

Jigyasa, Rajput JK (2022) Nanomaterial-based sensors as potential remedy for detection of biotoxins. Food Control 135:108686. https://doi.org/10.1016/j.foodcont.2021.108686

Johnston CT (2010) Probing the nanoscale architecture of clay minerals. Clay Miner 45:245–279

Joseph B et al (2020) Extraction of nanochitin from marine resources and fabrication of polymer nanocomposites: recent advances. Polymers 12:1664. https://doi.org/10.3390/POLYM1208 1664

Joudeh N, Linke D (2022) Nanoparticle classification, physicochemical properties, characterization, and applications: a comprehensive review for biologists. J Nanobiotechnol 20:262. https://doi.org/10.1186/s12951-022-01477-8

Katzschmann RK, DelPreto J, MacCurdy R, Rus D (2018) Exploration of underwater life with an acoustically controlled soft robotic fish. Sci Robot 3:eaar3449

Khan F et al (2022) Green nanotechnology: plant-mediated nanoparticle synthesis and application. Nanomaterials 12:673. https://doi.org/10.3390/nano12040673

Kim SK, Chojnacka K (2015) Marine algae extracts: processes, products, and applications vols 1–2

Kiuru P et al (2014) Exploring marine resources for bioactive compounds. Planta Med 80:1234–1246. https://doi.org/10.1055/s-0034-1383001

Kong X et al (2016) Chemical and biological sensing using diatom photonic crystal biosilica with in-situ growth plasmonic nanoparticles. IEEE Trans Nanobiosci 15:828–834

Krishna Mohan MV, Srinivasa Reddy P, Bhanu Prakash TVK, Mukheerjee A (2024) Impact of titanium and silicon metal oxide nanoparticles on surface coatings for marine vehicles to mitigate biofouling. Particulate Sci Technol 42:915–928

Liao L, Wang M, Li Z, Wang X, Zhou W (2023) Recent advances in black TiO_2 nanomaterials for solar energy conversion. Nanomaterials 13:468. https://doi.org/10.3390/nano13030468

Liu J, Bao B, Chen J, Wu Y, Wang Q (2023) Marine energy harvesting from tidal currents and offshore winds: A 2-DOF system based on flow-induced vibrations. Nano Energy 114:108664

Maliszewska I, Kocek D, Wanarska E (2020) Microbial synthesis of unique nanoscale minerals—Challenges and prospects. Physicochem Probl Mineral Process 56:244–256

Maltby KM et al (2022) Marine climate change risks to biodiversity and society in the ROPME Sea Area. Clim Risk Manag 35:100411

Mann KH, Lazier JRN (2005) Dynamics of marine ecosystems. https://doi.org/10.1002/978111868 7901

Martínez-Vázquez, R. M., Milán-García, J. & de Pablo Valenciano, J. Challenges of the Blue Economy: evidence and research trends. Environ Sci Eur 33, 61

Mathew J, Joy J, George SC (2019) Potential applications of nanotechnology in transportation: a review. J King Saud Univ Sci 31:586–594. https://doi.org/10.1016/j.jksus.2018.03.015

Mehmood A et al (2021) Carbon and polymer-based magnetic nanocomposites for oil-spill remediation—A comprehensive review. Environ Sci Pollut Res 28:54477–54496. https://doi.org/10.1007/s11356-021-16045-0

Mekuye B, Abera B (2023) Nanomaterials: An overview of synthesis, classification, characterization, and applications. Nano Select 4:486–501

Mgadi K, Ndaba B, Roopnarain A, Rama H, Adeleke R (2024) Nanoparticle applications in agriculture: overview and response of plant-associated microorganisms. Front Microbiol 15:1354440. https://doi.org/10.3389/fmicb.2024.1354440

Munyalo JM, Zhang X (2018) Particle size effect on thermophysical properties of nanofluid and nanofluid based phase change materials: a review. J Mol Liq 265:77–87. https://doi.org/10.1016/j.molliq.2018.05.129

Mustafa G, et al Water pollutants and nanosensors. In: Aquananotechnology: applications of nanomaterials for water purification. https://doi.org/10.1016/B978-0-12-821141-0.00018-5

Nasr-Eldahan S, Nabil-Adam A, Shreadah MA, Maher AM, El-Sayed Ali T (2021) A review article on nanotechnology in aquaculture sustainability as a novel tool in fish disease control. Aquaculture Int 29:1459–1480. https://doi.org/10.1007/s10499-021-00677-7

Nguyen DTC et al (2021) The sunflower plant family for bioenergy, environmental remediation, nanotechnology, medicine, food and agriculture: a review. Environ Chem Lett 19:3701–3726. https://doi.org/10.1007/s10311-021-01266-z

O'Dor RK, Fennel K, Berghe E, Vanden (2009) A one ocean model of biodiversity. Deep Sea Res 2 Top Stud Oceanogr 56:1816–1823

Ogunkalu OA (2019) Utilization of nanotechnology in aquaculture and seafood sectors. Eurasian J Food Sci Technol 3:26–33
Oshiro N, Gago-Martínez A, Tubaro A (2024) Chemistry, toxicology and etiology of marine biotoxins. J Marine Sci Eng 12:236. https://doi.org/10.3390/jmse12020236
Pace LA, Saritas O, Deidun A (2023) Exploring future research and innovation directions for a sustainable blue economy. Mar Policy 148:105433
Page TM, Lawley JW (2022) The next generation is here: a review of transcriptomic approaches in marine ecology. Front Mar Sci 9:757921. https://doi.org/10.3389/fmars.2022.757921
Parmar K, Bhande R (2022) Future of nanotechnology in solar energy and hydrogen energy. J Energy Res Rev. https://doi.org/10.9734/jenrr/2022/v10i230254
Patra JK, Baek KH (2014) Green nanobiotechnology: factors affecting synthesis and characterization techniques. J Nanomater 2014:417305. https://doi.org/10.1155/2014/417305
Pestovsky YS, Martínez-Antonio A (2017) The use of nanoparticles and nanoformulations in agriculture. J Nanosci Nanotechnol 17:8699–8730. https://doi.org/10.1166/jnn.2017.15041
Plank MJ, Law R (2012) Ecological drivers of stability and instability in marine ecosystems. Theor Ecol 5:465–480
Qu X, Alvarez PJJ, Li Q (2013) Applications of nanotechnology in water and wastewater treatment. Water Res 47:3931–3946
Radhakrishnan S et al (2022) Triboelectric nanogenerators for marine energy harvesting and sensing applications. Results in Engineering 15:100487
Rosa-Santos P, Taveira-Pinto F, López M, Rodríguez CA (2022) Hybrid systems for marine energy harvesting. J Marine Sci Eng 10:633. https://doi.org/10.3390/jmse10050633
Roy RB, Alahakoon S, Arachchillag SJ, Rahman S (2023) Optimizing dynamic electric ferry loads with intelligent power management. Energy Rep 9:5962–5963
Sajid M (2022) Nanomaterials: types, properties, recent advances, and toxicity concerns. Curr Opin Environ Sci Health 25:100319. https://doi.org/10.1016/j.coesh.2021.100319
Saliu F et al (2021) Evaluation of chitosan aggregates as pickering emulsifier for the remediation of marine sediments. Chemosphere 273:129733
Salvador-Morales C, Grodzinski P (2022) Nanotechnology tools enabling biological discovery. ACS Nano 16:5062–5084. https://doi.org/10.1021/acsnano.1c10635
Sangeetha N, Priyenka Devi KS, Saravanan K (2013) Synthesis and characterization of silver nanoparticles using the marine algae Ulva lactuca. Pollut Res 32:135–139
Sarangi U (2023) Blue economy, blue finance and ocean governance for achieving sustainable development goals. Nat Resour Forum 47:3–21
Segovia Ramírez I, Bernalte Sánchez PJ, Papaelias M, García Márquez FP (2021) Autonomous underwater vehicles and field of view in underwater operations. J Mar Sci Eng 9:277
Shams F et al (2022) Nanotechnology-based products for cancer immunotherapy. Mol Biol Rep 49:1349–1412. https://doi.org/10.1007/s11033-021-06876-y
Singh H, Bhardwaj N, Arya SK, Khatri M (2020b) Environmental impacts of oil spills and their remediation by magnetic nanomaterials. Environ Nanotechnol Monitor Manage 14:100305. https://doi.org/10.1016/j.enmm.2020.100305
Singh RP, Handa R, Manchanda G (2021) Nanoparticles in sustainable agriculture: an emerging opportunity. J Control Release 329:1234–1248. https://doi.org/10.1016/j.jconrel.2020.10.051
Singh I, Gautam HK, Dhawan G (2020) Nanobiotechnology: Current and future perspectives in combating microbial pathogenesis. In: Pathogenicity and drug resistance of human pathogens: mechanisms and novel approaches. https://doi.org/10.1007/978-981-32-9449-3_17
Sinha S et al (2023) Nanotechnology-based solutions for antibiofouling applications: an overview. ACS Appl Nano Mater 6:12828–12848
Sivanesan I et al (2021) Green synthesized chitosan/chitosan nanoforms/nanocomposites for drug delivery applications. Polymers 13:2256. https://doi.org/10.3390/polym13142256
Squire HJ, Tomatz S, Voke E, González-Grandío E, Landry M (2023) The emerging role of nanotechnology in plant genetic engineering. Nat Rev Bio 1:314–328

Suri SS, Fenniri H, Singh B (2007) Nanotechnology-based drug delivery systems. J Occup Med Toxicol 2:16

Vetrivel R, Navinselvakumar C, Samuel Ratna Kumar PS (2018) Carbon nanotubes and its applications—A review. Int J Mech Prod Eng Res Dev 8:288–293

Vigneron JP, Simonis P (2012) Natural photonic crystals. Physica B 407:4032–4036

Virginia Armbrust E, Palumbi SR (2015) Uncovering hidden worlds of ocean biodiversity. Science (1979) 348:865–867 (2015)

Vukusic P, Sambles JR (2003) Photonic structures in aquatic systems. Nature 424:516–537

Wallace JM et al (2003) Silica nanoarchitectures incorporating self-organized protein superstructures with gas-phase bioactivity. Nano Lett 3:1463–1467

Wang W-X (2023) Environmental toxicology of marine microplastic pollution. Cambridge Prisms: Plastics 1:e10

Wang Z, Carriveau R, Ting DSK, Xiong W, Wang Z (2019) A review of marine renewable energy storage. Int J Energy Res 43:6108–6150. https://doi.org/10.1002/er.4444

Wang Y et al (2023) Marine biomaterials in biomedical nano/micro-systems. J Nanobiotechnol 21:408. https://doi.org/10.1186/s12951-023-02112-w

Wu Z et al (2018) Microstructural characterization and in-situ sulfur isotopic analysis of silver-bearing sphalerite from the Edmond hydrothermal field, Central Indian Ridge. Ore Geol Rev 92:318–347

Wu Q, Zhang WS, Du Y, Gao HJ, Hui D (2020) Mechanical properties of nanomaterials: a review. Nanotechnol Rev 9:259–273

Youssef AM, Abdel-Aziz MS (2013) Preparation of polystyrene nanocomposites based on silver nanoparticles using marine bacterium for packaging. Polym Plast Technol Eng 52:607–613

Zäch M, Hägglund C, Chakarov D, Kasemo B (2006) Nanoscience and nanotechnology for advanced energy systems. Curr Opin Solid State Mater Sci 10:132–143

Zang L (2011) Energy efficiency and renewable energy through nanotechnology. In: Green energy and technology, vol 33

MIX
Papier aus verantwortungsvollen Quellen
Paper from responsible sources
FSC® C105338

If you have any concerns about our products,
you can contact us on
ProductSafety@springernature.com

In case Publisher is established outside the EU,
the EU authorized representative is:
Springer Nature Customer Service Center GmbH
Europaplatz 3, 69115 Heidelberg, Germany

Printed by Libri Plureos GmbH
in Hamburg, Germany